Capacity Development for Improved Water Management

Capacity Development for Improved Water Management

UNESCO-IHE Editors

M.W. Blokland
G.J. Alaerts
J.M. Kaspersma

UNW-DPC Editor

M. Hare

CRC Press
Taylor & Francis Group
Boca Raton London New York

CRC Press is an imprint of the
Taylor & Francis Group, an **informa** business

A BALKEMA BOOK

UNESCO-IHE Institute for Water Education
(UNESCO-IHE)

UN-Water Decade Programme on Capacity
Development (UNW-DPC)

Published by: CRC Press/Balkema
P.O. Box 447, 2300 AK Leiden, The Netherlands
e-mail: Pub.NL@taylorandfrancis.com
www.crcpress.com – www.taylorandfrancis.co.uk – www.balkema.nl

ISBN 13: 978-0-367-45237-7 (pbk)
ISBN 13: 978-0-415-57398-6 (hbk)

Visit the Taylor & Francis Web site at
http://www.taylorandfrancis.com

and the CRC Press Web site at
http://www.crcpress.com

Except: ECDPM, Chapter 7 "Capacity Change and Performance: Insights and implications for development cooperation", pp. 135–154, © ECDPM

Typeset by Vikatan Publishing Solutions (P) Ltd, Chennai, India.

Library of Congress Cataloging-in-Publication Data

Capacity development for improved water management / UNESCO-IHE editors,
M.W. Blokland ... [et al.].
p. cm.
Includes bibliographical references and index.
ISBN: 978-0-415-57398-6 (hardcover : alk. paper)
1. Water-supply--Management--Congresses. I. Blokland, M. W. II. Unesco-IHE
Institute for Water Education. III. World Water Symposium
(5th : 2009 : Istanbul, Turkey)
TD365.C42 2010
363.6'1--dc22   2009051400

Contents

Foreword *UNESCO-IHE*

Capacity development is important in any development process – and in many instances it is the fundamental prerequisite to make development possible. This is obviously true for developing countries, but equally holds for the processes of economic and social development in the industrialized economies. Our world will need to keep investing in human and financial resources as well as in the technological know-how if our objective is to improve the quality of the human experience. Without a commensurate investment in capacity development and the subsequent management of the new knowledge generated through such an investment process, any investment in infrastructure, or any effort to manage our natural resources and eco-systems is at risk of failure. In fact, as we are increasingly concerned about the most *efficient* use of our financial and other resources, decision-makers cannot be expected to make optimal decisions for achieving development goals without relying on knowledge management and capacity development – regardless of the level of investment or reach of the political support. Capacity development is emerging as the single best tool with which to create transparency and an informed citizenry to combat corruptive and other negative practices that taint the use of both private and public funds expended on behalf of the governments and the donor or lending communities.

Tertiary education is a vital link in this capacity development and knowledge management continuum. It helps ensure that a new generation of water leaders is prepared to assume technical and decision-making roles at each level of government and in political and social organizations. We, at UNESCO-IHE, take this responsibility very seriously and are continually adapting our technical specialties to become better able to address the true demands from the marketplace of 'development.' We are working with partners in such a way that our assistance is less needed the next time around, and they themselves are becoming capable to define their new challenges, and articulate the solutions for them. We must help build societies through investments in education if we are to help ensure a brighter future for all strata in civil society.

Informal water education – 'water literacy' – for citizens is an equally vital component of any capacity development process. It needs to ensure their informed participation in investment and management decisions. Women play a critical role in water resource management in many developing countries – in maintaining water sources, in agricultural production, when hauling water; and teaching sanitary habits to their children – therefore, they should be assigned a more important role in the management of that resource. Water is one of the most essential resources for any society to progress and live healthy lives, and building the capacity to manage that resource, must be a prime concern.

The 5th World Water Forum in Istanbul, March 2009, has brought together a wide range of stakeholders from all parts of civil society dedicated to improving the management of water. The Forum has catalyzed learning at all levels and among all corners of the world. No one region holds the key to success in capacity development and knowledge management. Workable solutions are partly generic and partly location-specific. So, there is much experience and knowledge that has been shared, on all continua – North-South, South-South and South-North. The World Water Forum has been a truly enriching experience.

This publication provides a wide range of cutting-edge papers on recent advances in capacity development. It describes the current conceptual approaches to knowledge and capacity development, coming from a broad yet representative variety of fields of social sciences, field practice and case studies. I welcome the reader to make use of the expertise displayed in this book and also build on the fruitful discussions at the 5th World Water Forum. I thank the authors for their time and efforts to contribute to this exciting book.

Richard A. Meganck, Rector
UNESCO-IHE Institute for Water Education

Foreword *UNW-DPC*

The broad mission of the *UN-Water Decade Programme on Capacity Development* (UNW-DPC) is to enhance the coherence and integrated effectiveness of the capacity development activities of UN-Water, an inter-agency mechanism involving the 26 UN members, and more than a dozen non-UN partners, working and collaborating on water-related issues. By doing so, it intends to strengthen the efforts of the members and partners of UN-Water in their quest to support Member States in meeting international goals, such as the Millennium Development Goals (MDGs) related to water. Consequently, since its establishment in August 2007, UNW-DPC has been collaborating with more than half a dozen UN-Water members (including UN-HABITAT, WMO, UNESCO, UNU, UN-DESA, UNCCD and FAO) and other partners on different capacity development activities as well as supporting the Office of the Chair of UN-Water and the UN-Water Task Forces in multilateral UN agency collaborations.

This joint publication represents further rewarding cooperation with UNESCO-IHE. It is aimed at promoting effective knowledge management and delivery by compiling existing expert knowledge on the state-of-the-art in water-related capacity development and making it easily accessible to members of the water community in the form of this high quality book. A pre-print was launched successfully at the 5th World Water Forum in Istanbul in March 2009, and the available copies were quickly snapped up. To augment the already substantial treasury of expert contributions, this new edition now includes a seventeenth chapter, provided by World Bank Institute authors, describing lessons learned from the last decade of capacity development in Africa.

I would like to express my gratitude to all the authors who have volunteered their time and expertise on this most exciting of collaborative projects. I welcome the interested reader to make use of the wide variety of expertise on capacity development on offer in this publication. Such expertise ranges from experiences in e-learning and networking, to community knowledge management and the running of training-of-trainers courses, in a book that maintains a global focus, providing case studies from Asia, Africa, the Americas and Europe.

Reza Ardakanian, PhD
Director, UN-Water Decade Programme
on Capacity Development,
Bonn, Germany

Acknowledgements

This publication would not have been possible without the authors' and organizations' original contributions, as well as those existing papers and briefs that have been kindly provided by authors and organizations for reproduction in this publication. Alongside the work of the publication copy editor, Adele Sanders, the involvement of all those who have supported the publication's editing and formatting, as well as the work of the anonymous reviewers, is gratefully acknowledged. Additionally, the generous sponsorship and support of the Women for Water Partnership in the development of Chapter 13 is recognized.

About the editors

GUY ALAERTS, PhD

is Principal Water Resources Specialist at the World Bank, and is an extraordinary Professor in Knowledge and Capacity Development at UNESCO-IHE. He advises on research on knowledge and capacity development of public water management organisations, and is working on sector policies and large scale sector reform processes in Asia and South Eastern Europe.

MAARTEN BLOKLAND, MSc

is Associate Professor in Water Services Management at UNESCO-IHE Institute for Water Education, a position he took up again in 2007, after working fulltime as the Institute's Deputy Director and Acting Director from August 2002 to October 2007. He has worked in the water supply and sanitation sector since 1975, and worked in more than 40 developing and newly industrialized countries. His current work is on benchmarking for pro-poor water services provision, managing water organisations and capacity building for water utilities.

MATT HARE, PhD

is Senior Programme Officer at the UN-Water Decade Programme on Capacity Development, in Bonn, Germany. He specializes in the areas of curriculum development, training-of-trainers programmes, climate change adaptation and stakeholder participation. Prior to UNW-DPC, he worked for EAWAG, in

Switzerland, before co-founding and managing a consultancy firm specialising in participatory water resources management.

JUDITH KASPERSMA, MSc

is working on her PhD research in knowledge and capacity development at UNESCO-IHE. She has been previously involved in consultancy activities in the field of institutional and organisational development of public organisations in the water sector in Asia and the Middle East.

About the organisations

UNESCO-IHE

UNESCO-IHE envisages a world in which people manage their water and environmental resources in a sustainable manner, and in which all sectors of society, particularly the poor, can enjoy the benefits of basic services.

UNESCO-IHE provides postgraduate education, training and research in the fields of water and the environment for mid-career and senior professionals from developing countries. The Institute is the largest water education facility in the world, and the only institution in the UN system authorised to confer accredited MSc degrees.

Since 1957, the Institute's community of partnerships includes public and private organisations. This community also includes over 13,500 alumni active in water sectors worldwide, representing an extensive network of international water professionals of yesterday, today and tomorrow.

UNESCO-IHE believes that partnerships and networks are of vital and strategic importance in improving access to, sharing and disseminating information. The Institute aims to strengthen the efforts of other universities and research centres to increase the knowledge and skills of professionals working in the water sector, and contributes to the achievement of synergy in the programmes and activities of existing knowledge networks.

UNW-DPC

The UN-Water Decade Programme on Capacity Development (UNW-DPC) is a joint programme of UN agencies and programmes cooperating within the framework of UN-Water. The broad mission of UNW-DPC is to enhance the coherence and integrated effectiveness of the capacity development activities of the more than two dozen UN organizations and programmes already cooperating within the interagency mechanism known as UN-Water and thereby support them in their efforts to achieve international goals such as the Millennium Development Goals (MDGs) related to water and sanitation.

About the authors

CHAPTER 1

GUY J. ALAERTS, PhD

is Principal Water Resources Specialist at the World Bank, and is an extraordinary Professor in Knowledge and Capacity Development at UNESCO-IHE. He advises on research on knowledge and capacity development of public water management organisations, and is working on sector policies and large scale sector reform processes in Asia and South Eastern Europe.

JUDITH M. KASPERSMA, MSc

is working on her PhD research in knowledge and capacity development at UNESCO-IHE. She has been previously involved in consultancy activities in the field of institutional and organisational development of public organisations in the water sector in Asia and the Middle East.

CHAPTER 2

VELMA I. GROVER, PhD

is the freshwater ecosystem programme and Water Virtual Learning Centre coordinator at UNU-INWEH. She has an extensive experience in capacity building in developing countries, particularly related to distance learning for integrated water resource management. She has published articles in journals on distance education and various environmental management issues, and is an editor (and contributor) to seven books.

COLIN MAYFIELD, PhD

is the Assistant Director of UNU-INWEH responsible for Freshwater Ecosystems. He is also a Professor of Microbiology at the University of Waterloo in Waterloo, Ontario, Canada where he was also Associate Dean of Computing. He has authored or co-authored over 200 papers in those areas and has worked as a consultant for many governments, international and national agencies and various companies. He has taught many microbiology and IWRM courses; many of them were also given through Distance Learning programmes at the University of Waterloo and UNU-INWEH.

CHAPTER 3

MATT HARE, PhD

is Senior Programme Officer at the UN-Water Decade Programme on Capacity Development, in Bonn, Germany. He specializes in the areas of curriculum development, training-of-trainers programmes, climate change adaptation and stakeholder participation. Prior to UNW-DPC, he worked for EAWAG, in Switzerland, before co-founding and managing a consultancy firm specialising in the area of participatory water resources management.

CATHARIEN TERWISSCHA VAN SCHELTINGA

works as a researcher and lecturer with the Earth System Science and Climate Change Group at Wageningen University and Research Centre, the Netherlands. She concentrates in her work on capacity development for water management and climate change adaptation.

CAROLINE VAN BERS

works with the Institute of Environmental Systems Research at the University of Osnabrück (Germany) and is a guest researcher with the Department of Water Management, University of Twente (Netherlands). Her current work focuses on training and education in Integrated Assessment and Adaptive Water Management.

CHAPTER 4

WOUTER LINCKLAEN ARRIËNS

is the Lead Water Resources Specialist of the Asian Development Bank (ADB). He has coordinated ADB's work to support its member countries in improving water governance through policies, sector reforms, investment projects, capacity development, and regional cooperation since 1994, and serves as spokesperson for ADB's water work.

JAN LUIJENDIJK, MSc

is Head, Scientific Department of UNESCO-IHE. He has over 30 years professional experience in Land and Water Engineering and Knowledge Management and water sector Capacity Building Programs. He has been Project Director of several long-term Capacity Building Programs in developing countries in Asia, East-Africa and the Middle East.

CHAPTER 5

PAOLA CHAVES

works as a journalist and social communicator and is a staff member of the Research and Development Institute on Water Supply, Environmental Sanitation and Water Resources Conservation, Cinara, of the Universidad del Valle.

MARIELA GARCÍA, MSc

is a sociologist working as a staff member of the Research and Development Institute on Water Supply, Environmental Sanitation and Water Resources Conservation – Cinara, and Associated Professor of the Universidad del Valle, Colombia. She has a Master's degree in development studies from the International Institute of Social Studies, ISS, the Netherlands.

CHAPTER 6

LÉNA SALAMÉ

is a jurist and trained mediator. She is the coordinator of *UNESCO's 'From Potential Conflict to Cooperation Potential' (PCCP)* programme (www.unesco.org/water/wwap/pccp/). She lectures on issues related to shared water resources and alternative dispute resolution within the framework of various educational programmes. She is involved in the development of, and research on, educational tools related to the same topics.

LARRY A. SWATUK, PhD

is Associate Professor and Director of the *Programme for Environment and International Development in the Faculty of Environment at the University of Waterloo*, Canada. From 1996-2007 he was a lecturer in the Department of Political and Administrative Studies at the University of Botswana and Associate Professor of Natural Resource Governance at the Harry Oppenheimer Okavango Research Centre in Maun, Botswana. His most recent publication is the co-edited collection (with Lars Wirkus) entitled Transboundary Water Governance in Southern Africa: exploring under-examined dimensions (Bonn: Nomos, 2009).

PIETER VAN DER ZAAG, PhD

is Professor of water resources management at the *UNESCO-IHE* Institute for Water Education in Delft, The Netherlands. He has a special interest in the management of catchment areas and transboundary river basins, as well as in curriculum development. He is involved in several multidisciplinary research and capacity building projects, mainly in Africa.

CHAPTER 7

ECDPM

The European Centre for Development Policy Management (**ECDPM**) works to improve relations between Europe and its partners in Africa, the Caribbean and the Pacific. The contributors to this Brief were core members of the study on Capacity, Change and Performance (www.ecdpm.org/capacitystudy). At the time of the study, Heather Baser, Volker Hauck, Niels Keijzer and Anje Kruiter were staff of ECDPM. Tony Land and Peter Morgan were ECDPM Programme Associates. The study was led by Heather Baser and Peter Morgan.

CHAPTER 8

JAN LUIJENDIJK, MSc

see Chapter 4.

IWAN NURSYIRWAN, MSc

graduated from Technical Institute Bandung (ITB) and UNESCO-IHE (the Netherlands), and has more than 30 years experience in Water Resources

Works in Indonesia. He currently is Director General for Water Resources at the Ministry for Public Works in Indonesia.

KLAAS SCHWARTZ, PhD

is Senior Lecturer in Water Services Management. He has been involved in research, consultancy and training activities in the field of water governance and public management of river basin organizations and water supply and sanitation utilities in various countries in Asia, Africa and Latin America.

AART VAN NES, MSc

holds an M.Sc. in civil engineering of the Delft University of Technology. He has more than 30 years experience in Asia and Africa mainly as team leader for multidisciplinary teams concerning technical, socio-economic and institutional aspects. He currently is involved in Jakarta Flood Management.

CHAPTER 9

SEYED HAMED ALEMOHAMMAD, MSc

holds a Master of Science in Water Resources Engineering from Sharif University of Technology, Iran. He also holds a Bachelor of Science in Civil Engineering. He has about 14 papers in international and national conferences and national journals. He will start his Ph.D. in Water Resources Management at Massachusetts Institute of Technology (MIT) from September 2009.

REZA ARDAKANIAN, PhD

is founding director of UN-Water Decade Programme on Capacity Development (UNW-DPC), former Vice-Minister in charge of water affairs of Iran and has a PhD in water resources management from McMaster university, Canada. He is a university faculty member and has more than 50 publications in scientific journals and conferences' proceeding.

MOJTABA NIKRAVESH, MBA

is the director general for Management Improvement & Renovation in Water Resources Management of the Ministry of Energy, I.R. Iran. He holds a Bachelor of Science in Industrial Engineering (System's Planning and Analysis) and Master of Business Administration (MBA). He has 8 publications in conferences and journals about Comprehensive Structure of Water Resources Management in Iran and Iran's Regional Water Authorities.

CHAPTER 10

SILVER MUGISHA, PhD

is Chief Manager, Institutional Development and External Services, at the **National Water and Sewerage Corporation (NWSC)** of Uganda. He joined NWSC in 1994 and has been a key champion in the performance programmes that have turned around NWSC. Mugisha has wide experience in utility management and institutional development.

CHAPTER 11

GRAHAM ALABASTER, PhD

is Section Chief responsible for Africa in Human Settlements Financing Division of UN-HABITAT. He has a first degree in Chemical Engineering and a PhD in Civil Engineering. He has over 25 years experience in water, sanitation and solid waste management working in over 30 countries. He joined UN-HABITAT in 1992 and has responsibility for all UN-HABITAT's operation projects on water, sanitation and waste management in Africa & Latin America in addition to global responsibility for policy issues relating to sanitation, pro-poor water and sanitation governance, solid and hazardous waste management, and monitoring water and sanitation MDGs.

CHAPTER 12

JOHN ETGEN, MSc

is Senior Vice President, Project WET Foundation. He has an M.S. in Science Education and has devoted more than 25 years to natural resources education for diverse audiences around the world. He is currently creating innovative WASH education materials for sub-Saharan African teachers and students. Through these efforts, he is able to follow his passion for improving sustainability through education.

AMY FULLER

is a graduate student at the University of Pennsylvania working on a dual Masters degree in Public Health and Environmental Studies. Ms. Fuller has been specializing in water and sanitation issues and their impact on the development of African communities. She is especially interested in transboundary water cooperation and the role that resource distribution plays in regional conflict.

DENNIS NELSON, MSc

is President and CEO of Project WET Foundation which he founded in 1984 growing it into a global leader in water resources education. Dennis has taught hundreds of courses and authored/contributed to more than 300 activities and 30 children's activity booklets about water. Through these efforts, he has become a recognized leader and advocate for water education worldwide.

TEDDY TINDAMANYIRE, MSc

is Principal Environment Officer, Uganda Ministry of Water and Environment. She is a trained female scientist and science teacher from Makerere University. Mrs. Tindamanyire holds a Bachelor of Science degree, a Diploma in Education, a Diploma in Human Resources Management and a M.S. in Environment and Natural Resources Management. Mrs. Tindamanyire has worked in secondary schools and with Wetlands Programme for 14 years as an educator and public awareness officer.

CHAPTER 13

PRABHA KHOSLA, MES

is an urban planner. She works on issues of equality and equity in urban governance including urban sustainability; women's rights and gender equality; water and sanitation; production of training materials and training. Her recent publication – *Gender in Local Government: A Sourcebook for Trainers* is available from UN-HABITAT.

CHAPTER 14

DANIEL VALENSUELA

is Deputy General Manager at the International Office for Water, Paris and senior assistant to the Technical Permanent Secretary of the International Network of Basin Organisation. Previously, he worked six years as Network officer in the Global Water Partnership Secretariat in Stockholm. With his agronomist background, he has also worked in the area of water for agriculture and impact of agriculture on, water resources, particularly at basin level in Sub-Saharan Africa.

CHAPTER 15

STÉPHAN VINCENT-LANCRIN, PhD

is a Senior Analyst at the OECD Centre for Educational Research and Innovation. Before joining the OECD, Stéphan worked for 7 years as lecturer and researcher in economics at the University of Paris-Nanterre and the London School of Economics. He holds a PhD in economics and master's degrees in business administration and in philosophy. His recent work has focused on tertiary education and on the relationship between innovation and education.

CHAPTER 16

MARK NELSON

is the team leader for governance diagnostics and capacity development at the World Bank Institute and co-chair of the Learning Network on Capacity Development, an international network of development practitioners. Mark also works on the role of the media in development, access to information and transparency, and advises country teams on how to incorporate these issues into overall development strategies.

AJAY TEJASVI

is a consultant working on capacity development policy issues at the World Bank Institute. An engineer by training, he is currently working towards a Ph.D. in Political Science from Claremont Graduate University.

CHAPTER 17

YEMILE MIZRAHI, PhD

is a political scientist specializing in democratic governance and anti-corruption. She has more than 15 years of experience conducting research and analysis; designing governance programs; training and advising government officials, political party leaders, journalists, and civil society organizations on strategies for increasing government effectiveness, build capacity, and promote transparency and accountability. Since 2005, Dr. Mizrahi has been a Senior Associate at Casals and Associates, Inc., a consulting firm recognized for its pioneering work in designing and implementing anti-corruption projects across the world. Before joining Casals, Dr. Mizrahi worked as an independent consultant for international donor organizations, including the World Bank and USAID. From 1999–2000, she was a Fellow at the Woodrow Wilson International Center for Scholars where she wrote the manuscript for her book, *From Martyrdom to Power: The Partido Acción Nacional in Mexico,* published by Notre Dame University Press, in 2004.

Part I

Setting the scene

Chapter 1

PROGRESS AND CHALLENGES IN KNOWLEDGE AND CAPACITY DEVELOPMENT

Guy J. Alaerts
World Bank and UNESCO-IHE Institute for Water Education

Judith M. Kaspersma
UNESCO-IHE Institute for Water Education

ABSTRACT

Knowledge and capacity development (KCD) involves something more than the strengthening of individual skills and abilities. Trained individuals need an appropriate environment, and the proper mix of opportunities and incentives to use their acquired knowledge. The article therefore discusses KCD for the water sector at three different levels, from individual to organizational to the institutional level and enabling environment. Secondly, the article describes the current conceptual approaches to KCD. They come from a wide variety of fields of social sciences, as well as from field practice and case studies, and are sometimes contradicting. Ideas about capacity originate from fields including organizational development, political economy, public administration, pedagogy, institutional economics and sociology. The most important views and fields, including some examples from the field that influence our thinking on KCD, are described and discussed. The article will discuss the link between knowledge and capacities, clarify that one of the basic capacities is to learn, and highlight the importance to create, share and manage the knowledge that results from learning, at the three levels. Extensive reference will be provided to the respective other chapters in the volume where more detail and case studies are provided, as well as examples of other approaches.

I.I INTRODUCTION

It is nowadays generally agreed that capacity enhancement involves more than the strengthening of individual skills and abilities. Trained individuals need an appropriate environment, and the proper mix of opportunities and incentives to apply their acquired knowledge. Understanding capacity development therefore requires a more comprehensive analytical framework that takes into account the individual, the organizational and the institutional levels of analysis (Alaerts, 1999, EuropeAid, 2005, McKinsey, 2007). The water sector is a sector of particular complexity, and therefore highly dependent on strong institutions and individual capacities. This complexity derives from the fact that daily decisions of each and every individual in society impact on water management – regarding water use, water pollution, sanitation, etc. – which is different from, say the roads sector. Also, water is a bulky and fugitive resource that has to be managed continuously in order for it to be available in the right quantities at the right time to sustain life – not too little to avoid drought and not too much to avoid flood – which means that it requires heavy investment and laborious operation and maintenance. The decisions and behaviour of mothers and farmers, thus, matter as much as those of the Minister of Water Resources or of Public Works. Not surprisingly, this sector was one of the first to identify the need for capacity development and introduce focused capacity development programs (Alaerts et al., 1991).

The first UN Conference on Water, in 1977 in Mar del Plata, sounded for the first time the alarm bell over the vulnerable and finite nature of water in light of the rapidly growing demands on the resource. The Mar del Plata Action Plan prioritized the provision of drinking water and sanitation – "drinking water and sanitation for all by 1990" – and the need to save water and protect it from wastage and pollution. The 1981–1990 Drinking Water Supply and Sanitation Decade managed to dramatically increase the coverage for water services but proved on many counts less effective. It was outpaced by the growth in population and demand for water. This led to the recognition that strong institutions and proper social behaviour are as important as the infrastructure itself. Also, it was found that development of this sector would require a higher level of pro-activity and effective strategies to deal with future challenges. The UN Conference on Environment and Development in Rio in 1992, and the 2002 UN Millennium Development Goals re-iterated the same priorities. The 2006 World Water Forum in Mexico City again highlighted the role of strong local capacity. After three decades of major investment efforts the world has much achievement to show for, yet serious challenges remain, both old and new.

The exponential growth in demand for water, the strong urbanization, and the persistent poverty have kept the coverage rates for *drinking water supply and sanitation* at modest levels in some regions, though other regions are well on their way to close the gap. Still, major challenges remain: the water use efficiency and service reliability are often unacceptably low; many water supplies are precariously vulnerable; and pollution taints water quality much faster than pollution control measures can be put in place. Weak performance of the institutions and of the users remains the bottleneck, especially in sanitation and in general *water service delivery*. Yet, it could be argued that the main future concern will be on *water resources sustainability*, as the large-scale and steady transformation of the earth's surface by human interventions as well as by climate variability is rapidly adding new stresses on the natural eco-hydrological systems that will have to support the continual generation of water resources to meet the demand for water–for food production, for drinking, for hydropower and navigation, flood management, etc. Competition for access to water will rise.

Part of the knowledge to address these challenges is available. However, gaps still do exist in our knowledge, for example regarding how the global changes are going to affect us and what the responses should be, and how the water service delivery and the resource should be managed more effectively. This represents a first key challenge. Equally important, one often observes that even when available, this knowledge does not get readily translated into proper planning or effective action. Weak institutions, especially at local levels of government, and in many communities, form a second key challenge, in particular in developing nations. This lagging or constraining effect is especially visible in countries that are developing into modern economies, but it is a challenge for all societies as they continuously must adjust their sector to new outside changes or to new internal demands.

1.2   KNOWLEDGE AND CAPACITY

1.2.1   CAPACITY CONSTRAINTS TO APPLY KNOWLEDGE – THE CASE OF DEVELOPING ECONOMIES

Many countries – that is, their governments as well as their civil society – are observed to have a weak "capacity": limited knowledge bases; small numbers of professionals with the right education and skills; and, in general, administrative and managerial arrangements, and laws and regulations (the "enabling environment") that eventually fail to facilitate the swift and effective actions that in their aggregate can deliver the desired outcomes and results on the ground. First, knowledge is required that can identify and describe the issue, challenge or problem that one desires to be addressed. A different knowledge is required to then articulate how to address this. Thereafter, this knowledge needs to be communicated, shared, refined and confirmed among experts, peers and decision-makers as prerequisite for action, after which implementation of the action necessitates a functional and capable organization and an enabling policy and administrative environment to do so and mobilize the matching financial and other resources. The implementation capacity, thus, also depends on the knowledge and skills of the implementing agency, and the incentives it responds to. This creates the potential to act. However, this potential will materialize only in presence of positive incentives (such as, financial or political incentives, personal motives to further one's career, etc.) which outweigh negative ones (such as, vested interests of an elite, lack of reward, opportunities to extract rents, etc.).

The action, therefore, comes as the aggregate of a series of sequential causal steps and decisions. The eventual outcome or impact from that action can be observed only much later, at substantial distance from the original knowledge. Hence, it is often tenuous to correlate the outcome with that knowledge, or with the capacity of the administration and the quality of the enabling environment. In addition, these processes take place in a dynamic and changing environment, and political contexts are continuously shifting. This difficulty to define unequivocal causalities is further compounded by the fact that similar outcomes can also be generated by several other sets of knowledge, capabilities and circumstances.

1.2.2   DEFINING CAPACITY

The concept of capacity refers to development in general, and several definitions have been proposed that reflect the theoretical (or political) frameworks from which the subject is approached (see Box 1.1).

1   Firstly, because of the original concern with effective government, the public administration science was one of the first disciplines to define "capacity" referring to the organizational structures and operational procedures of administrations. In this perspective, the public administration receives its capacity from the education and training of the civil servants, from proper administrative procedures, and appropriate incentives (e.g. Shafritz, 1985).

2   However, drawing experience and insight from the only modestly successful efforts in developing countries with the International Drinking Water Supply and Sanitation Decade spanning the 1980s, the water sector was early to devise a practical definition expressing its strong interest in making overall development programs more effective and sustainable, and to articulate in a coherent fashion the need for *knowledge and capacity development* (KCD) (Alaerts et al., 1991, 1999). This experience also highlighted the critical function of the "enabling environment" of the broader policy, legal and regulatory frameworks in which the public administration, and the investment projects funded by the international donor community, have to operate. For example, many regulations and procedures in other sectors one way or another were found to restrict the effectiveness of policies in the water sector: regulations on urban settlements, for example, make it often impossible to extend water services to "irregular" city quarters; human resource procedures in the civil service often preclude that incentives can be provided; and centralized administrations often have no

> **Box 1.1 What is capacity?**
>
> Shafritz (1985) approached capacity from the perspective of public administration sciences: "... any system, effort, or process ... which includes among its major objectives strengthening of elected chief executive officers, chief administrative officers, department and agency heads, and program managers in general purpose government to plan, implement, manage or evaluate policies, strategies, or programs designed to impact on social conditions in the community". The 1991 *Delft Declaration* (Alaerts et al., 1991) suggested "Capacity comprises well-developed institutions, their managerial systems, and their human resources, which in turn require favorable policy environments, so as to make the [water] sector effective and sustainable". UNDP states that "Capacity is the ability of individuals, groups, institutions and organizations to identify and solve problems over time" (Morgan, 1993, UNDP, 1993). Hildebrand and Grindle (1994) emphasize the dynamic nature of capacity: "Capacity is the ability to perform appropriate tasks effectively, efficiently, and sustainably. This implies that capacity is not a passive state – the extent of human resources development, for example – but part of an active process." More recently, the complexity or systems nature of capacity has been emphasized: "Capacity is ... the emergent combination of attributes that enables a human system to create developmental value" and "... the overall ability of a system to perform and to sustain itself: the coherent combination of competencies and capabilities"(Zinke, 2006). And: "Emergent properties, such as capacity, come from the dynamism of the interrelationships in the system. The challenge is not so much to build or enhance them as it is to unleash them or find ways to encourage their emergence" (Morgan, 2005).

place for a role by local water users. It became recognized that the water users, the consumer, the electorate and other distinct stakeholders in civil society have to play equally important roles in making things work at the local operational level of the household, the irrigation plot or the water catchment, and in providing the political foundation for decisions. This holds especially true in developing countries where the national government has often a very limited reach and capacity, and, thus, much depends on whether local users and communities are willing to take initiative, co-operate and contribute. This approach, in addition, also for the first time linked capacity with knowledge – as generated and disseminated by educational institutions and knowledge centres locally and at a global level. The water sector is increasingly considered a knowledge-intensive sector.

3 Thirdly, UNDP, and later OECD, the World Bank and EuropeAid took this definition to a higher level, relating capacity to the overarching goal of national development across the board. That definition, however, tends to dissociate the capacity and the knowledge components.

4 Further recent work emphasizes the complex and systems nature of capacity in development efforts, as described below. This approach argues that the capacity of an organization is both a distinct entity by itself and the resultant of the capacities of the individuals in that organization. It is also the resultant of a wide variety of inputs (types of knowledge that have been transferred, structure and procedures, leadership and managerial capabilities of the individuals, etc.). All these attributes tend to change over time and mutually influence each other. This complexity tends to blur the relationship between the capacity development input, and its outcome. This school of thought rejects any normative or deterministic approach and posits the pivotal role of process in which all stakeholders are involved in determining the objectives that are consistent with the capacity (e.g. Pahl-Wostl, 2002).

5 Finally, the behavioural and business management sciences have extensively researched behaviour of humans and of organizations. Sveiby (1997) for instance draws from this body of knowledge and also builds on Polanyi's dynamic concept of knowledge (see section 1.3.2). This approach helps to make the concepts more practical for the purposes of managing knowledge organizations, and for budgeting and results assessment, in contrast with the complexity approach which tends to dispute the notion that capacity can be managed.

Based on the above and for our purposes here, capacity can be defined as the capability of a society or a community to *identify* and *understand* its development issues, to *act* to address these, and to *learn* from experience and accumulate knowledge for the future. A country, then, has its "government" and public administration system as its formal tools to make this possible. Other less formal systems do exist as well. This definition emphasizes the linkage of capacity with a verifiable impact on-the-ground after the "act" as well as with the generation of fresh knowledge. It also points out that critical "extra" capacity is required to allocate the resources and incentive for continual learning and improvement that characterize the "learning organization'. Few governments or sector agencies, or non-governmental organizations for that matter, have provisions to allow for this learning.

This definition pertains equally to *individuals* – from technicians and community members up to ministers and politicians – and to the *institutions* in which they work and operate together and carry out their work. Overall performance thus depends on the simultaneous effect of capacity of the individuals as well as of the institutions. Developing countries currently tend to possess weak institutional and human capacity, with administrative systems that tend to be static and bureaucratic, and pre-occupied with technical aspects and standardized solutions. Especially local governments and local communities are at risk and may have little capacity to anticipate, and adapt to the changing demands and environment effectively. Notwithstanding, many of these local communities have generated over generations a body of traditional knowledge regarding the local conditions and how to cope with them. Tribal communities in the Andaman Islands in the Gulf of Bengal appeared better prepared to deal with erratic devastating events like the December 2003 tsunami than the urbanized and wealthier communities in west Indonesia and east India.

This weak capacity and knowledge impede the proper targeting and absorption of development funds and the sustainable operation and management of "feasible" investments. Feasible proposals are those that have been designed properly based on sound analysis and with respect to realistic outcomes, and that are embedded in a policy and administrative structure that is able to ensure sustained use and maintenance. Different from the situation up to the 1990s, the current experience in most developing countries suggests that it is the shortage of such "feasible" investment proposals that is impeding development, not any longer the shortage of funds *per se*.

1.3 SCHOOLS OF THOUGHT ON KNOWLEDGE AND CAPACITY

1.3.1 UNDERSTANDING KCD

KCD, and in a more general sense, the generation and dissemination of knowledge, take place through formal education, training, and institutional development. The KCD activity always implies a status change, and indeed, the capacity development in organizations and in the administrative and regulatory frameworks are irrevocably linked to change processes and reform (Alaerts et al., 1999, EuropeAid, 2005). The generation and dissemination of the knowledge can be carried out via different vehicles or modes, such as secondary education, tertiary education, learning-by-doing, learning from peers (mimicking), learning through formal and through social networks, purchase of patents, and the physical acquisition or import of individual specialists (and, in reverse, brain drain). However, these vehicles only concern the action of knowledge transfer itself, whereas in a policy or development context one should be equally concerned about the "indirect instruments" that help retain and improve knowledge and capacity. For example, financial and other career-related incentives in the Human Resources policy strongly influence the retainment and commitment of specialists and qualified staff, and make or break the organization's capacity. The appropriate definition of the objectives of a KCD initiative and the choice of the KCD process and instruments depend highly on the context, as the institutions they focus on are the exponents of a particular set of economic, social and cultural factors.

As argued in section 1.2, KCD has a broad remit, and several disciplines are studying parts of the subject, each highlighting an essential aspect and offering complementary and sometimes conflicting analysis and advice (Alaerts, 2009). One can distinguish a number of "schools of thought". For our case, with the aim of developing a well-performing water sector and extending advice to, amongst others, government, the insights of public administration sciences are of central importance. The formal water "sector" is largely embedded in, and is a part of the public administration (Figure 1.1). Still, as each individual in society must take decisions on water in his vicinity – from the mother in the household to the farmer – and elect his government, the achievement of any water sector policy depends highly on the individuals and communities in society, their representatives, the press, etc. Thus, many of the KCD processes equally pertain to the formal and informal organizations and processes in society. The way the organizations are functioning and are to be managed, is the subject of management and business administration sciences, and what has come to be known as knowledge management in the corporate sector. Several concepts are derived from political and institutional economics and sociology. Other contributing disciplines are educational sciences, pedagogy and didaxology. On a more fundamental level, the epistemology (the philosophy of knowledge and knowing) and the complex-systems analysis provide better understanding of the nature of KCD. These disciplines should be acknowledged by KCD practitioners as important sources of knowledge. Often, policy papers and working documents on KCD are not based on solid and disciplinary research but rather on prima facie observation and intuitive analysis.

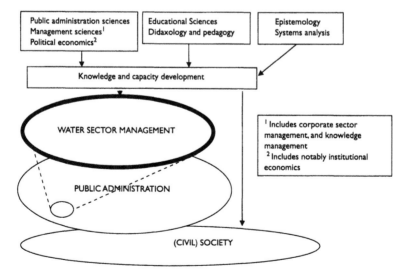

Figure 1.1 Disciplines that contribute to understanding of KCD in the water sector (Alaerts, 2009).

1.3.2   LINEAR APPROACH VS. COMPLEX SYSTEMS APPROACH TO KCD

A first important development in the understanding of KCD is the distinction that is made between a linear approach and a dynamic systems or complexity approach. In approaching complex dynamics in our daily life, and certainly for those complex societal processes, we generally tend to apply a reductionist approach. Complex problems become more easily understandable when they are cut up in isolated events that stand in "linear" cause-effect relationships to each other. However, the process of capacity development is very complex, and full of interconnections. The study of the nature of the parts does not necessarily lead to a correct understanding of the whole as the whole is more than the summation of the parts, since the parts interact. Furthermore, the relationships within the parts and among the parts may be causal under certain conditions but not under others. Some of the parts may have a strong influence on the internal relations within other parts. In addition, these interactions do not take place in a static environment and usually play out over time in sequences. With all interconnections between the components of the system, trying to obtain an overview of the system as a whole might be impossible because of the complexity but also because of increasing uncertainties associated with the interconnections. Complex systems are far more difficult to engineer successfully since potential outcomes can be difficult to identify, let alone measure (Parker and Stacey, 1995). Figure 1.2 illustrates the complexity of cause and effect for three individuals or organizations X, Y and Z. X as central actor discovers the actions and preferences of Y and Z, chooses how to respond appropriately and then acts. That action would have in turn consequences which Y and Z discover. This would lead each of them to choose a position or response which may differ from their original one, which, of course, has again consequences that X then needs to discover and respond to, and so on. This mutual interaction creates non-linear systems. X could be for example Water Users Organizations, and Y and Z a Ministry of Water Resources and local governments, that all are competing for authority and budgetary power. In the context of achieving a reform process of decentralization of management authority over water, it is in principle quite possible for the Organizations to predict what kind of knowledge and capacity development they will require. However, the Ministry and the local governments may have different perspectives and be subject to national policies, regulations and informal vested interests that may or may not preclude, obstruct or re-direct the reform. It is not difficult to imagine that in such non-linear system, it will be difficult to predict, and measure, what the outcome will be of capacity development interventions. In complex systems feedback can be used to monitor how incremental change takes place or what influence a certain measure is having. It can be described as an influence or message that conveys information about the outcome of a process or activity back to its source. Whereas prediction of outcomes in

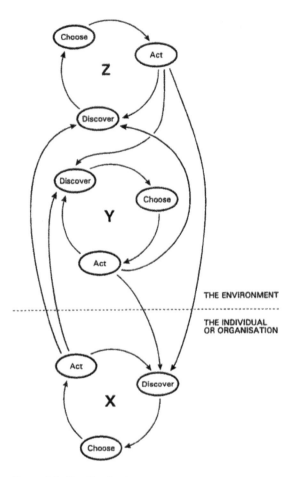

Figure 1.2 Feedback loops in complex systems (Parker and Stacey, 1995).

a complex and fuzzy system such as capacity development, thus, is difficult, "managed learning" as a way to improve systems capacity in an iterative way may be a more modest and realistic option. However, this "process approach" assumes continuous flexibility which conflicts with the desire for outcomes that are agreed upon up-front. For example, governments, and donors will be unwilling to allocate substantial funds in annual budgets for ventures where the outcomes cannot be defined well, and, by implication, where the efficiency and feasibility of the investment cannot be confirmed ex ante. This outcome-focus requires a certain amount of planning in the development processes. It is suggested that for government and accountability purposes a predefined goal and direction are needed, as well as a strategy to achieve this goal. This strategy should, however, be a flexible plan that can cope to a certain extent with unexpected influences and developments.

1.3.3 PUBLIC ADMINISTRATION

Water sector management is part of a nation's public administration. Proper understanding of this field is therefore a requirement to understand capacity development in the water sector. The field covers among others governance, public choice theory, management and human resources management. Farazmand (2004) describes what human resources management should consider in public sector organizations:

- Motivating people at work is of prime importance. Employees must see real purpose in public service, and have a sense of belonging to the organization.
- Compensation must be equitable to treat every citizen equally, offer equal opportunity and avoid any semblance of preferential treatment. The organization must be seen as impartial and objective. This objective is essential but in practice often hard to maintain. It also tends to become very inflexible and bureaucratic because the decision is determined by "the rule" rather than by considerations of overall effectiveness. The compensation should, however, also be efficient in order to prevent organizational brain drain and to attract the most competent talents to the public service.
- Organizational mobility and rotation are designed for strategic personnel to move periodically to new positions, to learn and apply their knowledge and experience. At the same time continuity in the organization should be guaranteed and skills mixes maintained.
- Personnel should have the chance to periodically refresh their knowledge and skills they need to manage, and function in, the organizations in this information age. This can be done by in-service training programs, seminars, conferences, and workshops.

The above objectives are laudable and appropriate, but implementation remains a major challenge in developing countries facing on one hand a dearth of qualified staff and budget constraints limiting the number of civil servants, and on the other hand a weak support in society for an impartial and meritocratic civil service that can be held accountable. Good governance implies transparency and accountability. Transparency of, say, budgets, plans and staffing arrangements, eliminates opportunities for corruption and closes the door to secrecy and abuse. Pro-active accountability mechanisms help serve as a strategic instrument in pursuit of excellence and trust, and hence legitimacy. The comparatively low compensation for employees in public sector organizations, and opaque structures and regulations in many developing nations are often an incentive to engage in corruptive practices.

1.3.4 KNOWLEDGE MANAGEMENT

A large proportion of the literature on "knowledge management" and "learning organizations" is geared at corporate businesses and firms. For the public sector the rationale for knowledge management is just as important, however, instead of the maximization of profit under conditions of competition there is a need for maximizing public service delivery at minimal cost, under pressure from society that demands a healthy public sector at low fiscal requirement.

For the water sector this would mean investing in the capacities that are needed for a well-performing sector, and working towards better governance, transparency and accountability. Knowledge management tools can be utilized drawn from human resources management and personnel management, and putting in place the objective and the administrative procedures that create a climate where people are encouraged to learn about their field and share their knowledge. A learning organization will have incentive systems that encourage knowledge, and internal procedures that facilitate, and require, open discussion, "peer reviews" and the valuation of knowledge. Information and Communication Systems (ICS – comprising both the technologies proper, and the interfaces and educated users who are able to utilize and work with these systems) are essential for example to develop databases and make information broadly available, and to facilitate communication and exchange of information. In reality, many water sector agencies notably in the developing world do not actively promote critical reviews and discussion, and staff careers are often based on performance in contract management for works and political relationships.

Drawing on epistemological analysis, knowledge can be both explicit (that what can be articulated in written form, for example, and formally taught) and tacit (such as the capability to ride a bike). Recognizing the value of the tacit knowledge, and figuring out how to use it, are the key challenges in a knowledge-creating company; they require extended communication, intensive and critical personal interactions among peers, and an knowledge enabling environment (Krogh et al., 2000). The enabling environment for bringing out tacit knowledge is a shared space that fosters emerging relationships. Tacit knowledge is to be considered more crucial and needs to be recreated from scratch by any new apprentice who is mimicking what the mentor demonstrates (Sveiby, 1997). As Tsoukas (2002) mentions, tacit knowledge can only be manifested and valued in what we do, and "that new knowledge comes about when our skilled performance is punctuated in new ways through social interaction".

What makes knowledge creation a fragile process is the fact that it requires individuals to share their personal beliefs with others about a particular situation. There is need for justification, explanation, persuasion and human connectedness. In organizations with a poor transparency or a culture of fear, knowledge creation will be difficult. Likewise, treating staff as conscripts or just implementers, and applying unnecessary bureaucratic burdens, significantly impede effective knowledge sharing (Schenk et al., 2006). For the water sector in different countries, a useful indicator for the knowledge facilitation is the presence of informal and formal networks among sector specialists and peers. In many countries, in particular developing ones, the absence or limited size and relevance of these networks is a barrier to knowledge generation and sharing. These networks are of great importance in the exchange of knowledge between government and other actors in the sector. Through formal and informal networks educational institutes are connected to the government and society to provide part of the "answers" through appropriate curricula and research. They also can have the overview of the human resources needs both in terms of quantity and skills mixes. Similarly, the informal and formal debates between the water sector and the civil society and its representatives and the press are critical to identify pressing issues in the sector and foster consensus on the action to be taken. In strongly hierarchic societies that tend to wield large bureaucracies and where the government basically takes all key decisions, knowledge creation and sharing is much less intensive.

1.3.5 EDUCATION AND TRAINING

Education and training is an obviously important aspect of knowledge and capacity development, especially at the level of individuals (see also Chapter 18 in this volume). Capacity development relies on the strengthening of individual capacity through training and learning, in order to raise the domestic stock of human capital, or the "social capital". Although some of the necessary skills would typically be acquired on-the-job or through learning-by-doing, developing countries characterized by less efficient organizations of work or by obsolete technologies might need to rely more on formal vocational education and training (Vincent-Lancrin, 2007).

There are four different types or levels of knowledge for which different methods of learning are appropriate:

- Factual knowledge ("water is boiling at 100°C").
- Understanding ("where does rain come from?").
- Skills (proficiency in a language, ability to work in a team).
- Attitudes (problem solving attitude, capability to approach a complex challenge, ambition, "gut feeling", and the drive to keep learning).

Skills generally can be acquired by training, whereas for the other three types of knowledge, education is more appropriate, partly through formal schooling. Training can be accomplished through apprenticeships and mentoring, seminars, workshops, classes, or through self-study. With training, a task analysis will yield a complete "step-by-step" list of what needs to be done to accomplish the skill being learned (Fabri, 2008). Training has therefore predefined content and is a closed system. Usually training needs a fixed amount of time – weeks or months.

Box 1.2 Effective training and education.

The World Bank evaluation on WBI training for capacity building (World Bank 2008) suggests a number of simple but essential success factors for *training*, namely:

- appropriate pedagogical tools and frameworks, classroom teaching, self-study, group work, etc.;
- support for the transfer of learning to the workplace, to consolidate what was learned by embedding it through on-the-job support; and
- targeting of training to organizational needs, anchored in a diagnosis of institutional and/or organizational capacity gaps, and formal assessment of participant training.

The early preparation to facilitate the application of what was learned in the workplace through action learning and practical exercises is of utmost importance.

Success factors for *education* include, to get the right people to become teachers and develop them into effective instructors (Fullan, 2007), (McKinsey, 2007). With regards to capacity needs for teachers, Fullan (2007) further states that a high level of success in the classroom requires personalization, precision and professional learning by teachers:

- Personalization is the capacity of a teacher to acknowledge what each individual student needs.
- Precision refers to the capacity to address these learning needs.
- Finally, learning by doing and mentoring by experienced colleagues are very effective educational and training methods.

Instructors require continuous learning and development, not only during their education, but especially during their working life. In general for training to be effective, it should be taken to the work floor, be integrated in operational practices and there should be room for learning-by-doing and for coaching by experienced colleagues.

1.3.6   INSTITUTIONAL CAPACITY BUILDING

The weak institutional capacity at the levels of the organizations and the enabling environment, are an equally important impediment to assimilate modern approaches in science, technology and management, essential to deal with the complex challenges in the water sector. The term institution is described by North (1990) as "the rules of the game in a society, the humanly devised constraints that shape human interaction". Institutions can take the forms of policies and objectives, laws, regulations, administrative rules, organizations, and norms and traditions. They determine how we think and what we do (North, 1990, 2005).

It is now conventional wisdom to say that strong institutions are the critical variable in development, not only in the sense of their quality, but also in their very capacity – for example as defined as their quantity of civil servants – to absorb initiatives and funds and prepare for future action. In the recent past several seminal studies have provided empirical documentation on the critical role of institutions (Fukuyama, 2004). Incremental change comes from the perceptions of the entrepreneurs in political and corporate organizations that they could do better by altering the existing institutional framework at some margin. But the perceptions crucially depend on both the information that the entrepreneurs receive and the way they process that information (North, 1990). One aspect of capacity development is to make sure the information received is as comprehensive as possible and to equip the entrepreneur with the right skills and mental constructs to process the information for the benefit of the organization and society. This will determine the transaction costs. Transactions costs are the costs of measuring the value of attributes of what is being exchanged and the costs of protecting rights and of policing and enforcing agreements. When the costs associated with an exchange are higher than the perceived benefits, there will likely be no transaction, meaning, things stay the way they are. Incomplete information, limited mental or institutional capacity by which to process the information and the risk of failure associated with this, determine the cost of transacting.

1.3.7 A SCHEMATIC OF KNOWLEDGE AND CAPACITY DEVELOPMENT

Figure 1.3 brings together the different elements of KCD. The diagram specifies in broad terms, for each of the three levels of action – the individual, the organization and the enabling environment – the sequence of what knowledge and capacity imply, by what means the knowledge and capacity development can take place, what the outcomes are and how these could be potentially assessed. These levels are "nested', that is, the individual operates embedded within his organizational context, and the organization operates within its broader context. The enabling environment itself is divided in that part that typically falls in the realm of the formal institutional frameworks, and in a second part referring to the broader context provided by society. (Civil) society itself is both part of this enabling framework and at the same time an actor in its own right, as within society numerous formal and informal non-governmental networks, associations and organizations take part in the broader game of water management.

The sector's performance derives from the effective action of individuals with the proper knowledge and capacity, who function in larger (sector) organizations (such as ministries, local governments, water user associations, civil

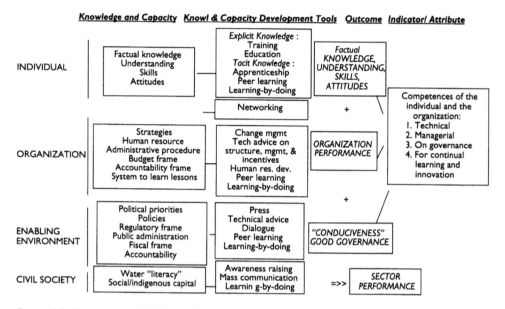

Figure 1.3 Schematic of KCD at different levels, indiating inputs and outcomes, and means for measurememt (after Alaerts, 2009).

society organizations, etc.). The effectiveness of these organizations depends both on the effectiveness of those individuals and on the typical features that shape the capacity of the organization itself through its skills mix, its internal operational and administrative procedures, etc. In turn, organizations with the right capacity and procedures still need an enabling environment to put in place the facilitating factors including an enabling legal and regulatory framework, financing and fiscal rules that stimulate proper action, and a broadly supportive political inclination in parliaments and among the voters and consumers. As mentioned above, KCD is essential to support and implement improvement of institutions or change in institutional arrangements. Often it is difficult to distinguish KCD proper from institutional development, and, indeed, KCD is embedded in and in effect helps shape, any institutional development and reform effort. KCD is part and parcel of change management.

Knowledge, understanding and skills are generally developed through typical knowledge transfer instruments such as education and training, however, whether the desired knowledge (or capacity) is explicit or tacit does make a difference in the choice of instrument. As Polanyi and Sveiby argued from their respective epistemological and practical management insights, tacit knowledge is eventually far more important as it shapes the skills and deeper attitudes. Tacit knowledge can best be transferred through one-on-one interaction between junior and senior, apprentice and teacher. Organizational capacity development is achieved by educating/ training the (staff) members and by helping the organization as such learn from the experiences from others. Technical assistance, management advice, learning experiences, comparison with peers and benchmarking, are important instruments. Both for individuals and for organizations, networks are playing an increasingly important role for generating, sharing, corroborating and improving knowledge and capacity. Networks – both formal and informal associations and "communities of practice" – are becoming the main mechanisms for professional improvement for many water professionals. ICS are powerful tools to support and intensify communication and open up new avenues for the dissemination of knowledge including best-practices.

At the level of the enabling environment, governments and other actors also "learn" and acquire capability to, in turn, become more enabling. Policy makers, governmental departments, and politicians, also increase their understanding about new challenges and solutions by, for example, drawing lessons from international "good practice". Technical advice, communication platforms and peer-learning are useful instruments for exchanges of information. In civil society, capacity already exists in the form of what is often called social capital or indigenous knowledge that resides in communities, and in water literacy. This can be further developed through comparison activities and peer-learning with other communities. The press and mass communication – such as Discovery Channel – are powerful vehicles to disseminate knowledge and develop attitudes, and reach many communities that otherwise receive little opportunity to communicate with the government. Non-governmental organizations and networks are important and often effective actors in this capacity development. Finally, the role of society is of course critically important as it shapes the nation's consensus about the priorities for the future (and the budget allocations) through electing its political representatives and holding its government accountable.

Eventually, "capable" individuals and organizations possess aggregate competences to act. Four types of aggregate competences can be distinguished. Firstly, technical competence is required to analyze and solve the problems that have a technical nature. For instance, all sector agencies need to regularly acquire new technical knowledge on an array of subjects from piling and construction techniques to climate change mitigation and adaptation. Second, organizations need to have an adequate pool of management competences embodied in their senior staff. In many developing countries sector agencies may score well on technical and civil engineering aspects, but often the competence to manage personnel and organizations, as well as the water resource itself, is modest. Third, an effective and performing water sector requires organizations that possess skills to foster and apply principles of good governance, such as dialogue and communication with stakeholders, resource allocation within policy frameworks that aim for equity and poverty alleviation, transparency and accountability. Finally, as mentioned before, capable individuals and organizations are those that manage, by deliberate decision, to keep learning and innovate. Learning and innovation do not come automatically but require financial resources and personal and managerial procedures to foster knowledge generation and sharing. This learning can emerge from an acquired attitude or natural inquisitiveness, but institutions may become interested in and coaxed into learning only after being held accountable for poor performance. Table 1.1 provides an overview of features and attributes related to these four competences.

Table 1.1 Examples of competences for each level.

	Individual level	Organizational level	Institutional level
Technical competence	Regularly updated knowledge and skills. Understanding of the broader technical context.	Appropriate knowledge and skills mixes for the services that are delivered, such as engineering, legal, financial, institutional knowledge. Knowledge on procurement and investmentprocedures.	Technical expertises and available skills mixes in a broader setting. Procedures for critical review and corroboration of knowledge and information.
Management competence	Project mgt skills. Financial mgt skills. Personnel and team mgmt skills. Mentoring skills. Understanding of political consensus building. Ability to 'deliver'. Leadership.	Leaders able to operate with goals and objectives as agreed with supervisory entities and main stakeholders. Ability to set goals, strategy. Financial management. People management. appropriate staff rotation; talent spotting, incentive systems, etc: Project management. Ability to 'deliver' timely.	Sound and workable task assignments of sector agencies. Minimal overlap between agencies, and size and task of agencies facilitate proper management and task execution. Sound financial, fiscal and budgeting systems. Facilitating proper management by organizations.
Governance competence	Understanding of procedures. Ability to engage with and listen to stakeholders. Ability to apply inclusiveness. Focus on results.	Transparent decision making processes. Procedures to consult with stakeholders, and provide empowerment to others. Procedures to be held accountable, including transparency in budgets and plans.	Distinction between 'operator' and 'regulator'. Procedures to ensure inclusiveness in particular regarding objectives, priorities and strategies. Procedures to ensure transparency and accountability.

(Continued)

Table 1.1 (Continued)

	Individual level	Organizational level	Institutional level
Learning competence	Desire to 'keep learning', readiness to critically reflect on own's performance. Availability for training and education in new skills and knowledge.	Readiness, and procedures, to critically review own's performance on a continuous basis, and revise if necessary. Goal, procedures and resources to support learning by staff, organization and if necessary other stakeholders. Support of 'communities of practice', and rewards for staff learning.	Procedures to promote open working atmosphere and critical reflection on performance. Openness to review sector performance on a continuous basis, and revise policies and arrangements if necessary. Foster inclusiveness.

1.4 TOOLS AND INSTRUMENTS FOR KCD

1.4.1 A STRONGER CASE

If sustainable development is our prime objective, then KCD should have the priority attention in developing and industrialized countries alike. Governments – for having legal authority and responsibility over the sector – can devise appropriate policies and take the main initiatives. Notwithstanding, an equally important role is to be played by civil society organizations and NGOs, and through the awareness and political vote of the electorate. The discussion below does not claim to be comprehensive, and focuses on the government role.

Investing in knowledge and capacity development pays off. Recent evaluations have demonstrated that development programs in the water sector are now more effective and sustainable than, say, before the mid-nineties. Many local communities have become less vulnerable to external upheaval or natural disaster. This higher effectiveness can be attributed for a large part to stronger institutions, better governance and more technical and managerial competence in the developing countries whose capacity has been strengthened. Several studies on irrigation, for example, have shown that nowadays the best return on the investment in canal improvement is achieved when a substantial effort is also placed in capacity enhancement, including empowerment, of irrigators and local-government officials. Of course, at a higher level of aggregation, the impact of education of the young generation and of communities on sector development and sustainability is without doubt. These impacts are being recognized. Although accurate figures are lacking, the amount of funds allocated to capacity development by donors seems to have increased about tenfold in the period 1995–2004 over the preceding decade. For instance, the World Bank provided for Sub-Saharan Africa between 1995 and 2004 about US$9 billion in lending and US$900 million in grants for the broader goal of capacity enhancement and education (World Bank, 2005). Across the globe, the Bank financed about $720 million annually over the past years for training activities for all sectors (World Bank, 2008).

1.4.2 STRATEGIES AND NEXT STEPS

But governments and other decision-makers have still a long way to go to align their administrative systems and sector policies to international best-practices, and put in place a knowledge-management system stimulating structured learning. The coordination between institutions must be deepened, and more structural capacity developed. Communication with the stakeholder groups from local communities to politicians is to become a priority, partly for awareness raising and education, but also to listen in and forge cooperation.

i   The fact that KCD is a formidable agenda by itself does not mean that meaningful KCD should always be carried out with such comprehensive ambition. KCD usually is more manageable and better targeted when carried out on a smaller working area or on a confined issue. Still, governments should at the same time start analyze their sector's knowledge and capacity weaknesses and outline a longer-time strategy comprising a series of steps.

ii As a first initiative to inform the policy and the next steps, the institutional and human capacities of the country's sector, or part thereof, should be assessed to define their strengths and weaknesses, and how robust the capacities are to deliver more effective services and prepare for future uncertainties. The assessment can cover a larger or smaller part of the sector (e.g. the management of river basins, basic sanitation, vulnerable communities, youth, etc.), or focus on a part of the overall institutional architecture and capacity (e.g. the education system, community management, or the legal framework). The SWOT analysis technique (Strengths-Weaknesses-Opportunities-Threats), results frameworks, risk analysis, stakeholder analysis, and similar techniques can be very useful to conduct the assessment (see Box 1.1–1.3). Such analysis always needs to be conducted together with all key stakeholders, including also non-governmental entities, and educational and training outfits besides academia. UNDP (2007) has reviewed and compiled the experience with capacity assessments and offers a rational framework for capacity assessment. It suggests that *core issues* to be assessed cover institutional development, leadership, knowledge, and mutual accountability. Critical *functional capacities* include, e.g., capacity to engage in multi-stakeholder dialogue, situational analysis, vision creation, policy and strategy formulation, budgeting, and monitoring and evaluation. It should be borne in mind that proper analysis requires substantial time and funds, as the assessment is specialistic and interactive and presumes adequate meeting and communication opportunities. Other best-practices are offered in a World Bank review of experiences in Africa (World Bank, 2005).

Box 1.3 **A check-list to assess the capacities at the sector level (Lopes and Theisohn, 2003).**

Human resources: Refers to the process of changing attitudes and behaviours–imparting knowledge and developing skills while maximizing the benefits of participation, knowledge exchange and ownership.

Job requirements and skill levels	Are jobs correctly defined and are the skills available?
Training/retraining	Is the appropriate learning taking place?
Career progression	Are individuals able to advance and develop professionally?
Accountability/ethics	Is responsibility effectively delegated and are individuals held accountable?
Access to information	Is there adequate access to needed information?
Personal/professional networking	Are individuals exchanging knowledge with peers?
Performance/conduct	Is performance effectively measured?
Incentives/security	Are these sufficient to promote excellence?
Values, integrity and attitudes	Are these in place and maintained?
Morale and motivation	Are these adequately maintained?
Work redeployment and job sharing	Are there alternatives to the existing arrangements?
Inter-relationships and teamwork	Do individuals interact and form functional teams?
Communication skills	Are these effective?

Organizational capacity: Focuses on the overall organizational performance and functioning capabilities as well as the ability of an organization to adapt to change.

Mission and strategy	Do the organizations have clearly defined mandates?
Culture/structure/competencies	Are organizations effectively structured and managed?
Process	Do institutional processes such as planning, quality management, monitoring and evaluation work effectively?

Human resources	Are the human resources adequate, skilled and developed?
Financial resources	Are financial resources managed effectively and allocated appropriately to enable effective operation?
Information resources	Is required information available and effectively distributed?
Infrastructure	Are offices, vehicles and computers managed effectively?

The enabling environment: Focuses on the overall policy framework in which individuals and organizations operate and interact with the external environment.

Policy framework	What are the strengths, weaknesses, opportunities and threats operating at the societal level?
Legal/regulatory framework	Is the appropriate legislation in place, and are these laws effectively enforced?
Management/accountability framework	Are institutional responsibilities clearly defined, and are responsible institutions held accountable?
Economic framework	Do markets function effectively and efficiently?
Systems-level framework	Are the required human, financial and information resources available?
Process and relationships	Do the different institutions and processes interact and work together effectively?

Lopes and Theisohn (2003) offer a useful check-list to help assess a sector's capacity by analyzing the three levels of individual (or human resources) capacity, organizations (or institutional) capacity and the capacity of the enabling environment (Box 1.3).

From this assessment, a strategy and action plan can be derived. The strategy should be shaped contextually through dialogue and stakeholder involvement. Because the environmental, social-economic and cultural contexts differ between countries and sub-sectors, there is no "one size fits all" strategy. Such process cannot be imposed from the outside, and requires a home-grown demand and political commitment. However, often the capacity assessment and its development help to make the case and demonstrate the benefits, and at the same time help develop capacity. The process often turns out to be slow, incremental and patchy. Addressing weak institutional environments is not a straightforward or "linear" process but often works best through "strategic incrementalism", i.e., pragmatic incremental reform steps that may not fully address all the current institutional performance problems but can alleviate some acute problems while at the same time creating the conditions for deeper and more favourable change in the longer run (World Bank, 2008). In other words, one should continually adapt the approach as specific new opportunities arise along the way.

iii The water sector is knowledge intensive. This calls for much investment in the creation of new knowledge, through research and innovation. On the other hand, a lot of knowledge to guide local action resides within the traditional knowledge of local communities, and is often untapped or dismissed. Knowledge on water tends to be available in fragmented form among a growing number of actors who each hold part of the solution. Communication, therefore, is becoming increasingly important in building the knowledge base and the institutional and human capacities; to disseminate and acquire knowledge from across the sector; and to forge political consensus in society. Precisely because of this complexity and the distributed nature of sector knowledge, a judicious balance needs to be sought between centralized sector management and collaborative arrangements among decentralized entities.

ICS is a powerful tool to support and intensify communication and open up new avenues for the dissemination of knowledge including best-practices. ICS is also becoming the key instrument to forecast with greater precision the future consequences of current decisions and policies in development scenarios, and to reduce the uncertainty related to climatic variability. The access to relevant data facilitates decision-making and in general increases transparency and governance. Thorkilsen (2001) and Abbott (2007) describe how ICS has been at the core of the broad knowledge-sharing effort that was necessary to create the political and societal support for one of Europe's largest and controversial infrastructure projects, namely the rail and road bridge and tunnel across the Øresund strait connecting Denmark and Sweden.

Decision-makers have nowadays an array of tools and "knowledge pools" such as data bases, research and educational centres, consultants, etc., at their disposal to develop or enhance capacity and facilitate knowledge generation in those areas and with those actors whose low capacity is considered a key constraint.

iv The strategy generally is implemented through a combination of education; training; technical advice for institutional strengthening and change; institution of appropriate incentives and procedures that encourage staff to seek innovation and learn; facilitation of research and innovation, and of communication and interaction. Given that much of the sector's education is actually carried out by the educational establishment – through polytechnics, and a variety of university-level studies – sectoral decision-makers would do well by engaging in a dialogue with these establishments to ensure that sectoral and educational perspectives are aligned. The generation of professionals is in many developing countries a prime concern. In many countries nominally adequate numbers of graduates are churned out by universities and polytechnics, but their specializations and skills may not be attuned to the modern challenges and the expectations from the societies they are supposed to serve. "Education" should also be understood to go well beyond tertiary education, and society at large and youth in particular need to be educated about water and how it impacts their future. Finally, "education and training" encompasses a wide gamut of instruments spanning from conventional class-room teaching to Objective-Based Learning, hands-on learning and mentoring. Recent comparative studies have shed new light on which approaches are likely to be most effective. In this fields too, ICS has opened up powerful new tools for real-time access to data and teachers across the globe, and communication among peers.

v Governments can encourage the development of systems to generate and share knowledge among "centres of knowledge", as well as between those who are in need of knowledge and those who possess it. As highlighted above, with the advent of ICS and globalization, governments can create and actively fund "communities of practice", networks of professionals and institutions, and databases, on the relevant subjects. In such networks, local governments and communities are not to be left out – they are always the first at risk, but they also often hold a lot of traditional wisdom. Some such networks preferably should also be international, both south-south and south-north.

vi A special challenge concerns the facilitation of institutional strengthening and KCD in civil society. Although many governments may not find this obvious or against their mandate, civil society needs to receive special attention based on the following considerations.

- With the large numbers of stakeholders in water management, governments will increasingly depend on informed and "capacitated" actors in civil society to play a growing role in water management, over and beyond the role they are already playing.
- As the technical agencies at national as well as at local government levels depend on budgets voted by Parliament or Councils, it is essential that their staff is better able to make the case with politicians to secure the budgets. Eventually, this also means that civil society too needs to feel more strongly about the priority for water, and lend its support to water sector initiatives.
- Climate and other changes need to be better forecasted, and remedial actions identified, agreed upon and taken. Much of this implies non-technical measures as well as some technical measures at local level to strengthen preparedness and resilience of local settlements and their

arrangements for agriculture and for natural resource management. This is of special relevance in light of climate variability and other environmental changes.

- Civil society has the right to hold government accountable for delivery on its policies. To enhance governance, civil society therefore should have suitable capacities, have access to relevant information and be able to engage with (local) government on service delivery.
- Civil society as "user", on the other hand, needs to be able to involve itself in some decision making at local, river basin and national level, to help decide on priorities in spending and water management, and in such a way that at the same time it is aware of the costs involved in the options, and agrees to provide for the finance.

vii Finally, it should be recognized that "learning" is a continual effort. Governments can put in place the procedures, institutions and incentives to ensure that lessons are learned, documented and disseminated. Each time an issue has been addressed, or a particular action carried out more effectively, this information should be fed back to stimulate further improvement.

1.5   MEASURING KCD

1.5.1   INTRODUCTION

Little agreement exists about how to identify and measure KCD. Some sources tend to focus on the measurement of components of KCD only, such as the individual capacity or the organizational capacity. Others prefer to focus on measuring performance as a proxy for capacity, probably because it is more appealing to look at apparent results than to look at the mechanism of KCD itself. Performance however, depends on many other factors beside

Table 1.2 Three types of evaluation (Hospes, 2008).

Evidence-oriented evaluation	Realistic evaluation	Complexity evaluation
Measuring effects	Investigating "black boxes"	Exploring complexity
No use of policy theory	Use of policy or program theories on what happens in "black boxes"	Starting point is that policymaking is dynamic and interactive
Programs stand at the beginning of a results chain	Programs are "black boxes"	Programs are adaptive systems
Input => output => outcome => impact	Mechanism + context => outcome	Outcomes are emerging and quite unpredictable
One-way and single cause-effect relationships	One-way and multiple cause-effect relationships	Two-way and multiple cause-effect relationships

the current capacities of individuals and organizations, and there is no direct causal relation with KCD; one should be cautious.

Many attempts have been undertaken to develop frameworks for the measurement of KCD, and numerous capacity assessment tools exist; UNDP summarizes 20 just for organizational capacity measurement (UNDP, 2005). Several web-sites are specialised in capacity assessment. Mizrahi (Chapter 20) concludes that not one study offers indicators, benchmarks or measurement tools that can be used across regions, or indicators that assess all relevant levels of KCD.

1.5.2   APPROACHES IN MONITORING AND EVALUATION (M&E)

The various ways in which KCD is assessed can usually be categorised in two broad schools. One school advocates approaches based on results-based management; the other favours evaluation based on the complex-systems theory and a participatory approach.

Hospes (2008) has created a comprehensive overview based on three approaches (Table 1.2). The most commonly used evaluation method in development cooperation is evidence-oriented, as is described in the left column. This approach focuses on the impact of a given intervention. It assumes a linear, causal relation from input to impact. This type of evaluation is convenient when working with the short-term project goals and time slots, as donors often do. Realistic evaluation however, believes that the relation between input, output, outcome and impact is not linear. What happens in the process from input to impact, in the so-called "black boxes" should be investigated. The art in realistic evaluation is to map the dynamics in a program: social interaction and institutional influences that determine the relation between input and output.

Critics of realistic evaluation point out that the context should not be perceived as something external, but as part of the process. As a reaction to realistic evaluation, complexity evaluation is evaluating how policymakers themselves respond to complex problems. It wants to capture the dynamics of the context as well. Hospes (2008) states that for the evaluation of complex systems different approaches need to be brought together, based on adaptive systems, institutional dynamics and assigning meaning. The evaluation of complex systems is rooted in various schools of thought, some of which are described earlier in this article, and will receive more elaborate attention elsewhere in this volume.

1.5.3   ADAPTIVE SYSTEM MANAGEMENT

One management approach that is working according to the principles of complexity evaluation and the institutional dynamics in complex systems, is adaptive management (AM). AM is defined as the "integration of design, management, and monitoring to systematically test assumptions in order to adapt and learn" (Salafsky et al., 2002). AM originates from ecosystems research and has combined this with research on the social and institutional dynamics in a system. The objective of AM is to better manage the impact of an intervention, by investigating what happens in the "black boxes", and at same time it takes along the social and institutional dynamics of the system. An important component is the involvement of stakeholders in the process from beginning to end, to capture their knowledge and use it to arrive at a set of management options. New knowledge about the connection and interaction between the social and cultural system and the ecological system has to be generated, through extensive consultation and participation of stakeholders. The next step is how to use that knowledge for policy making. An important advantage of AM is that the process consists of step-by-step learning, iteratively allowing to monitor what worked and what not, and to adjust the intervention process. This facilitates easier and early evaluation of mistakes and successes, albeit that this monitoring system would likely involve high expense for the repeated consultations. Chapter 5 will discuss the principles of AM more elaborately.

NOTE

The opinions expressed in this paper are those of the author and do not necessarily reflect those of the World Bank.

REFERENCES

Abbott, M. 2007. Managing the inner world of infrastructure. *Civil Engineering* 160: 26–32.

Alaerts, G.J. 1999. Capacity Building as Knowledge Management: Purpose, Definition and Instruments. In: Alaerts, G.J., Hartvelt, F.J.A., Patorni, F.-M. (eds) Second UNDP Symposium on Water Sector Capacity Building Balkema Rotterdam Brookfield, Delft.

Alaerts, G.J. 2009. Knowledge and capacity development (KCD) as tool for institutional strengthening and change. In: Alaerts, G.J., Dickinson, N. (eds) *Water for a changing world – Developing local knowledge and capacity,* Taylor & Francis, London.

Alaerts, G.J., Blair, T.L. and Hartvelt, F.J.A.(eds.) 1991. *A strategy for water sector capacity building: proceedings of the UNDP Symposium, Delft, 3–5 June 1991.* IHE UNDP, Delft, The Netherlands.

Alaerts, G.J., Hartvelt, F.J.A. and Patorni, F.-M.(eds.) 1999. *Water Sector Capacity Building: Concepts and Instruments.* Balkema, Rotterdam.

EuropeAid. 2007. *Institutional Assessment and Capacity Development. Why, what and how?* Tools and Methods Series, Office for Official Publications of the European Communities, Luxembourg.

Fabri, P.J. 2008. Is there a difference between education and training? http://www.facs.org/education/rap/fabri0408.html. Cited 19 August 2008.

Farazmand, A. 2004. Innovation in Strategic Human Resource Management: Building Capacity in the Age of Globalization. *Public Organization Review* 4: 3–24.

Fukuyama, F. 2004. *State-building. Governance and world order in the twenty-first century.* 1 edn. Cornell University Press.

Fullan, M. 2007. Change the term for teacher learning. *Journal of Staff Development* 28: 2.

Hildebrand, M.E. and Grindle, M.S. 1994. *Building sustainable capacity: challenges for the public sector.* Harvard Institute for International Development, Harvard Univerity, [Cambridge, Mass.]

Hospes, O. 2008. Evaluation Evolution? The Broker IDP, Leiden, p. 5.

Krogh, G.V., Ichijo, K. and Nonaka, I. 2000. *Enabling knowledge creation: how to unlock the mystery of tacit knowledge and release the power of innovation.* Oxford University Press, New York.

Lopes, C. and Theisohn, T. 2003. Ownership, leadership and transformation. Can we do better for capacity development? Earthscan publications, London, p. 355.

McKinsey. 2007. *How the best performing school systems come out on top.* Series. McKinsey. 56.

Morgan, P. 1993. Capacity Building: an overview. Paper presented at the Workshop on Capacity Development, Ottawa, November 22–23, 1993.

Morgan, P. 2005. *The idea and practice of systems thinking and their relevance for capacity development.* Series. 33.

North, D.C. 1990. *Institutions, Institutional Change and Economic Performance.* Cambridge University Press.

Pahl-Wostl, C. 2002. Towards sustainability in the water sector – The importance of human actors and processes of social learning *Aquatic Sciences* 64: 17.

Parker, D. and Stacey, R. 1995. *Chaos Management and Economics. The implications of Non-Linear Thinking.* 2 edn. Institute of Economic Affairs, London.

Salafsky, N., Margoluis, R., Redford, K.H. and Robinson, J.G. 2002. Improving the Practice of Conservation: a Conceptual Framework and Research Agenda for Conservation Science. *Conservation Biology* 16: 1469–1479.

Schenk, M., Callahan, S and Rixon, S. 2006. Our take on how to talk about knowledge management. http://www.anecdote.com.au/whitepapers.php?wpid=6. Cited 25 July 2008.

Shafritz, J.M. 1985. *The Facts on File dictionary of public administration.* Facts on File, New York, N.Y.

Sveiby, K.E. 1997. *The new organizational wealth: managing & measuring knowledge-based assets.* Berrett-Koehler, San Francisco, Calif.

Thorkilsen, M. and Dynesen, C. 2001. An owners view of hydroinformatics: its role in realising the bridge and tunnel connection between Denmark and Sweden. *Journal of Hydroinformatics* 3: 30.

Tsoukas, H. 2002. Do we really understand tacit knowledge? Paper presented at the Knowledge Economy and Society Seminar, London, 14 June 2002.

UNDP. 1993. *National Capacity Building: Report of the Administrator.* Series.

UNDP. 2005. *A brief review of 20 tools to assess capacity.* Series. UNDP. 12.

UNDP. 2007. *Capacity assessment methodology – user's guide.* Series. UNDP. 77.

Vincent-Lancrin, S. 2007. *Cross-border tertiary education: a way towards capacity development.* OECD; World Bank, Paris.

WorldBank-IEG. 2008. *Using Training to Build Capacity for Development – An Evaluation of the World Bank's Project-Based and WBI Training.* Series. WorldBank. 119.

WorldBank. 2005. *Capacity Building in Africa. An OED Evaluation of World Bank Support.* Series. 85.

Zinke, J. 2006. *Monitoring and Evaluation of capacity and capacity development.* Series. 38.

Part 2

Tools and techniques

Chapter 2

E-LEARNING FOR CAPACITY DEVELOPMENT

Colin Mayfield & Velma I. Grover
United Nations University – International Network on Water Environment and Health (UNU-INWEH), Canada

ABSTRACT

E-learning means that electronic media and methods are used to enhance teaching and learning but it is often used to describe Distance Learning because of the common use of E-learning techniques in Distance Learning. Although not a monolithic model, Distance Learning implies asynchronous learning during which the participants study at their own rate and at various convenient times; the delivery mechanism for the course materials is not the defining issue. The tools and instruments used for E-learning in capacity development in the water sector have ranged from the standard and very common short workshop training sessions using E-learning tools to full, web-enabled, interactive courses of various types. E-learning can be synchronous or asynchronous or a combination of the two. Whether or not E-learning methods are used to produce courses for traditional face-to-face instruction or for Distance Learning, the curriculum and courses must be carefully designed and managed to make the learning experience a successful one.

Technology has only a small part to play in this equation since the experience of the student taking courses is only partially affected by the technology; it is more important that the curriculum design and presentation be sound, the communication between and within the groups involved be effective, and the learning experience be satisfactory in the students' many different environments. This, in turn, implies that effective monitoring and evaluation are performed during or at the end of the courses. In many cases, only some of these expectations have been met in the various courses offered in the water sector. The effectiveness of many courses and training processes has not and, sometimes, cannot be measured in terms of student progress or development.

The development of regional and local expertise in capacity development, specifically designed to replace and improve upon the course delivery method using "visiting experts", has not been a strong focus of many capacity-development efforts. It is also apparent that capacity development in the water sector cannot be totally successful without equal attention being paid to all sectors: educational, training, institutional development, technical, scientific, administrative, management, legal, socio- economic, governance, cultural and political.

2.1 INTRODUCTION

E-learning has been variously defined over the years. A common definition is one that specifies that electronic media and methods are used to enhance teaching and learning. E-learning can be synchronous or asynchronous or a combination. Similarly, Distance Learning does not have to be one monolithic model, but does imply asynchronous learning during which the participants study at their own rate and at various convenient times; the delivery mechanism (electronic or traditional) for the course materials is not the defining issue. The evolution of Distance Learning reflects the increasing role of electronic media delivery in teaching in general; the first Distance Learning involved mailing printed materials to students (the "correspondence course"), then audio tapes of lectures with that printed material, then videotapes and graphic materials finally arriving in a completely electronic delivery mode via CD-ROM, DVD and online materials. The widespread use of E-learning methodologies in Distance Learning means that any study of E-learning necessarily involves examining Distance Learning methods, procedures, outcomes and assessments.

E-learning in the water sector has followed this same evolution (Figure 2.1): incorporation of electronic media in conventional teaching methods (simulations, models, PowerPoint presentations, etc), conversion of lecture materials to electronic formats for delivery to classroom or to Distance Learning students, conversion of these electronic materials to more organized formats in simple CD-ROM or web-based delivery systems, and finally, the development of complex presentation tools, student tracking, communication and assessment tools together with electronic repositories of reference materials and a social networking capability. Most also include a comprehensive course management system. There are examples of most of these stages in operation today using both commercial and open-source systems.

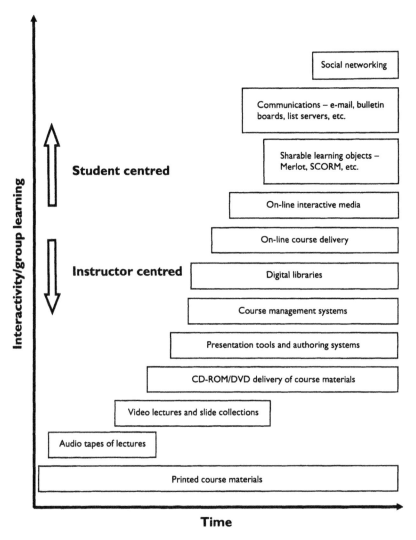

Figure 2.1 Evolution of E-learning: Time versus Interactivity/group learning.

The present models of E-learning are the most interactive, using electronic means to help with the delivery of course teaching and reference materials, assignments, tutorials, and with interactive assessments of progress. Cooperation and communications between students and with faculty members, establishing "communities of interest", encouraging knowledge sharing, producing electronic portfolios of student activities, and generally increasing the role and impact of social networking within the student population are other common functions. These functions are a close match to the observed uses of computer technologies by North American students – they want to be connected, entertained and able to present their own work or opinions. This is reflected in the most popular software sold in four U.S. campus bookstores studied by Zemsky and Massy (2004). After Microsoft Office and anti-virus software, the most popular titles purchased were photographic editing programs and web-development software. Social networking and messaging software is ubiquitous and free.

The history of E-learning is full of over-optimistic projections, false starts and failed systems. This is exemplified by assessments such as by Zemsky and Massy (2004) who address the issues of "Why the E-learning Boom Went Bust" and state that early attempts were often based on false premises such as, "If we build it they will come". Other assumptions were that students would inherently and immediately accept and like E-learning and that "E-learning will force a change in the way we teach". They point out that all of those premises were wrong: the range of E-learning software and materials is so variable that no dominant design type has taken root (including the idea of learning objects); the students are often ambivalent about E-learning – and many do not like it – and that ambivalence is shared by faculty members, even those who were early adopters; and, finally, the standard teaching function has proved remarkably resilient and resistant to change brought about by E-learning technologies. Others, such as Baggaley (2008) are now revisiting this somewhat pessimistic assessment. In an editorial in the International Review of Research into Open and Distance Learning entitled "Ideas whose time has come back", his assessment is that "a new world of Distance Education is coming together in the developing world as old media such as the radio and the telephone merge with each other and the Internet to form wholly original interactive partnerships".

This chapter will deal with the application of the various techniques and instruments or tools used in education and training in the water sector and will provide an overview and assessment of the application, uses and the relative success and problems associated with these methodologies or their application.

2.2   CURRICULUM AND CONTENT DEVELOPMENT

Curriculum development for courses or programmes is the complex of decisions that goes into selecting course topics and their organization into a coherent whole. This is often an individual set of choices by an instructor, but some programmes use a committee approach with some means of coming to a collective decision on these issues (Mayfield et al., 2004). Neither is necessarily superior in any given case, but some interesting interactions can occur between experts trying to fit their specialty into a broader course structure. This is especially true when dealing with the multidisciplinary or interdisciplinary issues that often occur in the water sector.

Developing excellent content for E-learning is the more intractable problem compared to the development of technologies for delivery of that content. Early efforts at repositories of reusable "learning objects" (e.g. Merlot at http://www.merlot.org/merlot/index.htm) have not developed the critical mass that has enabled them to set design standards for reusability. They are useful as simple repositories for simulations, animations, graphics materials, case studies, image banks and reference materials. A common criticism is that the learning objects are too specific and too closely linked to a particular "style" of teaching to be broadly useful. There are many exceptions to this, but it seems as if course authors still prefer a common "look and feel" to course or programme materials.

There have been attempts to set up standards for sharable content and the most prominent is SCORM or Sharable Content Object Reference Model. It is a "collection and harmonization of specifications and standards that defines the interrelationship of content objects, data models and protocols such that objects are sharable across systems that conform to the same model" (http://www.adlnet.gov/scorm/) and these standards and specifications for web-based E-learning define communications between client-side content and the host system. The latest version, SCORM 2004, introduces a complex idea called sequencing, which is a set of rules that specify the order in which a learner may experience content objects. In general, these sharable learning objects have been used when mandated but have not been widely used in other situations. It may be that course authors are hesitant to use materials from others unless they match the style of their own course presentation or are generally recognized as superior to other material. Other factors, of course, could be related to intellectual property rights in the course and simple job protection. A good example of excellent material available for sharing is the "Inner Life of the Cell" by Biovisions at Harvard University (http://multimedia.mcb.harvard.edu/), a superb animation of white blood cells and their cellular functions.

2.3   ASSESSMENT OF E-LEARNING

What has been missing from many E-learning efforts has been a rigorous appraisal of the outcomes and results of such programmes. The vast majority of programmes only offer anecdotal appraisals or assessments based on a selection of student responses to questionnaires or surveys. In a review of Distance and E-learning in the U.S., Mayes (2004) examined previous summaries of research (Merisotis and Phipps, 1999; Carnevale, 2000). He then continued with a review of more recent literature and concurs with the assessment by Phipps and Merisotis (1999) that "the vast majority of articles on distance education were opinion pieces, how-to articles and secondhand reports with no quality research basis". They identified only 40 articles that they could classify as original research, and they were primarily concerned with three broad measures of effectiveness: grades and test scores as measures of student outcomes, student attitudes, and overall student satisfaction with distance education. In general terms, from these studies the outcomes of Distance Learning were comparable with classroom-based instruction, but many of the studies failed to account for extraneous variables, did not randomly select respondents, and failed to control for attitudes and beliefs of students or faculty.

In an ongoing meta-analysis by Bernard et al. (2004) of 232 studies selected for content from studies on Distance Learning between 1985 and 2002, the achievement, attitude, and retention outcomes were analysed. Overall results indicated no significant differences between all three measures and wide variability in results. Synchronous applications favoured classroom instruction, while asynchronous applications favoured Distance Learning. However, there was still significant heterogeneity in the subsets. Many studies have found "no significant difference" between Distance Learning courses, E-learning courses and traditional ones (Russell, 1999). It seems that E-learning courses undergo more rigorous assessments of success than do equivalent classroom-based courses – there may be a (hidden?) systematic bias towards proving that classroom-based courses are superior.

Perhaps the appropriate question should be "what are the characteristics of a good course?" and not a question that implies there is some significant difference between E-learning and traditional modes of teaching and learning that has to be discovered.

2.4 WHAT ARE THE CHARACTERISTICS OF AN EXCELLENT COURSE OR PROGRAMME OF COURSES?

An excellent course or programme of courses should have proven excellence in the curriculum development, production, selection, quality, scope, organization and presentation of course material by experts in that field. These functions can be separated – presentation standards can be made uniform across the material by graphics or information technology specialists, and curriculum development may be improved by an individual expert or through consensus among a number of experts – but the basic curriculum and content production, selection, quality, scope and organization should remain unaltered.

An equally important criterion, and one that should precede or at least be coincident with the factors listed above, should be the "match" of the material and its delivery system to the needs and requirements of the intended audience. This means that some form of needs analysis should be performed and that rigorous testing of materials and technologies is needed. This is normally initially performed on subsets of the intended audience, but should lead to an ongoing review and assessment of each instance of the course or programme (see Figure 2.2).

For E-learning, the choice of delivery system (proprietary or Open Source) is the lowest priority, but often assumes primacy in the process and then dictates or influences the curriculum design and course production stages. This happens because of financial pressures to cut costs by using free software, to limit costs by using existing installations of particular software, or is driven by ideological factors such as a strong preference for Open Source software in an organization or group. All are valid reasons but should be carefully examined to ensure they do not overly influence the curriculum or content development.

There are many other factors that can play a role in the type of course/ programme production and dissemination mechanism chosen. They may be summarized as:

- Course or programme budget
- Style and intent of the course: e.g. group learning, knowledge sharing, community of practice, training or information delivery
- Central or distributed delivery and management of the course/programme and the allocation of central, regional or local responsibilities for course content, case studies or course customization
- Synchronous or asynchronous delivery (or both)

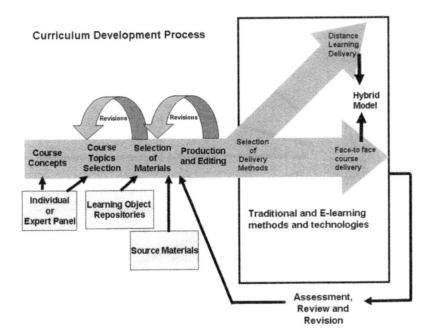

Figure 2.2 A typical curriculum development process.

- Use and re-use of pre-existing materials for course production
- Requirement for compliance with standards (SCORM, etc.)
- Need for digital library facilities or delivery of reference material
- Technologies available to the recipients at reasonable cost
- Degree of interactivity, communications and collaboration needed with the recipients, and the consequent need for instructors
- Assessments, examinations (distributed or centralized) and assignments required for the course
- Types of certification or qualification granted for successful completion of the course or programme
- Types and extent of course surveys and recipient course assessments needed.

Most distance education courses are taken for one of two reasons: to obtain certification or another type of qualification, or to improve knowledge and performance of the student. This is true in both developed and developing countries. One way to examine the characteristics of an excellent E-learning course is to consider the change in the dimensions of learning (Grunwald, 2007) between it and the traditional classroom model (Figure 2.3).

The excellent E-learning course will use student-centred methods, involve the students in the course in many different ways (e.g. interactive materials in the course content, list-servers, news feeds, social networking, e-mail, group

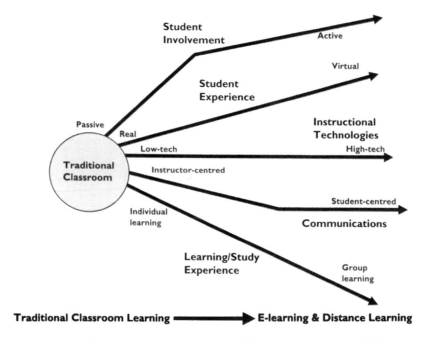

Figure 2.3 Dimensions of Learning (modified from Grunwald, 2007).

collaborations, student-instructor communications), and emphasize group learning versus the instructor-centric model. In E-learning, the course should provide appropriate technology for course delivery, taking into account the available technology and cost constraints for the students, and also provide a virtual environment with many interactive links to other students. Collaborative work is difficult for students in many surveyed courses and they often resort to individual efforts "pasted" together as collaboration

In the final analysis, the quality of the materials provided is a key factor and a high degree of interactivity (peer-to-peer and student-to-instructor) seems to help learning processes. Anything (technological, procedural, or academic) that impacts these, impacts the perceived course quality and outcomes.

2.5   SOFTWARE FOR E-LEARNING

The evolution of E-learning software has followed closely the capabilities and properties of larger and cheaper storage density (floppy disks, CD-ROM and DVD) and faster and cheaper communications technologies. This trend will no doubt continue as technology advances and becomes cheaper, communications costs (including VoIP and wireless) decline, interoperability improves and the convergence of all media (books, newspapers, magazines, journals, music, photographs, TV, and radio) to digital formats continues.

The software specifically designed for E-learning and Distance Learning includes both proprietary and Open Source (http://www.opensource.org/) solutions. Proprietary solutions include ANGEL (http://www.angellearning.com/), Blackboard (now incorporating WebCT – http://www.blackboard.com/us/index.bbb) and Desire2Learn (http://www.desire2learn.com/). Open Source solutions include Moodle (http://moodle.org/), Claroline (http://www.claroline.net/worldwide.htm), OLAT (www.olat.org) and ATutor (http://www.atutor.ca/). Most of these include some type of Learning Management System (LMS) to assist with the routine management procedures (marks, assignment delivery, examinations, etc). Other software solutions have been assembled from software for web delivery, messaging, bulletin boards, list-servers and databases normally using Open Source solutions such the Apache web server, MySQL and software from Sourceforge.net. A more recent trend is the use of "Web 2.0" components delivering web-based services and software and enhanced interactivity.

2.6   E-LEARNING IN THE WATER SECTOR

If we examine the water sector in particular, it appears that there have been, and continue to be, significant efforts to implement Distance and E-learning, but it is spread over many different operations from many different authoring and delivery groups. The courses range from very short technical courses presented on-site to small groups with face-to-face teaching using elements of E-learning, all the way to large multi-centred distance and E-learning programmes delivered through the Internet with complex, interactive group communications and social networks. Bourget (2008) has pointed out the particular difficulties in designing and operating a Masters Degree course in Integrated Water Resource Management (the extreme interdisciplinary nature, the conflicts between "hard" and "soft" sciences, the need for training guidelines from practitioners, the need for partnerships, etc). E-learning courses are delivered in both the developed and developing countries to a wide range of student types, aptitudes and professional backgrounds.

The challenge for E-Learning programmes increases when there are integration issues with a graduate programme in water (IWRM in particular). Major (1997) asked "Why is there still a need for an integrated approach ... when water professionals have been discussing the integration of disciplines since at least the 1960s." Some reasons are that the planning process has favoured "integrated multidisciplinary" approaches, which involved people working independently according to their disciplinary perspectives and then, only later, attempting to integrate the results (Kirshen et al., 2004); that there is a lack of national water policies (Stakhiv, 2003) and that academic institutions are traditionally organized in disciplines. Although multidisciplinary programmes are evolving, it is still a challenge to develop and offer them.

When the goal is to update the skills of people already working in the field, the challenges of capacity development are slightly different from the ones for degree or postgraduate courses. The aims should be to make learning and training resources accessible to all, to facilitate diligent, persistent, goal-based learning, to provide one-on-one tutor support for trainees and to evaluate and acknowledge the learning that has been done. Many capacity-development activities meet one or two of these challenges, but not all of them.

The development of regional and local expertise in capacity development, specifically designed to replace and improve upon the course delivery method using "visiting experts", has not been a strong focus of many E-learning capacity-development efforts until recently.

There has been a lack of real assessment of the outcomes of most such efforts in the water sector (and others) except for evaluations by students taking part in the programmes. Such assessments should ideally be designed by professionals in evaluation methodologies, be statistically valid and include a measure of real outcomes, i.e. changes that occurred because of taking the courses. Students should certainly be surveyed for their opinions of the courses, but a later survey conducted a few years after completion might reveal different views and assessments. The students' employers, supervisors or managers may also have some valuable information about the real outcomes of the courses.

If these assessments are performed, then future E-learning in the water sector can be properly aligned with the needs of the students, their abilities and backgrounds and their future career needs and goals. The choice of appropriate technologies to deliver these programmes is not critical as long as it provides the ease of use, information delivery, organization and degree of interactivity that is required. The course content (scope, organization, amount and quality) is the critical element deserving careful curriculum design and content production.

2.7   A CASE STUDY OF AN E-LEARNING PROGRAMME

The Water Virtual Learning Centre (WVLC – at http://wvlc.uwaterloo.ca) is an E-Learning and Distance Learning course programme developed by the United Nations University – International Network on Water, Environment and Health (UNU-INWEH), in conjunction with the United Nations, Department of Economic and Social Affairs (UN-DESA), in Integrated Water Resource Management (IWRM). The programme has attempted to incorporate some of the principles of E-learning discussed above. It is a ten-course, comprehensive programme leading to a Diploma in IWRM from the United Nations University (Mayfield et al., 2004).

The curriculum development process closely followed the outline in Figure 2.2. The first stage was an iterative process between experts in IWRM leading to an agreed-upon list of topics for each lecture in each of the ten courses. Experts in their respective area of specialization then produced and assembled the courses according to the curriculum outlines. Many agencies, universities, colleges and individuals supplied material for the course. The course materials were then edited to produce as much uniformity as possible in language, style and format. Each course is approximately equivalent to a university level course of 25 lecture-contact hours.

A platform for delivery was chosen (StudySpace™ software) that was specifically designed at the University of Waterloo, Ontario, Canada, to be simple to use, able to accept many different input formats, capable of tracking statistics of use and comprehension and, most importantly, capable of delivering very similar experiences to the students whether the courses were on CD-ROM, online on the Internet or in printed form. No specialized information technology knowledge is required to produce, edit or format a course for delivery (which is an essential requirement to deploy the programme in developing countries, where the faculty members can edit the courses as required). Students can individually track their progress through the course, rate their level of understanding of the materials on each page for future reference, make notes on any page, communicate with programme coordinators, faculty and each other, and test their knowledge with online quizzes.

Designated Regional Centres enroll students, administer courses, provide and mark assignments, add customized materials of specific interest to students in their region, provide case studies to enhance the teaching materials and deliver the courses. These Regional Centres are in Thailand (at the Asian Institute for Technology in Bangkok), in Kenya (at the University of Nairobi in Nairobi), in the South Pacific (at the University of the South Pacific at Fiji), in Ghana (at the University of Ghana in Accra), in Bahrain (at the Arabian Gulf University in Manama), and in Panama (at Cathalac in Panama City). The centres take ownership of the programme, administer the WVLC programme for their region, and deliver it as a standard core curriculum of IWRM for each student. They provide a high level of regional customization and enhancement to the basic curriculum and this added material/information is shared with other regional centres. The role of UNU-INWEH is to update and upgrade the basic curriculum in consultation with the regional centres and to disseminate and coordinate enhanced materials and case studies. Additional courses in related areas are produced after consultation with the regional centres.

The courses were initially tested, before final assembly and delivery, with graduate students from developing countries who were enrolled in Canadian Universities in Ontario but had undergraduate degrees from a developing country. Feedback from currently enrolled students is sought at intervals by the regional centres. The questionnaire covers the opinion/feedback of students on the entire programme, course content and the StudySpace™ software. Feedback received from the first two cohorts to graduate from the programme included that the students appreciated the combination of face-to-face and distance-based learning and that it was an effective way of delivering the programme. They also found the teaching/learning materials were convenient and easily transportable. The students found the programme useful and informative and the experience gained through this programme helped them to improve their knowledge and skills necessary to understand and apply IWRM concepts and principles in the planning and implementation of water resources management programmes and projects in their respective countries (Babel et al., 2008). In addition, the employers of the graduates of the programme will be surveyed after a short interval to determine their opinions on the value and outcomes of the WVLC IWRM Diploma programme.

2.8 CONCLUSIONS

In any review of the voluminous literature on E-learning, one must conclude that there is no definitive conclusion on the efficacy and efficiency of E-learning. Different investigators have found different outcomes and results from the use of E-learning techniques, but there does seem to be an emerging consensus that these differences between E-learning and traditional methods are not significant. There is a large degree of heterogeneity in the results and this might be because of the wide variety in types and quality of course materials, delivery systems, organization and communication methods used in the courses surveyed. There are simply too many variables in addition to the inherent differences in student attitudes, background, preparation and social and cultural issues.

Some general statements can be made concerning the characteristics of "successful" courses. In general, such courses have clear, well-organized and comprehensive materials, have a high degree of interaction and communication between students as well as between students and instructors, have well-defined goals, use appropriate learning technologies and the assessment methods are considered fair by the students. The role and importance of particular technologies seem less important than whether the technology is approachable and appropriate to the course and the student's environment. These conclusions are not surprising or unexpected, they are well known by good teachers in any field. Perhaps a new goal should be to provide the courses in a "malleable" format that is easily adjusted by the student to suit his or her own needs, preferences and environment (including their computing environment).

In addition to the considerations above, there is a need in the water sector for a more rigorous assessment of outcomes, as opposed to simply measuring outputs such as successful course delivery and student assessments of the courses. A more long-term assessment of the outcomes, in terms of changes in practice or successful implementation of programmes by the students at their place of employment, may provide one such metric. Typically, even very well-funded programmes are rarely able to get funding to carry out such in-depth assessments, often because the timeframe required stretches beyond the official end of the funding. It might be useful to start and implement a generic process to compare outputs and outcomes of courses and programmes in the water sector in a consistent and comprehensive manner. Successes might then be replicated.

REFERENCES

Babel, M.S., Grover, V.I., Sharma, D. and Wahid, S.M. 2008. Capacity Development in IWRM through E-learning – Initial Experiences of Water Virtual Learning Centre at AIT, Thailand. 13th World Water Congress, Montpellier, France, September 2008, Proceedings paper.

Baggaley, J. 2008. Editorial: Ideas whose time has come back. *International Review of Research in Open and Distance Learning.* 9(2): 1–2.

Bernard, R.M., Abrami, P.C., Lou, Y., Borokhovski, E., Wade, A., Wozney, L., Wallet, P.A., Fiset, M., and Huang, B. 2004. How Does Distance Education Compare With Classroom Instruction? A Meta-Analysis of the Empirical Literature. *Review of Educational Research,* 74(3): 379–439.

Bourget, P.G. 2008. Key Lessons Learned from the Masters Degree Program in Water Resources Planning and Management. *J. Contemporary Water Research and Education.* 139: 55–57.

Carnevale, D. 2000. *Quality on the Line: Benchmarks for Success in Internet-based Distance Education.* Washington D.C.: The Institute for Higher Education Policy.

Grunwald, S. 2007. E-delivery Methods of Learning Materials (U.S. Perspective). Indo-US Workshop on Innovative E-technologies for Distance Education and Extension/Outreach for Efficient Water Management. ICRISAT, Patancheru/Hyderabad. Andrha Pradesh, India. Proceedings paper.

Kirshen, P.H., Vogel, R.M. and Rogers, B.L. 2004. Challenges in Graduate Education in Integrated Water Resources Management. *J. Water Resour. Plng. and Mgmt.* 130(3): 185–186.

Major, D.C. 1997. Multiple objective water resource planning. *Water Resources Monograph 4*, Washington, D.C.: American Geophysical Union.

Mayes, R. 2004. *Review of Distance Education Literature.* Occasional Paper No. 6. Appalachian Collaborative Center for Learning, Assessment and Instruction in Mathematics.

Mayfield, C.I., Grover, V.I. and Daley, R.J. 2004. The United Nations Water Virtual Learning Centre: a flexible distance learning programme for integrated water resource management. *Global Environmental Change – Part A,* 13(4): 313–318.

Merisotis, J. and Phipps, R. 1999. What's the difference? Outcomes of distance vs. traditional classroom-based learning. *Change,* 31(2): 12–17.

Phipps, R. and Merisotis, J. 1999. *What's the difference? A review of contemporary research on the effectiveness of distance learning in higher education.* Washington D.C.: The Institute for Higher Education Policy.

Russell, T.L. 1999. *The No Significant Difference Phenomenon.* Chapel Hill, Office of Instructional Telecommunications, North Carolina State University.

Stakhiv, E. 2003. Disintegrated water resources management in the US: Union of Sisyphus and Pandora. *Journal of Water Resources Planning and Management,* 129(3): 146–154.

Zemsky, R. and Massy, W.F. 2004. Why the E-Learning Boom Went Bust. *The Chronicle Review.* 50(44): B6.

Chapter 3

LEARNING SYSTEMS FOR ADAPTIVE WATER MANAGEMENT:
Experiences with the development of *opencourseware* and training of trainers

Catharien Terwisscha van Scheltinga
Wageningen University and Research Centre, The Netherlands

Caroline van Bers
University of Osnabrück, Germany

Matt Hare
UN-Water Decade Programme on Capacity Development, Germany

ABSTRACT

Given the unprecedented global changes where "business-as-usual" is no longer an option in water management, teaching water management can also no longer be business-as-usual. In recent years the concept of adaptive water management (AWM) has been developed, giving managers guidance on how to respond adaptively to increasing risks of droughts and flooding. The most important implications of climate change for the water manager are learning to deal with uncertainty, measuring vulnerability, managing participative processes involving stakeholders, and linking water management to societal and social learning processes. AWM ultimately provides a framework in which individuals and organizations can perpetuate learning systems that enhance the capacity of society to deal with increasing risks and uncertainties in the water sector.

While today's water managers are already facing the challenge of addressing these issues, changes in university curricula need to be made in order to equip the next generation of water managers. Ideally, a teaching programme for AWM addresses the three aspects of learning: knowledge, skills and attitude. Effective teaching of AWM to tap these three aspects incorporates the use of a diversity of working forms, including new educational approaches like role-playing and participatory modelling. For this purpose, *opencourseware*[1] for adaptive water management has been created within the framework of the EU-funded NeWater research project, and a complementary international

[1] *Opencourseware* is a term for educational resources that are freely available on the Internet.

training-of-trainers programme has been developed and implemented. Tracking the subsequent developments in the curricula used by course participants in their home institutions is being undertaken.

This chapter explains and reviews the approach adopted in the NeWater project to develop an online AWM curriculum and to supplement this with face-to-face, training-of-trainers courses to create a multiplier effect in the dissemination of AWM teaching throughout university curricula within developing and developed countries. It also outlines the advances in individual, organizational and societal learning required if adaptive water management is to be realized.

3.1   INTRODUCTION

Water is essential for life. Internationally, access to clean water is recognized as a basic human right by the UN Committee on Economic, Cultural and Social Rights (UN CESCR, 2003). Action is being undertaken worldwide to increase access to safe water for drinking, sanitation, food and the environment, for instance by governments striving to implement the Millennium Development Goals (MDGs) and, in particular, MDG 7, one of the targets of which stipulates a 50 per cent reduction in the number of people without sustainable access to safe drinking water (UNDP, 2008).

Meanwhile, managing water resources is increasingly difficult: natural disasters involving water, such as floods and droughts, are increasing due to global environmental changes; international trade leads to virtual transfers of water from one region or country to another, and numerous conflicts arise over access to water.

At all levels and scales, water managers are facing more uncertainties and complex situations that require integrated solutions. As a result, business-as-usual is no longer an option for effective water management. However, it is not easy to quantify the changes observed or the uncertainties involved. How should our approach to water management change and in which direction should it go? Integrated Water Resources Management (IWRM) as an integrated management approach has been heavily promoted, although practical changes in the sector are often not yet in evidence.

Meanwhile, the next generation of water managers is being educated now. Their education cannot wait until water managers have solutions to our current problems. If water management cannot be approached in the "business-as-usual" way, then education in water management cannot continue in the same way either.

Higher education is based upon accredited course curricula that, in the case of water management, should reflect current trends in water management approaches and practices. As long as the direction to be taken by water managers is not clear and credible, and as long as appropriate teaching materials are not available, instructors will not be inclined to alter their existing curricula in favor of a new approach to water management. In order to enable instructors to move in a new direction, a special effort, including suitable training, support and capacity development, is required.

Adaptive Water Management (AWM) is one such new approach that is gaining recognition in the water management sector, and needs to be introduced into higher education curricula if global changes are to be effectively addressed by future managers. AWM ultimately provides a framework in which individuals and water management organizations can perpetuate learning systems that enhance the adaptive capacity of a society to deal with increasing risks and uncertainties in the water sector.

The provision of the training, support and capacity needed to alter curricula and implement the teaching of adaptive water management in the higher education sector is the subject of this chapter. First, however, AWM needs to be explained in more detail. Subsequently, an approach to teaching AWM and a complementary training course for teachers are elaborated. This experience is then translated into a multi-level approach to learning for AWM within society. The authors close with some conclusions about the attitudinal and institutional changes needed to support new didactic approaches.

3.2 ADAPTIVE WATER MANAGEMENT DEFINED

In order to address the problems of complexity and uncertainty in water management, the concept of adaptive water management has been developed (see, for example, Pahl-Wostl, et al., 2007). In this approach, management is seen as an ongoing, adaptive, and experimental process (Walters, 1986) in which it is important to learn continuously and systematically from any step in the policy or management process. More colloquially, adaptive management has been referred to as "learning to manage, by managing to learn" (Bormann et al., 1993). Learning is important since, in order to maintain high levels of adaptive capacity, one must be able to plan and build according to different scenarios so that flexibility can be maintained in both the planning and implementation phases to allow for adaptation as and when circumstances change. Adaptive management can be seen as "a systematic process for improving management policies and practices by learning from the outcomes of implemented management strategies" (Pahl-Wostl, et al., 2007, p. 3).

AWM is an intuitive response to dealing rationally with considerable uncertainty in the face of what might be perilous change. If one were to ask oneself what one would do in uncertain and potentially dangerous circumstances, one might instinctively consider continuously monitoring the situation, coming up with more than one plan, remaining flexible and ready to act, and seeking the support of others. One might also start to mentally prepare oneself for dealing with difficulties and change, and to learn quickly from initial mistakes.

Table 3.1 In the event of uncertainty.

What would you do if you had to act whilst facing significant uncertainty about a real and present threat to yourself ... and doing nothing was not an option?	
An intuitive response	*Adaptive management*
• Keep monitoring the situation	• Carry out real-time monitoring
• Take out insurance in some way	• Adopt a portfolio approach to risk management planning
• Develop more than one plan • Prepare to react quickly • Stay flexible	• Avoid lock-ins to particular management decisions
• Get support from those around you and find a common purpose	• Initiate stakeholder participation
• Prepare yourself mentally for upheaval	• Stimulate attitude change
• Consider learning from your mistakes	• Employ learning cycles
• Think about becoming resilient to the "slings and arrows of outrageous fortune"?	• Develop adaptive capacity

The AWM approach prescribes the following types of responses to managing water resources in the face of uncertain global change (Table 3.1): carry out real-time monitoring of resources; adopt a portfolio approach to risk management planning involving the implementation of a range of strategies whose implementation can reduce the overall risk to be faced; avoid lock-ins to particular technologies or institutions by choosing solutions that can be altered if necessary; support and encourage stakeholder and public participation in management decisions; change attitudes of professionals and society at large; and encourage social learning. All this will develop a society's adaptive capacity.

3.2.1   AWM AND IWRM

In general terms, as with IWRM, adaptive management is achieved by following a cyclic approach in which one plans – acts – monitors – evaluates – and adjusts (Jeffrey and Gearey, 2006). AWM can be seen (Figure 3.1) as an augmentation to the classic policy cycle as developed for IWRM (GWP-TEC, 2004). This augmentation takes the form of social learning (see Pahl-Wostl & Hare, 2004) within the water sector and society.

In order to make it an effective social learning cycle, the adaptive management process includes:

A   the identification of ambiguities, frames and paradigms within the sector and society, whose discussion and development are used to build commitment to reform,

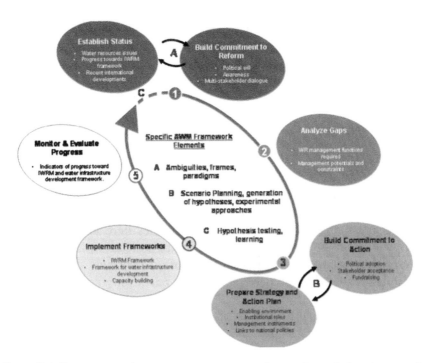

Figure 3.1 Five steps in the water management policy cycle, including the specific AWM framework elements (Source: Le Corre et al., 2009).

B   scenario planning, the generation of hypotheses and development of experimental approaches in preparation of strategies and action plans in order to help build commitment to actions,

C   hypothesis creation and testing in practice, so that plans and their implementation can be adapted.

The elements A, B and C are essential in the process of developing adaptive solutions, and are shown in Figure 3.1 where they may logically occur as part of the IWRM cycle. However, in practice, they can certainly also occur at other steps in the cycle (Le Corre et al., 2009).

The following steps are included in the AWM/IWRM policy cycle:

1   Establish status and build commitment to reform

The starting point of the adaptive water management cycle involves identification of the critical water resource issues that need to be tackled. This means that progress towards a management framework in which the issues can be addressed needs to be charted. To sustain this progress, political will is necessary, as is building awareness through a multi-stakeholder dialogue. The dialogue needs to be based on knowledge of the subject matter, and awareness-raising is one of the tools to establish this knowledge base and encourage participation of the broader population.

Critical elements of the participatory process that have to be considered are: developing stakeholder commitment, carrying out stakeholder and institutional analyses, and dealing with multiple actors, ambiguous issues and diverging perspectives.

2   Analyze gaps

The gaps in the adaptive water management cycle can be analyzed based on current development trends, policies, legislations, institutional situations, and capacity to reach the set goals. Important themes here are indicator development, establishment of monitoring and data collection, and participatory processes. This last one helps managers gain an overview of the issues and tools that are required when dealing with the parties involved.

3   Prepare strategy and action plan and build commitment
   to action

Application of the framework for water resources management requires a strategy and action plan. This means establishing an environment in which institutional roles and management instruments can be applied to set up relevant measures.

As with the first step in the cycle, commitment by others towards the necessary actions will be required. Plans need to be integrated into a political agenda; stakeholder acceptance is needed and funding should be committed to in order to achieve this acceptance. The essential theme here is participatory integrated assessment to develop scenarios with the support of stakeholders.

4   Implement frameworks

Having built commitment and developed plans, the actual implementation poses challenges. It is likely that changes will have to be made in the present management structures that will most likely require building capacity to implement the plans. The relevant themes here are building implementation capacity and the use of adaptive and flexible implementation plans to anticipate the uncertainties.

5   Monitor and evaluate progress

At the end of the first cycle, the monitoring and evaluation of progress will serve as input for adjusting or fine-tuning the course of action. It is important to choose indicators that describe the progress towards adaptive IWRM and towards the development of the infrastructure. In this step, as in each of the previous steps, the most relevant themes are monitoring the process and carrying out participatory evaluation.

It is also important to note that water management processes are iterative and that the cycle may be repeated a number of times; every time managers complete step 5 and return to step 1, they take into account what they have learned from the monitoring and evaluation processes and incorporate that knowledge in further fine-tuning of the decision-making process (Le Corre et al., 2009).

3.3   TEACHING AWM

Teaching adaptive water management requires a solid curriculum with a strong conceptual foundation, together with an array of commensurate approaches and methods for assessment, analysis and decision-making. In addition, such a curriculum needs to prepare students for the complexity and uncertainty inherent in water management, to be better equipped to address sudden changes, and to contribute to the development of adaptive approaches in water management as needed. To an extent, this also entails a shift in attitude or mindset. In this section, the curriculum developed within the NeWater project is described and suggestions are made for the effective learning of adaptive management through a range of teaching approaches.

3.3.1   CURRICULUM DEVELOPMENT

When developing a curriculum, both the content as well as the design of a learning process are important. In the design, the following questions are addressed: *why* is this particular stream of education important, *what* will be taught and *how* (under what conditions) will it take place?

Teaching curricula are generally underpinned with a description of the learning objectives (knowledge and skills to be imparted), an indication of the level of those being taught, together with prerequisite knowledge/skills, and an overview of the training content. Within academic institutions, an indication of the credits or value of the curriculum within the overall programme is normally provided. In some cases it may also be appropriate to indicate where the teaching should take place and what facilities and/or equipment are required. In Table 3.2 the general elements of curriculum development are outlined. Once a concrete direction is provided (the why, for whom and what), an outline of the specific courses or sessions can be developed.

Within the course outline, learning goals should then be formulated for each topic/session. Subsequent development of a session plan provides specific teaching approaches and materials to be used for a given topic. A sample session plan for an Introduction to Adaptive Water Management is provided in Table 3.3.

Table 3.2 Basic elements of curriculum development.

Why?	Specify learning objectives (knowledge – skills – desired attitude shift)
For whom?	Indicate the target group
What?	Formulate and organize the teaching content
How?	Prepare an outline of the courses or sessions; select teaching methods/forms of teaching
Where?	Indicate where the teaching will take place, and under what circumstances; describe facility and equipment requirements

Table 3.3 Sample session plan.

Title of Session: Introduction to Adaptive Water Management
27 November 2008

Introduction: AWM is an emerging concept that responds to the need for adaptation in the face of increased complexity and uncertainty, ... etc ...

Objectives: to teach participants about the AWM concept and the types of approaches and methods it entails, ... etc.	Material required: flipchart paper, post-it notes, markers
Expected outputs: students will demonstrate understanding of AWM concept and approaches through successful completion of group exercise and contribution to discussion	Time required: 3 hours 1.5 hours lecture, 1 hour group exercise, 30 min. discussion
Preparation required: provide background reading for students; prepare group exercise and discussion questions	Prerequisite: Introduction to Global Change, including understanding global cycles and their influence on water resources

Procedure (steps): Provide background reading at least two weeks in advance of teaching session

Information/Guidelines for (technical) preparation:
– NeWater Project: www.newater.info
– Online curriculum: www.newatereducation.nl

Questions for discussion: What are the characteristics of a regime that is adaptive? What is the influence of the context? Identify some barriers to change.

3.3.2   CHOICE OF TEACHING METHODS FOR AWM

In the design of the session, one also establishes the types of teaching methods to be used. Various teaching methods, most commonly lecturing, assignments, brainstorming, discussion groups, role play, and practical demonstrations are possible (examples that are applicable to teaching or training in AWM may be found in Snellen et al., 2005). There is a direct link between the goal (dissemination of knowledge, skills development, and attitudinal shift) and the choice of a teaching method. For example, to instruct students on the impacts of climate change, a lecture on the Fourth Assessment Report of the Intergovernmental Panel on Climate Change (IPCC) is an effective approach. But to begin to instill in the students an ability to deal with uncertainty, a lecture will not suffice. In this case, a role-playing game or series of games involving an iterative learning process can be very effective.

In order to address the interests of stakeholders, students will also require more than just lectures. An exercise in holding interviews using cognitive mapping approaches (see Vennix, 1996) will demonstrate to students not only how to hone in on the issues that are important to stakeholders, but also what the practical difficulties of doing so are. Another example is group model building – a process that is promoted for use in operations and management research by Vennix (1996). He developed such processes to allow managers and staff to better understand business problems and their solutions as a group. The idea is to involve the group in developing a system dynamics model of the management system and then to use the model, together, to investigate solutions to the identified problems. By doing so, the group becomes involved in a learning process that leads to a better understanding of the system in all its complexity and to a sharing of that understanding among the members of the group: the individuals' mental models are improved and harmonized. As with any participatory technique, greater acceptance of management decisions is an expected goal. In terms of training, only by carrying out such processes in practice can the student more readily understand their range of applicability.

3.3.3   DEVELOPMENT OF AN ONLINE CURRICULUM FOR AWM

In order to share knowledge and approaches developed or advanced in the NeWater project (www.newater.info) with young researchers and practitioners, three summer schools were organized in collaboration with the Global Water System Project (GWSP) in 2006, 2007 and 2008. In the summer schools, global change and its effects on water management were presented. Furthermore, the postgraduate researchers were introduced to the concept of adaptive and integrated water management as well as to methods and tools for adaptive water management and for managing the shift to adaptive management. The numerous instructors of the individual sessions were selected for their specific expertise as well as teaching skills.

These summer schools provided the teaching materials and approaches for an educational curriculum in adaptive water management that would be made freely available as *opencourseware*. Through their introduction and use at the summer schools, these teaching materials have been tested to ensure a high level of quality control. This applies as well to the method of training in this subject; over the years, the training approach has been adapted to become increasingly project-based, constructivist and hands-on, in response to the needs of the trainees. A wide variety of exercises and assignments have been developed in order to provide students with adequate opportunities for hands-on learning.

The materials developed for the summer schools have been compiled, edited, formatted, and organized in modules and made available online. The materials are primarily for use by university lecturers interested in including adaptive water management in their graduate level curricula (i.e. Master's and PhD level). The individually downloadable sections of each module include annotated presentations, exercises, discussion questions, relevant literature, and links to databases and tools.

A user-friendly open source e-learning platform, Moodle, has been used for this curriculum. Moodle is, in fact, designed for interactive use and, in principle, can be used by students directly for distance learning when administered by an educational institution. Users can interact with each other and the instructor(s) in discussion fora, and tests can be administered and grades posted. However, the AWM curriculum is not established as an interactive teaching course. It is, as stated earlier, a set of teaching materials for instructors to incorporate into their own courses or programmes. The curriculum may also be used directly by PhD and postgraduate researchers, as well as practitioners who may wish to supplement their knowledge and skills. The curriculum is available at http://www.newatereducation.nl (see Figure 3.2). For easy reference, a stand-alone version is available on CD-ROM for those who have limited or slower Internet access. A sample session slide (in this case

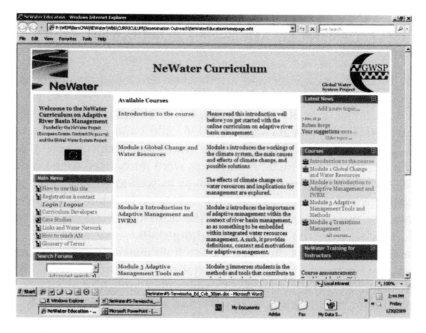

Figure 3.2 Homepage of NeWater-GWSP online curriculum AWM.

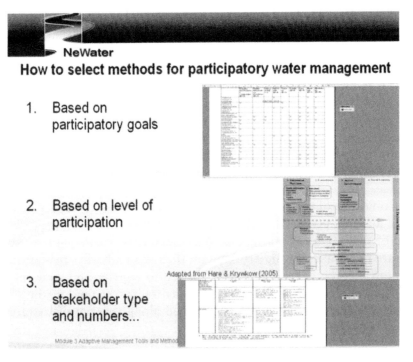

Figure 3.3 Sample session slide from NeWater GWSP online curriculum (reproduced from Hare, 2007).

"Participatory Management and Social Learning in Resource Management") is provided in Figure 3.3.

The copyright for the materials is held by the original authors, and therefore all teaching materials must be appropriately cited by users. For this reason, a citation reference has been provided for each downloadable document in the online curriculum. All other information on the site has been assigned a "Creative Commons non-commercial share alike" license[2], meaning that users can freely use the information and adjust it for their own purpose on a non-commercial basis (i.e. not-for-profit), as long as the creators of the website are duly credited.

3.4   TRAINING-OF-TRAINERS COURSE FOR TEACHING AWM

During the NeWater project, two training-of-trainers courses have been provided for university lecturers, one in Osnabrück, Germany in April 2008, and

[2] www.creativecommons.org

a second in New Delhi, India in October 2008. The course in Osnabrück was organized in close collaboration with the Global Water Systems Project (GWSP) and in New Delhi with the UN-Water Decade Programme for Capacity Development (UNW-DPC) at the International Human Dimensions Workshop 2008 of the International Human Dimensions Programme (IHDP). Extra funding was made available from a variety of sources to permit the participation of instructors from developing countries. Two central criteria underpinned the selection of participants: experience in teaching water management or moving in this direction, and a statement that clearly demonstrated that the applicant had sufficient understanding of the subject matter (e.g. the related field of IWRM) and the importance of it, in order to benefit significantly from the course.

The objective of this training course is to familiarize university instructors with the material provided in the online curriculum, with respect to both the content of the modules and the use of the materials. Participants were asked to prepare in advance by undertaking on their own a basic orientation of the curriculum (i.e. the participants were invited to register and view the online curriculum). Articles on AWM were provided as background reading.

More specifically, the course is designed along five parallel strands that alternate over the duration of the course:

1   content teaching, focusing on knowledge needed for AWM involving seminars;
2   skills training, involving hands-on experience with particular methods;
3   attitude and awareness-raising, involving role-playing and group discussions;
4   didactical development, in which the trainers are taught through theory and example the didactical approaches for AWM; and finally,
5   group project work, in which teams of trainers (i.e. participants) work throughout the course on developing a curriculum on AWM that will fit to their home courses.

The course begins with an introduction to the conceptual foundations of adaptive water management and then moves on to the implications for teaching in water management (see Box 3.1 for a sample discussion question). An orientation is provided of the AWM *opencourseware*. Practical assignments focus on using materials and methods available in the online material. For example, to improve skills in gathering or eliciting local and expert knowledge, the participants practise cognitive mapping; in order to improve skills in allowing participation in the establishment of status, gaps analysis or preparing strategies, the participants are taught group model building and its best practice. Participants work directly with the AWM curriculum during this training session and develop their own course outline (including learning goals, knowledge and skills to be acquired, and attitudinal shifts to be achieved). The participants can then experience how attitudes could be investigated and shifted through the use and playing of a specially designed role-playing game called *Floodplain*, which demonstrates the effects of uncertainty and learning upon management behaviours and shows players the gains to be made by applying AWM approaches.

Box 3.1 Sample discussion question and response.

Sample question:

When teaching water management, our curriculum normally includes elements of knowledge, skills and attitude. Working in groups of 4–5, discuss how you include skills and attitude in your teaching (15–20 mins.).

Sample responses:

- Ability to initiate dialogue among the students
- Policy-relevant teaching
- Attitude: to be tolerant of different perspectives, engagement with students
- Multi-dimensional channels for transmitting knowledge
- Presentation skills, IT, software, multimedia
- Practical experience – to develop skills and shift attitudes
- Discussion of various points of view
- Go beyond conventional way of thinking and beyond the boundaries (thinking "out of the box")
- Preparing students to affect change
- Instructors need to improve assessment of educational needs
- Present scenarios and case studies
- Focus on group dynamics and diversify roles among students
- Interactive learning

Thus far, the training course has been positively evaluated. Participants rank the course high with respect to relevance of the training to their own needs, comprehensiveness of the course and effectiveness of the assignments. Furthermore, the role-playing game was very well received as a teaching tool. Nonetheless, there have been several suggestions for improvement including:

- presentation by participants at outset on their current teaching activities;
- a lecture on climate change science prior to the introduction to AWM;
- expansion of the AWM teaching materials to include the teaching of methods;
- discussions given more time.

All of these recommendations are being considered for subsequent training courses, and though they are potentially valuable supplements to the course, including them would mean extending the course by at least two days. In order to evaluate the ultimate effectiveness of the course and the uptake of the online curriculum, participants are sent a short questionnaire four to six months after each course to assess the extent to which they have used what they learned in the training course (i.e. the value of the course and the use of the online curriculum in supplementing or expanding the participants' teaching programmes).

3.5   A LEARNING SYSTEM FOR AWM

The promotion and adoption of AWM within the professions and policy-making requires a multi-level approach to learning within society. It cannot just be left to higher education or to isolated training or training-of-trainers courses. What is needed are professional, societal, educational and didactic levels of learning that work in concert. Such a learning system forms a critical foundation for the changes required to shift to more adaptive water management.

3.5.1   PROFESSIONAL LEVEL: DEVELOPING INDIVIDUAL AND ORGANIZATIONAL CAPACITY IN THE PROFESSIONS

Capacity development in adaptive water management means developing the knowledge, skills and attitudes among individuals and the professional organizations so that they can increase their own adaptive capacity and create flexible and responsive institutions to support their work in the face of long-term uncertainties. This infers an enabling environment with adequate funding, professional support, cooperative networks, legislation and agreements.

3.5.2   SOCIETAL LEVEL: SOCIAL LEARNING – CREATING AN OPPORTUNITY TO TRY AND LEARN

Capacity development in AWM also means creating an enabling environment in which individuals and organizations can move from single- to double- and triple-loop learning (Argyris and Schön, 1978): that is to say, shifting from a simple update of their existing mental models to reframing their long-held perspectives about water management, and then orchestrating a paradigm shift within the profession and society about what constitutes effective water management. It means allowing new structures and institutions to develop, thereby linking groups of individuals and organizations that promote social learning. An effective transition requires creating an opportunity (a space or a transition area) in which innovators can learn, experiment, and work freely before channeling their new approaches into the mainstream (see Module 3:8.2.1 of the AWM online curriculum at www.newatereducation.nl).

3.5.3 EDUCATIONAL LEVEL: TEACHING AND TRAINING

Teaching also has to adapt. The sender-receiver paradigm of traditional training is no longer appropriate. New attitudes and skills need to be passed on, and this can, for example, be in the form of constructivist learning – learning by doing, or as the HarmoniCOP project declared: "learning together to manage together" (Ridder et al., 2005). The technical engineering skills of water resource management have to be harnessed to the social skills of negotiation, knowledge elicitation, group decision-making, effective communication and collaborative action. However, these skills can only be effectively taught through learning by doing. This requires that time be allocated in the curriculum for project work and more hands-on experiential learning in classes. It also requires that teachers be able to master the hands-on skills. Thus, there is a need for the training of trainers, not only to better disseminate the concepts and skills but also to develop new didactical skills in the teachers.

3.5.4 DIDACTIC LEVEL: TRAINING OF TRAINERS

New capacity, structures and institutions are needed that will allow trainers to develop and experiment with new approaches to teaching, if AWM is to be passed on to the next generation of water managers and policy-makers. Not only will the didactics of teaching have to develop, but there will also have to be a willingness, resources and organizational capacity to collaboratively adapt current (and in some cases long-established) technical curricula.

3.6 CONCLUSIONS

The promotion and adoption of AWM implies the activation of many different levels of a learning system in water management. At the professional level, it requires support for learning in terms of the knowledge and skills of the individual and of organizations, and it requires attitudinal change from individuals as well as society. At the societal level, it requires the development of an enabling environment, in the form of appropriate institutions that can encourage social learning with the aim of creating the new networks, management collaborations and stakeholder participation, in order to shift policy and management paradigms (triple-loop learning). At the educational level, this means that both technical and soft skills have to be taught to the next generation of water managers and policy-makers. The soft skills, such as organizing participatory planning sessions, facilitating intersectoral negotiation, coordinating communication and outreach programs or eliciting local knowledge, require hands-on learning-by-doing approaches, if they are to be effectively learnt. This necessarily changes the teaching paradigm, which leads to the final level of change in the learning system: the development of the training-of-trainers to support educators in the development of new curricula and new capacity in the education system adopting them.

This chapter has illustrated an approach to supporting change in curriculum development in higher education establishments in the developed and developing world. This approach has been a long-term and multi-pronged process. First, the development of course material by experts in their fields has been facilitated through running a series of postgraduate summer schools in AWM. These summer schools have allowed the material to not only be created, but over the years also user-tested in order to maintain a high level of quality control. This applies as well to the method of training in this subject; over the years, the training approach has been adapted to become increasingly project-based, constructivist and hands-on, in response to the needs of the trainees.

The materials and methods of teaching have then, with permission of the experts who developed them, been converted into *opencourseware*. This enables dissemination of the materials and methods as such free of charge, as well as of the *opencourseware* as a whole, and for materials to be altered and made use of by educators unhindered by copyright concerns, so long as the original materials are referenced. The final steps have been to move from training courses at the summer schools, to training-of-trainer courses designed to support the educators in using the *opencourseware* to develop their own curricula and specialized materials. A process of tracking the development of curricula in the participants' home organizations has been initiated to assess the effectiveness of the training-of-trainer courses and the application of the *opencourseware*.

3.6.1   A CRITIQUE

There are some notable criticisms that can be made of this approach. First of all, the training material and didactic approaches have come from experts primarily based in developed countries, largely in the European Union. The perspectives of the experts have been shaped, for the most part, by working with AWM in developed countries. Additionally, the choice of experts was not based on the quality of their expertise alone, but also on their connection with the project network and their willingness to put time and effort into developing course material (and subsequently teaching it – in most cases pro-bono) for use by all. Didactical approaches have also been allowed to evolve through trial and error, rather than through prior design. To continue the evolution analogy, what remains in terms of course material within the GWSP/NeWater Online Curriculum has also been designed through the forces of natural selection, rather than prior design. However, in light of increasing uncertainties, one may ponder the extent to which a project can be designed in advance – and the emerging results are also worthy of evaluation.

The extent to which educators in developing countries can develop their curricula on their own limited local resources remains questionable.

It is valuable to follow-up on individual efforts, and vital to support this development over a longer period of time. This is even more the case for those educators who work with organizations that might be in the middle of a five- to eight-year curriculum cycle and will not be planning to redevelop curricula for the next three to four years.

3.6.2   A STEP FORWARD ON A LONG ROAD

Support needs to be given not only to individuals but also to the organizations they belong to, in order to provide them with the capacity to plan and implement change in curricula. The training-of-trainer courses, therefore, have to be seen as the first step in a longer process of individual and organizational capacity development to effectively alter curricula and the way of teaching. In fact, the introduction or offer of *opencourseware* in a much-needed subject such as AWM can and should be used as a catalyst for strengthening the range of didactic approaches that an educational organization would adopt. The introduction of new course material via online fora necessitates, in most cases, not only new ways of teaching, it also provides an opportunity to bring in new teaching methods without directly challenging standard approaches.

ACKNOWLEDGEMENTS

The authors wish to express their appreciation for the fruitful discussions held with instructors and participants of the various training courses, as well as colleagues and project partners within the framework of the NeWater project (www.newater.info). In particular, the authors are grateful for the time and effort invested by the instructors/researchers who provided training and teaching materials for the online curriculum. The funding provided by the European Commission for the NeWater project (Contract No. 511179) and the Dutch Ministry of Agriculture, Nature and Food Quality are also kindly acknowledged.

Finally we would like to thank the Global Water System Project, The Integrated Assessment Society (TIAS), UNU-IHDP and the UNW-DPC for their support in setting up the various training courses.

REFERENCES

Argyris, C. and Schön, D. 1978, *Organizational Learning: A Theory of Action Perspective*. New York: McGraw-Hill.
Bormann B.T., Cunningham, P.G., Brookes, M.H., Manning, V.W. and Collopy, M.W. 1993, *Adaptive ecosystem management in the Pacific Northwest*. USDA Forestry Services General Technical Report PNW-GTR-341.
GWP-TEC (Global Water Partnership – Technical Advisory Committee) 2004, *Integrated Water Resources Management (IWRM) and Water Efficiency Plans by 2005. Why, What and How?* TEC Background Papers No. 10. Stockholm, Sweden: GWP.

Hare, M. 2007, Participatory Management and Social Learning in Resource Management. *NeWater online curriculum on Adaptive River Basin Management Version 1.0.* http://www.newatereducation.nl/

Jeffrey, P. and Gearey, M. 2006, Integrated water resources management: lost on the road from ambition to realisation?, *Water Science And Technology*, 53: 1–8.

Le Corre, K., Poolman, M. and van der Keur, P. (eds), 2009. *Training and Guidance Booklet for Adaptive Water Management.* Cranfield: 'New Approaches to Water Management under Uncertainty' Project.

Pahl-Wostl. C. and Hare, M.P. 2004, Processes of social learning in integrated resources management. *Journal of Community and Applied Social Psychology.* 14: 193–206.

Pahl-Wostl, C., Sendzimir, J., Jeffrey, P., Aerts, J., Bergkamp, G. and Cross, K. 2007, Managing change toward adaptive water management through social learning. *Ecology and Society* 12(2): 30.

Ridder, D., Mostert E. and Wolters, H. 2005, *Learning together to manage together – Improving participation in water management.* Handbook of the European 'Harmonizing Collaborative Planning' Project, Osnabrück: University of Osnabrück.

Snellen, W.B., Terwisscha van Scheltinga, C.T.H.M. and Schrevel, A. 2005, *Working with farmers. Towards a service approach in irrigation. Training. The socio-economic base line survey.* Alterra Report 1096, Wageningen.

UN Committee on Economic, Social and Cultural Rights (CESCR), 2003. Report of the UN Committee on Economic, Social and Cultural Rights, Twenty-eighth and Twenty-ninth Sessions (29 April – 17 May 2002, 11–29 November 2002), 23 June 2003, E/2003/22; E/C.12/2002/13. http://www.unhcr.org/refworld/docid/3f6b10ea4.html

UNDP 2008, *The Millennium Development Goals Report.* http://www.undp.org/mdg/

Vennix, J.A.M. 1996, *Group model building: facilitating team learning using systems dynamics.* New York: John Wiley & Sons.

Walters, C.J. 1986, *Adaptive management of renewable resources.* New York: McMillan.

Chapter 4

BRIDGING THE KNOWLEDGE GAP:
The value of knowledge networks

Jan Luijendijk
UNESCO-IHE Institute for Water Education, The Netherlands

Wouter T. Lincklaen Arriëns
Asian Development Bank, Philippines

ABSTRACT

Changes inside and outside the water sector are creating new demands for knowledge and capacity, but water institutions have considerable inertia and do not easily adjust to changing demands. The need for change and innovation has become the main driver for people and organizations to create new social structures like Communities of Practice and Knowledge Networks aiming to generate more value through information and knowledge sharing. In the past decades, ICT has further spurred the creation of these networks. In this paper, a framework for developing and improving knowledge networks in the water arena is proposed that focuses on 1) catering to demand, 2) delivering results, and 3) managing the networking process. Specific attention is given to determining and measuring outputs and outcomes, and to identifying success factors and overcoming constraints for effective networking. The performance of some selected water knowledge networks is assessed to define the most important issues for knowledge networking.

4.1 MANAGING WATER FOR A CHANGING WORLD – AN INTRODUCTION

The world community is facing an increasingly daunting challenge of how to manage its water. The term "water security" was used at the World

Water Forum in The Hague in 2000 to underline the vulnerability of people, economies and ecosystems to ever-growing pressures on the world's limited water resources (World Water Council, 2000).

Awareness is now increasing internationally about the need to mitigate the effects of climate change, and it is becoming clear that developing countries and poor communities, which find themselves least equipped with knowledge, capacity and infrastructure to adapt, may be hit hardest by water scarcity, floods, environmental degradation and rising sea levels.

Rapidly growing cities around the world have become the prime engines of economic growth, yet their capacity to manage their water and waste is being increasingly stretched. At the World Water Forum in 2003 in Kyoto, the international community already signalled an urgent need to double investments in water infrastructure, and to build the capacity of cities for managing their water services (World Water Council, 2003).

The year 2007 marked a turning point in history with half the world's population now living in cities. An estimated 180,000 people move into cities each day, adding more than 60 million city dwellers each year, the vast majority of whom are in developing countries. This rate is expected to continue for the next 30 years (UN-HABITAT, 2006). It is also expected that the larger and mega-cities will have the economic power to attract qualified people and build organizations to manage their water and waste.

The largest increase in population, however, is expected in small to medium-sized cities, and it is these that will face the greatest constraints in knowledge and capacity to manage their public services, including water and waste. Recognizing this challenge, the 2006 World Water Forum in Mexico City focused on local actions to address the global water challenge (World Water Council, 2006).

Similarly, it is expected that people and organizations working in the large river basins of the world will find ways to collaborate on improving water resources management in a more integrated manner.

At one end of the scale, the Yellow River Conservancy Commission in the People's Republic of China employs tens of thousands of people in nine provinces to carry out its challenging mandate. The Commission is already a knowledge leader in its field, and its success is primarily determined by the willingness of riparian stakeholders to collaborate on effective actions (Yellow River Conservancy Commission, 2007).

At the other end of the scale, newly established river basin committees, like the one in Thailand's Bang Pakong River Basin, may rely on just a handful of trained specialists to deliver results, starting with knowledge on particular problems and solutions in the basin. Such smaller river basins will definitely face a big challenge in building their local knowledge and capacity for water resources management (ADB, 2006).

While international attention in recent years has focused on the need to increase investments in water services and better management of water resources, there is a growing consensus that the world's water crisis is primarily one of water governance.

Investments in water infrastructure need to be complemented with much higher investments in knowledge and capacity development. There is ample evidence that this makes good economic sense. Drawing on the example of Japan's economic success since 1945, the Asia-Pacific presentation to the World Water Forum in 2006 in Mexico City recommended that for every dollar invested in infrastructure, another seventy cents should be invested in the "soft" side, including education and capacity building (Japan Water Forum, 2006).

The Asia Water Watch 2015 study, commissioned by the Asian Development Bank and its partners, reported that every dollar invested in clean water supply and sanitation might produce up to six dollars in economic benefit (ADB, 2005). At local levels, experience shows that the economic rate of return (ERR) of conventional water infrastructure rehabilitation projects can be doubled or even tripled when significant investment components are added for capacity development (van Hofwegen, 2004).

When the Asian Development Bank adopted its policy for supporting agriculture and natural resources research in 1995, it showed that the ERR of investments in such research could be up to 90 per cent higher than from regular development projects (ADB, 1995a). More than ten years later, an ADB Eminent Persons Group recommended that the bank should focus on combining higher levels of infrastructure investments with development of knowledge and research through regional networks (ADB, 2007a).

The case for making better use of local knowledge and capacity becomes even more compelling when we consider that, despite international awareness of the need to increase water investments, official development assistance (ODA) for water has declined since the middle of the 1990s. OECD figures show that the share of water supply and sanitation investments declined from 8 per cent in 1999/2000 to 6 per cent in 2001/2002, and remained constant at 6 per cent in 2003/2004. And of all private funding for infrastructure, only 5 per cent went into water supply and sanitation services (World Bank, 2007).

The water challenge remains enormous. Globally, 1.1 billion people lack safe drinking water and 2.6 billion lack improved sanitation. While the vast majority of unserved people live in the Asia-Pacific region (70 per cent of those without safe drinking water and 75 per cent of those without improved sanitation), the service levels are lowest in Sub-Saharan Africa.

Achieving the millennium development goal (MDG) target Number 10 globally would require a doubling of the current investment levels in the water supply and sanitation sector in developing countries: from 15 billion USD per annum to 30 billion USD per annum (World Bank, 2007). For the Asia-Pacific region, Asia Water Watch 2015 estimated that a minimum total annual investment of $8 billion was needed, not including any upgrading from basic service levels and wastewater treatment (ADB, 2005).

Apart from investments in water supply and sanitation, significant additional investments are also needed for irrigation services, river basin management, flood management and mitigation, hydropower and multipurpose infrastructure, and wastewater management.

While annual investment needs around the world for water services and better management of water resources keep rising, the key challenge is for people to deliver better results. If there is agreement that the world's water crisis is primarily one of water governance, the central challenge is to empower people and organizations with a vision, mission, and values, and to look for ways to enhance their knowledge and capacity to improve their performance. Luijendijk and Mejia-Velez (2005) concluded that the social capital of the water sector is at "the heart of the matter".

4.2   ABOUT KNOWLEDGE AND CAPACITY – SOME DEFINITIONS

Knowledge and capacity are like two sides of the same coin. The same is true of good governance and capacity development (ADB, 1995b).

Capacity development is understood to involve three levels: the individual, the institution, and the enabling environment. Figure 4.1 shows the three levels with resulting outputs and goals. Successful reforms in the water sector are often the result of leadership by one or more champions. Traditionally, capacity development has focused mainly on education and training to increase individual knowledge and skills. Motivating and empowering individuals through coaching to improve their attitude and develop leadership skills has received less attention to date.

Practitioners of knowledge management now refer to knowledge as "the capacity for effective action". Boom quotes Karl Erik Sveiby as saying, "I define knowledge as a capacity to act." According to Boom, knowledge is about "know how" or "what works" in a given context, while information in general is "know what" or "what is" (Boom, 2007). Luijendijk and Mejia-Velez (2005) suggest that the best definition might be the one provided by Ikujiro Nonaka, who said that knowledge is the "justified belief that increases an entity's capacity for effective action".

"Explicit" knowledge is distinguished from "tacit" knowledge. The first refers to knowledge that "can be expressed in facts and numbers and can be easily communicated and shared in the form of hard data, scientific formulae, codified procedures, or universal principles" (Luijendijk and Mejia-Velez, 2005: 115).

Tacit knowledge, however, is "highly personalized and hard to formalize. Subjective insights, intuitions and hunches fall into this category of knowledge." Boom refers to tacit knowledge as undocumented knowledge

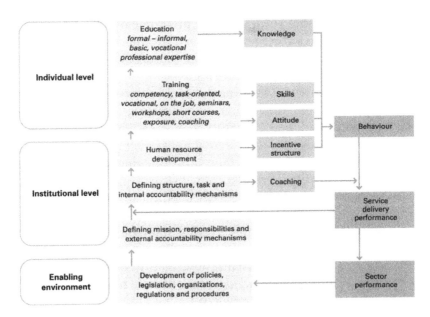

Figure 4.1 Capacity development: Levels, activities, outputs and goals (van Hofwegen, 2004).

(Boom, 2007). "It is embodied in people (human capital), or embedded in informal work processes (structural capital), or earned through working relationships outside (stakeholder capital)". Boom further suggests that tacit knowledge is more important than explicit knowledge for three reasons: 1) it accounts for an estimated 75–95 per cent of the total organizational knowledge; 2) expertise and mastery, the highest forms of knowledge, are mostly tacit; and, 3) innovation processes depend, for the most part, on tacit knowledge to get started.

In this age of information and communications technology (ICT)-driven knowledge applications, Boom warns that knowledge managed through ICT is often no more than a fraction of the total knowledge in an organization. Boom refers to claims that employees tend to contribute only 5 per cent of their personal intellectual capital of tacit knowledge to their organization's knowledge repository. Since ICT is limited to encoded or explicit knowledge, Boom contends that ICT "tends to distort policy and decision-makers' attention away from the greater fraction of what creates value for the organization, and ultimately for the economy" (Boom, 2007).

The process of knowledge management is understood to accommodate the development and use of both explicit and tacit knowledge. Management scholars have begun to proclaim the possibility of knowledge management becoming "the most universal management concept in history" (Takeuchi, 2001 as reported in Zhu, 2004).

Nonaka and Takeuchi (1995) adopted an epistemological dimension in their model, distinguishing between tacit knowledge and explicit knowledge, both of which are being continuously converted in a social learning process. The interplay between the two types of knowledge leads to processes of knowledge conversion, expansion, and innovation.

The starting point is the well-known cyclical model of knowledge generation (socialization, externalization, combination, internalization) shown in Figure 4.2. Socialization is the process of creating new tacit knowledge out of existing knowledge through shared experiences, for example in informal social gatherings or networks. Socialization, therefore, leads to shared knowledge. Externalization is the process of converting tacit knowledge into explicit knowledge, for example concept creation in new product development. Externalization leads to conceptual knowledge. Combination converts explicit knowledge into more complex and systematic sets of explicit knowledge, called systemic knowledge. This is where databases and computer-supported analyses come in. Internalization, finally, is the process of turning explicit knowledge into tacit knowledge, for example by training. This type of knowledge is called operational knowledge.

In this way, tacit knowledge, supported by explicit knowledge, becomes a synonym for "capacity to act" or a competence to solve problems (Luijendijk and Mejia-Velez, 2005). Tacit knowledge is highly contextual, and it recognizes the importance of local, traditional, or indigenous knowledge.

Knowledge management and capacity building are closely related. Knowledge management operates in an environment with relatively strong

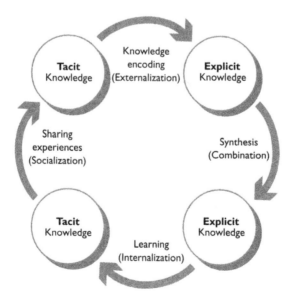

Figure 4.2 Types of knowledge and the knowledge creating process (Nonaka and Takeuchi, 1995).

institutions and where knowledge is considered a constraining factor for further development and improved efficiency (Luijendijk and Mejia-Velez, 2005: 117). On the other hand, capacity building primarily deals with environments of weak institutions and poor governance. Both knowledge management and capacity development deal with the same basic questions, i.e. how the decision-making process can be improved, and how tasks can be executed properly, by applying appropriate knowledge that has been collected in raw form or acquired through sharing in networks.

4.3   ABOUT KNOWLEDGE NETWORKING – SOME BASICS

It needs no explanation that rapid advancements in ICT in recent decades have enabled and catalyzed communication and collaboration among people around the globe. Luijendijk and Mejia-Velez (2005: 117) point out that "globalization, which is seen as a threat to many local communities, owes its pervasiveness to the worldwide networks that provide instant communication, including Internet." Most notably, ICT has caused the "death of distance" as a determining factor in human communications (Cairncross, 1997). ICT has, of course, not yet spread evenly around the globe, with millions of people in poor countries yet to be "connected" to this revolution and its opportunities. However, many poor countries are leapfrogging technological steps that took decades or longer in developed countries (e.g. cell phone rather than landline networks). And, applications of ICT for poor countries are increasing rapidly, e.g. GPS-based systems using solar energy, IT kiosks in local communities, and cell phones used for micro-credit and for issuing flood and typhoon warnings.

The introduction of networking itself has affected modalities of thinking and learning. Engel (1997) reports that one of the main problems constraining the development of sustainable solutions is the one-sidedness of many social and institutional learning processes. That is, people are told what to think rather than being invited to contribute their own ideas and insights, while the importance lies in people being able to adapt themselves effectively to rapidly changing circumstances. Networking has provided space for the latter to happen, and adaptation will occur. Importantly, networking has also diminished the predominance of traditional expert-to-counterpart models in favour of two-way communication and partnerships that focus on what each side can "bring to the table" (Fukuda-Parr et al., 2002).

Knowledge networks can exist in formal institutional arrangements between organizations (like in the Global Water Partnership) or in Communities of Practice, (CoP) where the focus is on individuals and their tight, informal relations within and also around organizations. As such, Luijendijk and Mejia-Velez (2005) explain that knowledge networks: 1) emphasize joint value creation by all the members within the network and thereby seek to move beyond the sharing of information to the aggregation and creation of new knowledge; 2) strengthen capacity for research and communication by all members in the network; and, 3) identify and implement strategies to engage decision-makers more directly, thereby linking to appropriate processes, and moving the network's knowledge into policy and practice (p. 121).

They continue to explain that

knowledge networks tend to be more focused and more narrowly-based than information networks; more cross-sectoral and cross-regional than internal knowledge management networks; more outward-looking than communities of practice; and, they involve more partners than some strategic alliances (p. 121).

Knowledge networks bring together providers and consumers of knowledge. Communities of practice (CoP) are seen as the primary building blocks of knowledge networks. CoPs are "groups of people that gather around a common interest or theme, and deepen their knowledge by interacting on an ongoing basis" (Wenger et al., 2002). CoPs serve to generate, share and disseminate knowledge. They often comprise 20–30 people and are cultivated by organizations as a contributing factor to knowledge management. CoPs normally use both face-to-face meetings and ICT applications. "Distributed" CoPs are those that link people across time zones, countries, organizational units, languages and cultures, and rely solely on ICT. For these CoPs, it is more difficult to build trust and personal relationships, which are key factors for success. The essential differences between a CoP and a knowledge network are in the scale/size, the relationships, and in the connectivity among members. A community of practice is smaller, more personal, more narrowly focused,

---

Box 4.1 Knowledge network approach (Luijendijk and Mejia-Velez, 2005).

Networks stimulate faster flow of knowledge to end-users.
   The network:

- Identifies, mobilizes and activates individual and organizational capacities in the different organizations and countries,
- Facilitates the process of sharing knowledge and experiences between people not only from the region but also with experts outside,
- Creates and supports opportunities for knowledge dissemination (training, education, workshops, seminars, etc.),
- Guides people to become involved in the application of knowledge in the real, knowledge-driven world, where quality is the key to success.

more informal and more tightly connected. Networks have less identity and practice. However, networks are important in developing relationships and are more about sharing information (rather than knowledge) among individuals and organizations.

Apart from CoPs, other building blocks for knowledge networks are Internet-based learning and education, and Internet-based interactive platforms offering functionalities and services to the network members.

From a broader perspective that regards knowledge and capacity as part of a continuum, the family of partnerships, networks, and communities of practice that support cooperation in knowledge and capacity development in the water sector can all be considered water knowledge networks. Below, we see that there can also be networks within networks.

Luijendijk and Mejia-Velez (2005) argue that the strength of knowledge networking is that it starts where the knowledge exists, namely, with the professionals who form the backbone of knowledge networks by collaborating in CoPs. They recommend that knowledge networks can best be initiated, therefore, through a bottom-up approach involving the professionals, and later seek a more formal institutional embedding through agreements with sector institutions. This process is illustrated in Figure 4.3.

In this constellation, partnerships form the enabling environment that serves to promote and support knowledge networks. And knowledge networks can rely on CoPs, nodes or knowledge hubs, which are sub-groups or

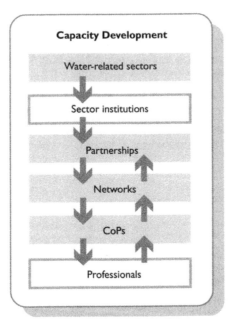

Figure 4.3 Institutional framework for capacity building (Luijendijk and Mejia-Velez, 2005).

sub-networks of the larger network that have accepted the role of focal point for a particular topic of interest or for a specific geographic region.

4.4 WATER KNOWLEDGE NETWORKING – A FRAMEWORK FOR FURTHER ANALYSIS

Further analysis and improvement of water knowledge networking can benefit from a three-pronged approach focusing on needs, results, and management. The first prong reviews how a water knowledge network caters to critical water sector needs. The second assesses what results are delivered by the network. The third reviews how well the process of networking is organized and managed. Key dimensions of each of these areas are explored in the following sections.

4.4.1 DEMAND-SIDE – WHAT THE WATER SECTOR NEEDS FROM NETWORKS

Key question: How can networks respond to priority water sector needs?

4.4.1.1 Water services and resources

Over the past decade, the needs for knowledge and capacity in the water sector have been better understood and articulated as countries and institutions around the world have carried out sector assessments and formulated policies, strategies and action plans on a more comprehensive basis. UNESCO-IHE itself has made a significant contribution to this area through its symposia on water sector capacity building in 1991, 1996, 2001 and 2007. The most recent symposium marked the Institute's 50th anniversary. Water policies increasingly distinguish between the governance of water as a service and water as a resource. There is also better understanding of the necessary national level superstructure of the sector, often referred to as national water reforms, or the "enabling environment"

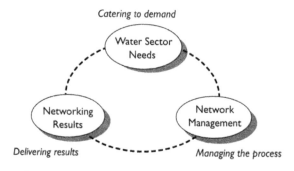

Figure 4.4 Analyzing water knowledge networks.

to oversee and facilitate the delivery of water services and the management of water resources. It is understood that all three areas (i.e. enabling environment, water services and water resources management) are in need of better knowledge and capacity development, and that dedicated knowledge networks can help to improve the performance of sector institutions in these three areas.

4.4.1.2 Involvement of civil society

The increasing involvement of civil society in the water sector is both a driver for change and an area in which more knowledge and capacity are needed. Civil society covers a multitude of different organizations. The Asian Development Bank regards civil society as including non-government organizations engaged in development and/or advocacy work, the academic and international research organizations, and parliamentarians. The media are either considered part of civil society or regarded as a separate group; however, the lines are increasingly blurred as media and communications modalities change and new formats like weblogs become popular.

4.4.1.3 Private sector participation

Participation of the private sector in delivering water services, and in supporting water resources management, continues to be a major challenge (issue) in both developing and developed countries around the world. What is clear is that government budgets alone will not suffice to finance the necessary investments. More knowledge is needed based on local experience and lessons learned, without ideological prejudice against, or in favour of, the merits of private sector involvement. Recent changes include the involvement of small entrepreneurs and medium-scale domestic firms to support water services, especially to poor communities (ADB, 2007b).

4.4.1.4 Changes abound

Much more work is needed to compile, analyse and compare experiences of water management in a decentralized environment, including the roles of local governments in water services and water resources management, and to determine their priority needs for information, knowledge and capacity.

Many of these needs reflect changes in the sector over the past decade. Water security, environmental integrity, urbanization, governance and the use of information and communications systems are some of the drivers for these changes. Policy- and decision-makers at national and local levels urgently need information and advice on how their jurisdiction will be affected by climate change, and what adaptive strategies and measures they can adopt.

Managers of coastal cities faced with increasing shortages of raw water resources and rising costs of developing new supplies want to know if large-scale desalination is becoming a feasible alternative to expand their water services. National and local decision-makers and river basin organizations are looking for advice to introduce explicit water-user rights through licensing.

Meanwhile, the challenge of restructuring urban water utilities to expand their services to poor communities and other customers remains as important as before, and mayors want to know what models of public-private partnership can be adopted to make this happen with the involvement of small entrepreneurs as well as medium and large private firms.

The paradigm of fixing environmental degradation only after economic development needs have been satisfied looks increasingly out of date, and knowledge is in high demand for proactive approaches that involve cost sharing by consumers, e.g. through payment for environmental services, adoption of environmental flows and sustainable wetland management.

The 2006 Stockholm Water Prize winner Asit Biswas predicted that "the world of water will change more in the next twenty years compared to the past 2,000. We must anticipate events and not follow them." Most of the changes will be driven from outside the water sector by population dynamics, trade policy and the growth of small cities, amongst others (SIWI, 2006).

4.4.1.5 Knowledge and ICT

The need for better knowledge networking in the water sector should also be seen in the broader context of countries taking deliberate steps to introduce knowledge-based economies (KBE) and knowledge-based development (KBD). This includes building ICT-supported networks that can benefit networking among water sector professionals. Some of these networks can address common areas of interest such as improving governance and enhancing civil society, which can have direct application and benefits for the water sector. Among developing countries, the recent emphasis on initiatives for managing for development results is one such example (Boom, 2007).

4.4.1.6 Regional development

Regional differences may, of course, also give rise to different demands for knowledge networking. For economies with highly developed institutional capacities, the demand for water knowledge networking would be different than for countries with low levels of capacity. Regions might also express demand for networks that amplify the identity, issues, solutions and social capital of the region. For example, in 2006, the countries of the Asia-Pacific region acted on their desire to establish a water network that would encompass all of the region, rather than only the sub-regions for which water networks (partnerships) were already established. The Asia-Pacific Water Forum was, thus, announced at the 4th World Water Forum and launched later that year in Manila. Similarly, countries in Africa had earlier moved to establish water sector networking, which helped to articulate the needs and strategies of that region.

4.4.2   SUPPLY-SIDE – WHAT NETWORKS SHOULD DELIVER

Key question: How can networks focus on delivering results?

4.4.2.1   Outputs and outcomes

The merits of outcome-based approaches and managing for development results are widely recognized, yet their introduction is still relatively recent. Many organizations in the world are still using planning and reporting systems that focus predominantly on inputs, activities and sometimes outputs. A quick review of websites of water knowledge networks currently in existence reveals that many appear to be focused primarily on implementing activities under a broad umbrella of lofty objectives. Simple outputs are often mentioned, like the number of people trained. It is still rare to find descriptions of specific network services and products, and their intended outcomes. And rarely is there mention of expected outcomes in terms of measured performance improvement in network member organizations.

There is no doubt that a focus on results helps networks (and organizations) continually verify (regularly review) their relevance, effectiveness and efficiency. In contrast, an activity-driven approach risks missing out on their significance, effectiveness and efficiency, particularly in a sector that is subject to profound and rapid changes. Networks that do not focus on delivering specific results tend to rely on the satisfaction of their members for their continued existence. This satisfaction is derived from activities the members are used to and comfortable with, at least for as long as they can raise the necessary resources to continue the operation of the network.

An equally important challenge for networks is to keep a practical focus and remain grounded in reality. There is a need to guard against focusing on the virtual concepts of models (like IWRM) such that they replace the actual realities and needs on the ground. In their technical paper on water rights and allocation, Bird et al. (2009) urge members of the Network of Asian River Basin Organizations to keep a practical focus on solving and avoiding problems in their river basins while continuing to help build the enabling environment for IWRM over a longer term.

4.4.2.2   Determination of results

If the case for focusing on the delivery of results is acknowledged, the question arises: What types of results could a water knowledge network set out to achieve? The definition of knowledge as a "capacity to act" offers us some direction. A focus on results can be strategic (what to do), operational (how best to do it), and related to performance improvement (how to do it better).

Policy advice might be seen as a strategic result of networking. Toolkits, on the other hand, are useful for organizations and individuals to know how best to carry out activities. And, benchmarks are important for setting standards for better organizational performance.

The distinction between explicit and tacit knowledge is also helpful. The process of making tacit knowledge explicit and usable for sharing is an important network challenge. Referring back to Nonaka's cycle, networks could consider which tacit knowledge is externalized and socialized, and when and how this conversion would take place.

If knowledge is defined as the capacity to act, knowledge networking could be focused on a wide array of results. If the network focuses on producing knowledge products and services (strategic, operational or performance oriented), questions could be asked about who benefits from these services, and which network's clients' needs should be satisfied. If the network focuses on research, questions could be asked about the clients, topics and application. Networks that are focused on capacity development can be asked who their clients are, if the clients are in the driver's seat, and if capacity development activities are based on thorough and participatory diagnostic assessments. Networks can also focus on advocacy and the promotion of good practice, in which case results are linked to clear messages and their adoption by targeted audiences or client groups.

Lank (2006) suggests that there are eight principal reasons for organizations to collaborate, and these can be used to provide additional insight into what could be desired results of networking: 1) more effective research; 2) greater influence; 3) increased probability of winning business; 4) faster, better or cheaper development of products, services or markets; 5) faster, better or cheaper delivery of products or services; 6) in-depth learning; 7) meeting an external requirement; and, 8) saving costs. Reasons 1, 2 and 6 seem clearly relevant to water knowledge networking, while reason 8 might provide a result for networking in areas where human and financial resources in the water sector are severely constrained, for example in the Pacific island countries.

Lank also suggests that collaboration is concerned with the process of working together to achieve one or more specific outcomes, a focus on actions not just words, and collaborative leadership and consensus building. These distinctions provide further thought towards assessing the effectiveness and results of a network. Are outcomes achieved through collaboration? Do the results go beyond words? Has the networking produced collaborative leadership and consensus building?

Depending on specific objectives and circumstances of water knowledge networks, a number of other dimensions may also be important in measuring their success. For example, one might ask how the network 1) supports the MDGs by improving the water security of the poor; 2) focuses on increasing local capacity, ability and skills (the capacity to act); 3) facilitates a division of responsibilities by agreeing which organization is responsible for focusing on what topic; 4) provides advice to planners and decision-makers outside the water sector; 5) involves civil society, the private sector and the public; and 6) helps local organizations enhance their knowledge and capacity.

4.4.2.3   Measurement of results

Current practice in water networks seems often limited to measuring outputs, including the number of members, events organized, materials and tools produced, and people trained (including trainers and capacity builders). More consideration is needed to develop a framework for determining and measuring outcomes. Not all can be measured quantitatively (e.g. the extent of empowerment, or the use of policy advice); hence, the framework should include qualitative and narrative indicators. A breakdown of results by a target group will also be important, including the results that are considered to be relevant to enhancing local knowledge and capacity.

4.4.3   MANAGING THE PROCESS – HOW NETWORKS CAN WORK BETTER

Key question: How can networks organize for success?

In this section we focus on the "how-to" questions of effective partnering and organizing for success. We also refer to some recent developments that networks take advantage of, and recall some elements of earlier wisdom that continue to be relevant. A number of perspectives are offered for consideration.

4.4.3.1   Success factors

Are there common factors that will guarantee a successful network? The answer is both yes and no. There is enough knowledge about networks to point to some factors that are necessary for success in most cases. Networking, however, remains a fast-developing phenomenon, and new insights are emerging continuously.

"Trust is the basic lubricant for networking and sharing knowledge" (GTZ, 2006). Successful networks tend to operate informally with only a few rules. For example, there may be no rules for non-disclosure of important findings of a network member. Boom (2007) therefore contends that there are three important success factors: trust, a common goal, and the need to know other members personally (not only through cyberspace). GTZ (2006) asserts that good network management, transparency and trust are preconditions to involve decision-makers in networks. Gloor (2006) points out that effective networks for innovation are marked by high degrees of connectivity, interactivity and sharing.

Initiative, leadership and vision by members are also important for network success. Networks need champions, leadership, multipliers and standards (for example, benchmarks for performance of member organizations). Networks can promote organic and incremental growth of knowledge and capacity among their members. However, in the case of paradigm shifts being introduced, champions are needed, followed by leaders to internalize the changes and push through change against inertia.

Some networks, like professional associations and those focused on research, expect sustained operation over a longer term. However, such sustainability may not be necessary or even possible for networks focused on innovation. Permanence is therefore an option to be selected, with success factors changing accordingly. For networks seeking sustained operation, financial contributions from members will be more important than for temporary networks that might reach their objective with time-bound support from one or more sponsors.

Wenger et al. (2002) point out that the level of energy and visibility in communities of practice often varies, in accordance with five stages of development, with corresponding developmental tensions. The first stage is marked by the discovery or imagination of potential, and is followed by coalescence with a choice of incubation or delivery of immediate value. During the next stage of maturing, the community will need to choose between focusing and expanding. At its peak stage of stewarding, questions of ownership versus openness arise. And, in the final stage of transformation of the community, the issue may be to let go or to live on.

4.4.3.2   Constraints to success

When the success factors for effective networking are not achieved, problems are bound to occur, and they need addressing. However, further analysis of networking constraints is needed since not a great deal of information is available from water knowledge networks about factors that hold back their performance. In addition to the commonly raised issues of insufficient budgets and networking hardware, two factors seem particularly important.

The first is the ability of networks to reach decision-makers at both national and local levels with their products and services. This, for example, is a concern heard about networking among research organizations. The networking itself may be experienced as successful because it satisfies the professional interests of the participating researchers. However, achieving the outcome of the research depends on "clients" outside the network, and remains a challenge.

A second concern is how water organizations can be persuaded to spend more time and effort on networking. Ask a cross-section of staff in national and local governments and in development organizations, including the multilateral development banks, how much time they spend as a consumer or member of networks (by regular web surfing and reading), and the result is likely to be much lower than expected. This is partly because many organizations have not yet recognized the benefits of networking for their own work, lack a corporate policy to promote networking, have yet to put in place staff incentives to spend time on networking, and have no organizational focal point to coordinate this.

4.4.3.3 Digital divide

While ICT is driving development and networking forward with ever greater connectivity and speed, the prevailing digital divide in the world continues to impede access to such networking by many local practitioners and poor communities in the developing world, and also by the elderly who feel unable to participate. Unless water knowledge networks can find ways to bridge these divides, the risk is that the social capital of local traditional and indigenous water knowledge will be marginalized into extinction.

Also, it can be noted that modern ICT-supported networks are often premised on models, prescriptive solutions, and innovation through (perceived) paradigm shifts, while traditional low-tech communities in developing countries have a heritage of cultivating consultative approaches and incremental improvements. Modern ICT-supported water networks could, therefore, be encouraged to reach out and accommodate knowledge and social capital from low-tech local communities in developing countries. To enhance knowledge and capacity development at local levels, a combination of high-tech and low-tech networks, or a cross-fertilization among such networks, might be needed, with proper interfaces that need developing.

4.4.3.4 People for networking

Knowledge networks are networks of people and they therefore rely on them. However, reviews of networking experience and performance tend to overlook this obvious fact. By comparison, much more attention has been devoted to the analysis of the organization and ICT application for networking. The empowerment of individuals is, however, a key ingredient to capacity building of organizations and networks, together with the cultivation of vision, mission and values. Coaching and mentoring have a high return on investment for the enhancement of networking for knowledge and capacity building.

Lank (2006) distinguishes some typical functions and roles of individuals in collaborative ventures such as networking. She identifies organizational sponsors, gatekeepers (relationship and partnership managers), partnership coordinators, advisory partnership facilitators and project (network) managers. She also recommends that organizations consider appointing a chief relationship officer, who would help with partnering and networking. Commenting on the specific leadership qualities required for working collaboratively, she quotes Doz and Hamel (1998) who wrote that "executives do not wake up one morning with an unexplained urge to collaborate. It is not in their nature."

4.4.3.5 Networks within networks

Networks are not homogeneous. They often comprise groups within the network. Gloor (2006) identifies three types of networks within each other, which are similar to those described by GTZ (2006). The larger network is seen as a collaborative *insight* network which helps people with a shared interest. GTZ refers to these people as the lurkers in the network. Within that larger network, some members (people and/or organizations) take a more active role in sharing knowledge and act as a collaborative *learning* network that focuses on stewarding of best practices. Finally, at the core of the other two, an even smaller group works as a collaborative *innovation* network with total dedication to generating fundamentally new insights. These groups within the network are shown in Figure 4.5.

Gloor describes a number of conditions for collaborative innovation networks (COINS) to be successful, including 1) being a learning network, 2) having an ethical code, 3) being based on trust and self-organization, 4) making knowledge accessible to anyone, and 5) operating with internal honesty and transparency. He claims that the combination of the three networks creates a strong ripple effect.

4.4.3.6 Cutting across boundaries

Today's ICT-supported knowledge networks can cut across boundaries and levels in hierarchies. Network members, however, are still tied to professions, position levels and hierarchies within their organizations, and to the position of their organization in the constellations of local, regional, national and international levels. This raises the question of who should be linked to whom in effective knowledge networks, and this question is particularly relevant

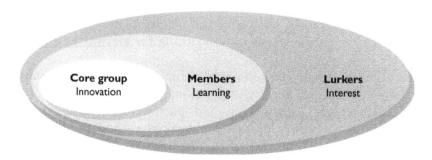

Figure 4.5 Groups within networks (after GTZ, 2006 and Gloor, 2006).

when we focus on enhancing knowledge and capacity at the local government level. Assuming that more networks will emerge in the coming years to link practitioners at local levels with each other, more knowledge is needed on how this can best be done. With the wide use of the Internet and ICT, open access networks are expected to increase, and individuals can take part in several networks at the same time.

Language is another issue to be considered. Most local government practitioners will want to communicate in their country's language, or even in local languages. Products and services at the local level will need to be disseminated in the appropriate language. Most of the ICT-supported water knowledge networks seem, however, to operate in one language or a few languages, and the questions arise as to who will network with whom in what language, and how language will be factored into a hierarchy or constellation of intersecting networks.

Another dimension of language in networking concerns the use of technical and scientific jargon versus language for consumption by decision-makers and the general public. Networks need to consider this for effective operation and to reach their intended clients.

4.4.3.7 Assessments to do

Many knowledge networks engage in regular surveys or assessments, and there seems to be scope for improvement by sharing knowledge and experience on how this can best be done. In a departure from past practice, more surveys and assessments will need to focus on the needs of local practitioners and decision-makers.

Gloor and Cooper (2007) identifies three tools: 1) the knowledge map that focuses on "what is" and provides global or local knowledge about a topic, 2) the talent map that focuses on "who is" and shows who has the expertise on the topic, and 3) the trend map that focuses on "what will be" and presents global or local trends about the topic.

Capacity-building activities, including those in the water sector, are all too often designed based on training needs assessments rather than more comprehensive diagnostic assessments that take into account the individual and institutional levels and their enabling environment (refer to Figure 4.1). Networking that aims to enhance capacity will therefore need to engage more in diagnostic assessments with the full ownership of the organizations concerned.

4.4.3.8 Stories that work

Why is it worth remembering that one of the most effective tools of knowledge management is the sharing of stories and anecdotes? It is because people remember and easily identify with them. The majority of management, leadership and self-enrichment literature published in the past decade has made extensive use of stories and anecdotes to get its message across to audiences. Water knowledge networks might consider adopting a similar approach in delivering some of their key messages through stories and anecdotes during networking activities and on their websites, and in newsletters and publications. This may also help in reaching out to local level practitioners and decision-makers, and to externalize tacit knowledge.

4.4.3.9   Networking push and pull

Successful networking requires the delivery of products and services (push) that add value and thereby attract (pull) existing and new members. More research seems needed on the types of products and services that consistently produce added value and create "pull". For example, anecdotal feedback to the authors suggests that network members might value receiving a regular synthesis of good practices in their topic of interest, for example an annual overview of what is happening in countries around the world on the topic of water legislation.

4.4.3.10   Networking everywhere

Recognizing the need to make the best use of time and financial resources, the organizers of large water events are increasingly supportive of organizing side events, and these provide a cost-effective opportunity for networks to conduct face-to-face meetings with their members. With some advance planning on the part of the host organization, even more use could be made of water events by creating opportunities for country delegations to meet and forge partnerships, ranging from a straightforward exchange of information to setting up exchange visits, staff exchange programs and developing joint programmes of collaboration or twinning. Establishing such partnerships could also be recognized as one of the many objectives of water events.

4.4.3.11   Incentives for change

Of all the products and services that networks can offer their clients, which ones are most likely to trigger change and improvements in performance? The ADB's recent experience in helping regional water networks introduce performance benchmarking and peer reviews has been welcomed by participating member organizations, some of which have given feedback that it has already changed the mindset of their staff. The ADB is supporting separate networks for water utilities, river basin organizations and national water sector apex bodies, and all are now benefiting from performance benchmarking, and the latter two from peer reviews (ADB, 2007c).

4.5   KEY MESSAGES AND PROPOSED ACTIONS

During an international symposium organized by UNESCO-IHE in June 2007 under the title "Water for a Changing World: Enhancing Local Knowledge and Capacity", participants unanimously agreed that networks are proving their value in disseminating and sharing knowledge. Networks are increasingly effective in sharing tacit knowledge, identifying common problems, building attitudes and confidence, and generating new knowledge. Water knowledge networks are becoming important vehicles for knowledge and capacity development in the sector.

The following key messages and short-term actions were agreed upon and served in this chapter as a reference for the discussions leading up to and during World Water Forum 5 in March 2009.

Key Messages:

- Water knowledge networks help water leaders and practitioners in addressing the complex challenges in today's fast-changing world.
- These networks can play an important role in shuttling information and knowledge among local, regional and global levels.
- Successful water knowledge networks have a clearly defined purpose, operate with transparency and trust among members, and deliver value to members and clients.
- Access to knowledge networks can empower local communities and vulnerable groups to take action to improve their water security.
- Since effective solutions are increasingly generated at local levels, knowledge networks should increase vertical connectivity to involve local stakeholders and tap their experience.
- Development partners and funding agencies should facilitate water knowledge networks among practitioners and support their activities, including face-to-face and video meetings and project visits, as these are critical in sharing important tacit knowledge in developing and adapting solutions.
- Funding is also needed to mobilize a critical mass of trainers in developing and managing effective knowledge networks that capitalize on the benefits of knowledge sharing.
- Decision-makers and managers should support their knowledge workers to engage actively in knowledge networks as a core part of their daily tasks.
- These networks will then play an important role in meeting the large and growing demand for training, innovation and research in water

organizations at all levels, by facilitating collaboration and removing barriers for sharing experiences and knowledge.

- Educational institutions should also capitalize on knowledge networking, as they have as much to gain from collaboration as from competition.
- Networking among educational institutions can involve matchmaking services for collaborative research projects, joint curriculum development, as well as staff exchanges.

Short-term Actions:

- Build regional water knowledge networks with lead organizations and knowledge hubs, using global experience in effective knowledge networking.
- Agree on common but differentiated responsibilities of water knowledge networks (knowledge, research, capacity, advocacy).
- Develop interfaces between local, country, and international networking (including the use of multiple languages).
- Develop coaching programs to promote leadership in networking.
- Build effective CoPs within water organizations.
- Develop incentive schemes for water organizations' staff to contribute and share knowledge through CoPs and networks.
- Equip and retool water organizations for better networking and knowledge sharing skills.
- Invest to bridge the digital divide (poverty, elderly).
- Develop and disseminate knowledge networking etiquette, values and standards.

4.6 OVERSEEING THE SCENE – NETWORKING GALORE

In this paper, new developments and directions have been explored in knowledge networking for the water sector. Finally, some relevant water-related networks of various types that are already in operation are listed below. In Section 4.7 some of these networks have been described in more detail.

4.6.1 GLOBAL NETWORKS

Water networks at the global level include the Global Water Partnership (GWP), the CAP-Net network for capacity development in integrated water resources management (IWRM), the International Network of Basin Organizations (INBO), the Partnership for Water Education and Research (PoWER), Streams of Knowledge (STREAMS), the Gender and Water Alliance (GWA), the Water Integrity Network, UNESCO's International Hydrology Program (IHP), the Global Environment Facility's International Waters Learning Exchange and Resource Network (IW:LEARN). Professional networks include the International Water Association (IWA), the International Water Resources Association (IWRA), and the American Water Works Association (AWWA). Global knowledge networks of interest to the water sector include the World Bank's Global Development Learning Network, the Managing for Development Results Community, and others.

4.6.2   REGIONAL NETWORKS

Examples of water networks at the regional level include the Asia-Pacific Water Forum (APWF) and its Network of Regional Water Knowledge Hubs (APWF *KnowledgeHubs*), the River Network (North America), the European Water Partnership (EWP), Nile-Net, the Nile Basin Capacity Building Network for River Engineering (NBCBN-RE), WaterNet (Southern Africa), the Network of Asian River Basin Organizations (NARBO), the Latin American Network of Basin Organizations (LANBO), the Southeast Asian Water Utilities Network (SEAWUN), the South Asian Water Utilities Network (SAWUN), the Central Asia and South Caucasus Water Utilities Network (CASCWUN), the Africa Water Utilities Partnership, Water for African Cities, Water for Asian Cities, Latin American Network for Water Education and Training (LA-WETNet), Arab Integrated Water Resources Management Network (AWARENET), and networking among national water sector apex bodies in Asia.

4.6.3   COUNTRY NETWORKS

A number of country water partnerships have been established around the world in response to the Global Water Partnership and its regional water partnerships. Consider, for example, the Indonesia Water Partnership and the Netherlands Water Partnership, and also the Collaborative Knowledge Network Indonesia (CKNet-INA), the Partnership for Capacity Building in IWRM Indonesia, and various country Cap-Nets, e.g. the Cap-Net in Brazil.

4.7   EXAMPLES OF WATER NETWORKS

4.7.1   GLOBAL NETWORKS

Cap-Net (http://www.cap-net.org) is an international network made up of autonomous international, regional and country networks and institutions committed to capacity building in integrated water resources management (IWRM). The network's mission is to enhance human resources development for IWRM by means of establishing or strengthening regional capacity-building networks. Cap-Net was established in 2002 and is a UNDP project, funded by the Netherlands Ministry of Foreign Affairs. The Global Water Partnership (GWP) has adopted the Cap-Net project as an associated programme and considers it one of its flagshi p projects. Cap-Net is hosted by UNESCO-IHE Institute for Water Education in Delft (The Netherlands). Cap-Net has three major interlinked lines of activity: 1) Networks: support the establishment, facilitate access to information and exchange of experience between regions; 2) Capacity building: analyse capacity-building needs, improve capacity-building materials, and assist in the development of capacity-building strategies; 3) Website to disseminate information on training programmes and courses: make training materials available, provide information on national, regional and global networks. Results include 1) 550 trainers who have in turn impacted thousands of decision-makers, water managers, and fellow capacity builders, thereby exponentially increasing capacity in IWRM; 2) establishment or strengthening of 20 geographical and four thematic networks affiliated with Cap-Net; 3) over 1,000 member institutions organized in regional and country partnerships; 4) network training events and education programmes; and, 5) nine operational topic or geographic e-discussion groups on capacity building in IWRM.

Global Water Partnership

The **Global Water Partnership (GWP)** (http://www.gwpforum.org) was formed in 1996 with the vision of developing a worldwide network that could pull together financial, technical, policy and human resources to address the critical issues of sustainable water management. It sees itself as a worldwide partnership for all those involved in water management: government agencies, public institutions, private companies, professional organizations, multilateral development agencies and others who are committed to the Dublin-Rio principles. GWP identifies critical knowledge needs at global, regional and national levels, helps design programmes to meet these needs, and serves as a mechanism for alliance building and information exchange on integrated water resources management. The mission of the Global Water Partnership is to *"support countries in the sustainable management of their water resources"*. GWP's objectives are to clearly establish the principles of sustainable water resources management, to identify gaps and stimulate partners to meet critical needs within their available human and financial resources, to support action at the local, national, regional or river basin level that follows principles of sustainable water resources management, and to help match needs to available resources. GWP's architecture comprises twelve regional water partnerships across the world, an annual meeting of consulting partners (the members), a steering committee, a technical committee of up to 12 internationally recognized professionals, a financial partners group, and the Secretariat in Stockholm which is supported by advisory centres in Denmark, the UK and Sri Lanka. GWP has fostered the establishment of many country water partnerships in developing and developed countries across the world, for example the Indonesia Water Partnership (http://www.inawater. com/) and the Netherlands Water Partnership (http://www.nwp.nl).

STREAMS of Knowledge (Global Coalition of Water and Sanitation Resource Centres) (http://www.streams.net/) (STREAMS) describes itself as an autonomous, self-sustained, recognized and credible leader organization of resource centres, uniquely positioned locally and globally, and effective in capacitating the sector in supporting the development of pro-poor policies for the sustainable implementation of water, sanitation, health and hygiene.

The mission of the **Gender and Water Alliance** (GWA, http://www.genderandwater.org/) is to promote women's and men's equitable access to and management of safe and adequate water for domestic supply, sanitation, food security and environmental sustainability. GWA is active in 56 countries in Sub-Saharan Africa, Asia, Eastern Europe, Europe, Latin America and the Caribbean, Middle East and North Africa, North America and Oceania. GWA operates as an associated programme of the Global Water Partnership. Outputs include electronic conferences, gender ambassadors' reports of major events, an annual gender and development report, a best practices and lessons learnt booklet, case study compilation, gender policy development guide and advocacy material (brochures, posters, postcards, checklists, video), monitoring and evaluation frameworks, training of trainers manual for gender ambassadors, training-of-trainers programmes, a manual for gender and water, and pilot-project documentation and reports.

PoWER.

The **Partnership for Water Education and Research** (PoWER) is a network of 18 educational institutions in the field of water that aims to build capacity in developing countries by training people to become qualified professionals in the field of integrated water resources management (http://www.unesco-ihe.org/power/about.htm). The network was formed in 2001 in response to urgent appeals to support the millennium development goals (MDGs) agreed on at the World Summit on Sustainable Development in Johannesburg. PoWER's mission is to combine the strengths of all partners to enhance the capacity of each partner.

It mobilizes and shares knowledge globally to deliver it locally. PoWER enables people to acquire the specialist knowledge they need in the fields of water management and environmental resources, helps strengthen the capacities of its partner institutions and stimulates the use of innovative educational methods and techniques. PoWER builds up the capacity of institutions and communities, enhancing a quality system for water education. In the process of combining strengths and levelling the capacities of the individual partners, joint products in the field of education, training and collaborative research are developed in a multi-disciplinary manner. These products are demand-responsive, validated to meet mutually agreed upon standards. PoWER's annual results include 1) delivering 1,000 postgraduates; 2) training 3,000 participants through short courses; 3) reaching 800 decision-makers through seminars; and, 4) connecting 3,000 professional participants in e-seminars, videoconferences, and face-to-face learning sessions. PoWER partners are involved in 130 research projects, including research at PhD level, and have an alumni community of more than 18,000 professionals. PoWER is guided by a task force, technical committee, panel of rectors, vicechancellors and directors, academic working groups for joint products, and its alumni community.

4.7.2   REGIONAL NETWORKS

Nile Basin Capacity Building Network
for River Engineering

The **Nile Basin Capacity Building Network for River Engineering** (http://www. nbcbn.com/Home.asp) was established in 2002 as a regional knowledge network to strengthen the human and institutional capacity of the riparian countries managing the water resources of the Nile River Basin. The network's long-term objective is to contribute to the establishment of an overall knowledge network for the Nile region (Nile-Net) as a means to support stability and solidarity, and to support the activities of the Nile Basin Initiative in building strategic partnerships between water professionals, research and government authorities in a sensitive environment in Africa. Members are water sector practitioners and institutions from all ten riparian countries, with over 200 water professionals collaborating in 13 communities of practice. The network focuses on building an environment for stimulating and supporting collaborative applied research. It supports the development activities of in-country "nodes" as well as joint regional research "clusters" in which all countries participate. Six country nodes serve as host institutions for regional research clusters: Egypt for GIS and modelling, Ethiopia for river structures, Sudan for river morphology, Tanzania for hydropower, Uganda for environmental management, and Kenya for flood management. Regular face-to-face research cluster events organized in each of these hosting countries are complemented by interaction through the website, which offers the members a virtual meeting space to work collaboratively regardless of time and space. (Luijendijk et al., 2000, NBCBN-RE 2005).

The mission of the **River Network** (http://www.rivernetwork.org) in North America is to help people understand, protect and restore rivers and their watersheds. Established in 1988, it supports grassroots river and watershed conservation groups. With 19 staff members working in three offices across the USA, the network connects over 4,000 organizations. It cooperates with local watershed protection groups, state river conservation organizations, tribes and schools. The network operates a river source information centre, provides publications, training and consultation, offers a grant programme, and provides referrals to other service organizations and networking opportunities.

Information on global and regional **river basin networks** can be found at International Network of Basin Organizations (INBO) (www.inbo-news.org/), Latin American Basin Organization (www.inbo-news.org/relob/Lanbogwp.PDF) and the Network of Asian River Basin Organizations (NARBO, www.narbo.jp). NARBO has pioneered a process for performance benchmarking of river basin organizations (www.adb.org/Water/NARBO.asp) supported by peer reviews.

Building Capacity for Water Resources Managment in Southern Africa

WaterNet (http://www.waternetonline.ihe.nl/) is a UNESCO-IHE supported regional network of university departments and research and training institutes specializing in water. Its mission is to enhance regional capacity in IWRM through training, education, research and outreach by sharing the complementary expertise of its members, based in Botswana, Kenya, Lesotho, Mozambique,

Namibia, South Africa, Tanzania, Zambia and Zimbabwe. The network's strategy is to 1) raise awareness for IWRM, 2) stimulate regional cooperation, 3) increase access to training and education in IWRM, and 4) stimulate and strengthen research in IWRM in Southern Africa. Among its results to date are 1) scientific papers published in *Physics and Chemistry of the Earth*, an Elsevier international peer-reviewed science journal, 2) a modular master's degree programme with five specializations (water resources management, hydrology, water and environment, water for people, and water and society), with full scholarships to pursue the programme, 3) English language courses to ease access to the courses delivered to participants from non-English speaking countries of the area, and 4) collaboration with other networks (such as the Water and Sanitation Program Southern Africa and Cap-Net) to facilitate training for country water partnerships in compiling national IWRM plans. Research is stimulated through Water-Net's symposia, staff exchanges between institutions in the region, the master's theses research projects supported by the WaterNet Fellowship Fund, and collaboration with the Water Research Fund for Southern Africa resulting in some synergies in research support.

The **Asia-Pacific Water Forum** (http://apwf.org/) was launched in 2006 as an independent and inclusive network for sustainable water management in the Asia-Pacific region. The network aims to capitalize on the region's diversity and rich history in dealing with water as a fundamental part of human existence. Specifically, the APWF will help boost investments, build capacity and enhance cooperation. The network is organized to deliver results under three priority themes (water financing, disaster management and ecosystems) and five key result areas (developing knowledge and lessons, increasing local capacity, increasing public outreach, monitoring investments and results, and supporting the forum and regular summits for leaders of government, private sector and civil society). Each theme and key result area has a lead organization for coordinating work planning and implementation. PUB Singapore and UNESCO are co-lead organizations for the first key result area of developing knowledge and lessons, and the Asian Development Bank and UNESCO-IHE have facilitated the launch of a network of regional water knowledge hubs to address priority water sector topics (http://www.apwf-knowledgehubs.net). The network was announced at the first Asia-Pacific Water Summit in December 2007 in Beppu, Japan, and launched during the Singapore International Water Week in June 2008. Japan Water Forum serves as the Secretariat of APWF, while Singapore chairs the APWF Governing Council. The secretariat of APWF *KnowledgeHubs* is hosted by PUB Singapore.

Networking of water sector apex bodies in Asia (http://www.adb.org/Water/ NWSAB/). National water sector apex bodies in Asia, such as national water resources councils and boards, started networking in 2004 with the support of the Asian Development Bank. To date, ten countries in the region have established a national water sector apex body, and more are expected to do so in the near future. No formal network has yet been established, but regional meetings have resulted in agreements to work on priority activities, starting with the introduction of performance benchmarking and peer reviews. The apex bodies of the Philippines and Thailand have already completed a peer review process.

The **South East Asia Water Utilities Network** (SEAWUN) (http://www.seawun. org/) was established in 2002, with the support of the Asian Development Bank, to help utilities improve their performance in the delivery of water supply and sanitation services for all, including operation and management efficiency, achieving financial viability, and advocating for sector reforms for improved policy environment, contributing to realize the goal "Water for All". The vision is to develop a strong regional, non-profit making and self-sustainable organization, which is demand driven, by focusing its activities on the key issues agreed upon by the members of the organization. The network's five priority programmes are benchmarking, cost recovery, human resources development, unaccounted-for-water reduction, and regular member needs surveys. The Secretariat of SEAWUN is located in Hanoi, Viet Nam. Building on the experience of SEAWUN, the **South Asia Water Utilities Network** (SAWUN, http://www. adb.org/water/SAWUN/) was established in 2007.

M-POWER

Mekong Program on Water
Environment and Resilience

The **Mekong Program on Water Environment and Resilience** (M-Power, http:// www.mpowernet.org) works as a network of organizations to help democratise water governance and support sustainable livelihoods in the Mekong Region through action research. Activities are undertaken throughout mainland Southeast Asia including major river basins such as the Irrawaddy, Salween, Chao Phraya, Mekong, and Red as well as other smaller basins. Water governance involves negotiating decisions about how water resources are used. Benefits and risks – both voluntary and involuntary – are redistributed by such decisions.

4.7.3   COUNTRY NETWORKS

The **Collaborative Knowledge Network (CKNet-INA)** aims to increase the capacity of leading Indonesian universities in delivering capacity building services for water professionals and institutions. The network was established in 2005 with ten member universities working together in capacity building in the field of infrastructure, water and environmental management. The objective of the network is to strengthen water sector performance and support water sector reform activities. The network focuses on building capacity in Indonesian universities to deliver demand-based training courses in water resources management at national, regional and local levels for both individuals and institutions. Development of decentralized capacity building and training applications are planned, including an e-learning system for online education and training. Planned topics include water quality management, and gender and water management. Partners collaborate using a communications network for knowledge sharing. The network recently joined the Platform for Capacity Building in Integrated Water Resources Management in Indonesia (http://www.cbiwrm.ihe.nl/) which aims to facilitate communication and collaboration between the networks, institutions, and their individual members involved in capacity building for IWRM in Indonesia.

NOTE

The views expressed in this article are those of the authors and do not necessarily reflect the views and policies of UNESCO-IHE, the Asian Development Bank or its Board of Governors, or the governments they represent.

REFERENCES

Asian Development Bank, 1995a. The Bank's Policy on Agriculture and Natural Resources Research. Manila: ADB. http://www.adb.org/Documents/Policies/Agriculture-Natural-Resources/agri-natural-resources.pdf

Asian Development Bank, 1995b. Governance: sound development management. Manila: ADB. http://www.adb.org/Documents/Policies/Governance/default.asp?p= policies

Asian Development Bank 2005. Asia Water Watch 2015: Are Countries in Asia on Track to Meet Target 10 of the Millennium Development Goals? Manila: ADB.

Asian Development Bank 2006. Final Report of the Pilot and Demonstration Activity: Bang Pakong Dialogue Initiative.

Asian Development Bank 2007a. Toward a New Asian Development Bank in a New Asia: Report of the Eminent Persons Group to the President of the Asian Development Bank. Manila: ADB. http://www.adb.org/Documents/Reports/ EPG-report.pdf

Asian Development Bank 2007b. Water and small pipes: what a slum wants, what a slum needs. *Water for All e-Newsletter* May 2007. Available at http://www.adb. org/Water/actions/phi/water-small-pipes.asp

Asian Development Bank 2007c. Information on networks and partnerships. http:// www.adb.org/water/operations/partnerships

Bird, J., Lincklaen Arriëns, W. and Custodio, D. 2009. Water rights and water allocation: issues and challenges for Asia. Manila: Asian Development Bank.

Boom, D. 2007. Unpublished draft paper on knowledge economies in Asia for a seminar at the ADB Institute. Manila: Asian Development Bank.

Cairncross, F. 1997. *The death of distance.* Boston: Harvard Business School Press.

Doz, Y. and Hamel, G. 1998. *Alliance Advantage: The art of creating value through partnering.* Boston: Harvard Business School Press.

Fukuda-Parr, S., Lopes, C. and Malik, K. 2002. *Capacity for development. New solutions to old problems.* London: UNDP, Earthscan. http://capacity.undp.org/ books/book1.htm

Gloor, P. 2006. *Swarm creativity: competitive advantage through collaborative innovation networks.* Oxford: Oxford University Press.

Gloor, P. and Cooper, S. 2007. *Coolhunting: chasing down the next big thing.* New York: Amacom.

GTZ 2006. Work the Net – A management guide for formal networks. New Delhi, March 2006.

Japan Water Forum 2006. Asia-Pacific Regional Document for the 4th World Water Forum Final Report. Mexico. http://www.apwf.org/archive/documents/ APSynthesis_PartI.pdf

Lank, E. 2006. *Collaborative advantage – how organizations win by working together.* New York & Basingstoke, New Hampshire: Palgrave Macmillan.

Luijendijk, J., Boeriu, P. and Saad, M.B. 2000. Capacity building in the water sector in Africa. Proposal for a Nile river network. In *Symposium proceedings "The learning society and the water environment", Paris, 2–4 June 1999.*

Luijendijk, J. and Mejia-Velez, D. 2005. Knowledge networks for capacity building: a tool for achieving the MDGs? In *Workshop proceedings on design and implementation of capacity development strategies, Beijing, China, September 2005.* Rome: IPTRID Secretariat, FAO.

NBCBN-RE 2005. *Knowledge networks for the Nile basin: Using the innovative potential of Knowledge Networks and CoP's in strengthening human and institutional research capacity in the Nile region.* Delft: NBCBN-RE.

Nonaka, I. and Takeuchi, H. 1995. *The knowledge-creating company; how Japanese companies create the dynamics of innovation.* New York: Oxford University Press.

Stockholm International Water Institute (SIWI) 2006. World Water Week 2006 Synthesis Report. http://www.siwi.org/downloads/WWW-Symp/2006_synthesis_web.pdf

Takeuchi, H. 2001. Towards a universal management concept of knowledge. In I. Nonaka & D. Teece (eds.) *Managing Industrial Knowledge*: 315–329. London: Sage.

UN-HABITAT 2006. *State of the world's cities report 2006/7*. Nairobi: UN-HABITAT.

van Hofwegen, P. 2004. Capacity-building for water and irrigation sector management with application in Indonesia. In Capacity development in irrigation and drainage; *Proc. of the International Workshop during the International Commission on Irrigation and Drainage, Montpellier, France, 16 September 2003*. FAO Water Reports 26. Rome: FAO.

Wenger, E., McDermott, R. and Snyder, W. 2002. *Cultivating communities of practice: A guide to managing knowledge*. Boston: Harvard Business School Press.

World Bank 2007. *Staff analysis of water investments*. Washington DC: World Bank.

World Water Council 2000. Ministerial Declaration of The Hague on Water Security in the 21st Century. The Hague: The Netherlands. <http://www.waternunc.com/gb/secwwf12.htm>

World Water Council 2003. The 3rd World Water Forum Final Report. Financing Water Infrastructure Statement. Kyoto Shiga & Osaka: World Water Council. http://210.169.251.146/html/en/finalreport_pdf/FinalReport.pdf

World Water Council 2006. Final Report of the 4th World Water Forum: Local Actions for Global Challenge. Mexico. http://www.worldwaterforum4.org.mx/files/report/FinalReport.pdf

Yellow River Conservancy Commission 2007. Information on the work of the commission. http://www.yellowriver.gov.cn/eng/introduction/

Zhu, Z. 2004. Knowledge management: towards a universal concept or cross-cultural contexts? *Knowledge Management Research & Practice* 2(2): 67–79.

Chapter 5

KNOWLEDGE MANAGEMENT AT THE COMMUNITY LEVEL IN COLOMBIA

Paola Chaves & Mariela García
Cinara, Universidad del Valle, Colombia

ABSTRACT

The global water and sanitation sector is focused on reaching the Millennium Development Goals[1]. To accomplish this task, it is necessary to seek strategies which involve social actors from different backgrounds. They should be included in processes involving participation and negotiation, processes in which knowledge from all disciplines and realities are respected and recognized. This involves taking into account the needs and capacities of communities, involving the citizens, and sharing and improving the flow of information and knowledge. For regions such as Latin America, where rural water and sanitation services are in the hands of community based organizations, community management plays a key role in achieving both the sustainability of investments and improvement in the quality of service.

Similarly, combining community management with the use of resources for information and communication is a strategy for promoting knowledge management in the water and sanitation sector, and thus for allowing the efforts and investments in this sector to have the desired results.

This article presents the case of the Colombian Association of Community Water Supply and Sanitation Public Service Providers, or AQUACOL. This organization has been developing a knowledge management process at

---

[1] Target 10 of the Millennium Development Goal specifically addresses the water and sanitation sector: "Halve, by 2015, the proportion of people without sustainable access to safe drinking water and basic sanitation."

the community level, supported by processes of community management and using the media as well as information and communication technologies.

5.1   INTRODUCTION

In Colombia, as in most Latin American countries, the level of coverage of water supply and sanitation services has increased. Even so, the problems in this sector continue to be complex, especially in rural and marginal urban areas.

While major progress in coverage and management was achieved in the urban areas of the country in 2003, the rural areas, according to the National Administrative Department of Statistics DANE[2], still showed a significant gap, with an average coverage of 66 per cent in water supply (of which only 12 per cent is treated). There is only 34 per cent coverage in wastewater and sewage systems, according to the Rural Sanitation Inventory (*Inventario Sanitario Rural*)[3]. Moreover, there are deficiencies in the management of these services and in their level of technical development, as well as other issues which reflect disparity between rural and urban areas.

The main advances at the national level for improving these services have been regulatory, and a great change for rural areas was decentralization. In the Constitution of 1991, municipalities were mandated to manage their own services, and later, in 1994, Act 142 opened up opportunities for rural and marginal urban communities to take over providing services. Nevertheless, "decentralization means, above all, strengthening the regional regulatory agencies so that they are able to carry out their function ..." (García et al., 2000), and in this sense the decentralization process has not succeeded so far in making the necessary transfer of knowledge.

According to data from the Ministry of Environment, Housing and Development in 2005, of the 11,552 organizations which provide water supply services in the rural part of the country, 90 per cent are community organizations, and of these only 17 per cent are legally registered, which means that at least 74 per cent of the organizations (community or not) are not officially established.

The main reasons that public services providers in small towns have not registered officially are: legal requirements unsuited to this type of enterprise, the small number of advantages associated with legalization, the high direct and indirect costs of being legally constituted or of handling the transaction, the complicated procedures for legalization, a lack of enforcement, and the ease with which the process can be evaded (Pérez, 2001).

[2] Quality of life survey developed in 2003.
[3] This Inventory was carried out in 12,813 localities from 2000–2002 by the Ministry of Development.

As a consequence, a high percentage of organizations that manage local waterworks remain extra-legal and informal, thereby losing the opportunity to be recognized and to receive financial assistance and training.

On the other hand, some community organizations offer only water supply services, because so many communities have obsolete or incomplete sewage systems; and, upon administering these, the community would lose money and would have to invest their surplus from the water supply system in maintaining the sanitation systems. Some communities are waiting to assume management of sewage systems until municipal governments have invested in their improvement.

5.2 CONCEPTUAL FRAMEWORK

5.2.1 COMMUNITY MANAGEMENT

The focus of community management in water and sanitation should be understood as

the highest expression of participation, involving a decision-making process through which the future of a locality, in terms of the development of its water supply and sanitation, is determined. Administrative criteria are involved in community management because they have recourse to an institutional structure consisting of regulations, competencies, procedures, and administrative, economic and human resource structures, whose coordination allows the organizations which finance them to meet the demands and needs of its users (García and Bastidas, 2000).

That is to say, community management in the water and sanitation sector cannot be fully functional while remaining on the fringes of the legal processes and institutions that make up the sector.

Another factor that should be taken into account is the use of influence and power: *"community management is not only the ability to administer resources, but also the capacity to negotiate decisions and to resolve conflicts"* (García and Bastidas, 2000). This ability rests on two concepts: *empowerment* and *appropriation*. The first involves control, authority, responsibility and the planning needed for the provision of services, as well as the establishment of horizontal relationships with governmental and non-governmental agencies which support the work of the community based organizations. The second assures sustainability, because only when community members feel that projects are contributing to the improvement of their living conditions, when they take part in constructing and adapting the projects, and when they are capable of maintaining them, are they committed to their effective operation and administration.

One of the central ideas of this report is to demonstrate that community management and the use of different forms of communication to replicate and disseminate experiences in this field are fundamental tools for generating knowledge management in the water and sanitation sector.

5.2.2   KNOWLEDGE MANAGEMENT

Knowledge management can be understood as a cluster of strategies for improving the overall capacity of a sector through knowledge exchange. The focus is on access to and the use of information. Some of these strategies are: documentation of processes, community participation, capacity development and creation and strengthening of networks, among others.

Knowledge management can give greater potential and visibility to processes developed within an institution or a sector in such a way that new processes include the lessons learned in the past. For this reason, knowledge management has enormous potential to improve the water and sanitation sector, and also to increase the sustainability of services.

5.2.2.1   Information as a decision-making tool

There is a close relationship between the need for information and the ability to make decisions. However, there are different ways of ensuring that information is distributed, that it reaches its audience and that it has the desired effect. Some ways are more efficient than others.

In many research projects, it is assumed that communication and information resources can easily be aimed at a general audience, and that this audience can easily understand and apply it. For this reason, we find that, as it is now, institutions create websites and publish leaflets, bulletins, videos and so on; but their plans do not include conducting any follow-up on the use of their products, let alone evaluating their impact. Generally, they use these products only as publicity for the institution.

It is important to recognize that creating a communication product requires a research process which is directed at understanding the context in which it originates, as well as the audience at whom it is aimed. The product itself is no guarantee of the true communication of knowledge.

In the case of the water and sanitation sector, the production of resources for information and communication has taken place in a superficial manner. There is a flow of information among academic peers through publications and events; there are also commendable community participation processes within the projects. But the management of media or of resources for communication and information that would allow the generation of grass roots knowledge management is largely unexplored territory.

5.3 COMMUNITY MANAGEMENT – CASE STUDIES

5.3.1 THE CASE OF AQUACOL

The situation of the community management organizations in Colombia has led to the growth of networks for mutual support and for keeping an eye out for their interests in the face of government requests. This is the case of AQUACOL, an association made up of thirty-three organizations in the region of Valle del Cauca and Cauca, which provide services to more than 70,000 inhabitants.

The advantages of associating are related to having representation and lobbying ability. The more community organizations become official and group together, the more they can improve their level of representation in the water and sanitation sector, thus participating more actively in the related political decisions, and making more visible their needs, their problems and their strengths.

AQUACOL has been functioning since 2000, and the services it offers are related to helping their associates in everything to do with providing public services and offering technical, administrative and legal assistance. The guiding principles for growth in the organization are related to work among equals, the capacity to influence state policies, the reduction of operating and maintenance costs through economies of scale, and the generation of larger-scale projects by the associated organizations.

The Institute for Research and Development in Water Supply, Environmental Sanitation and Conservation of Water Resources – or *Cinara* – has, for over fifteen years, incorporated community expertise into its teaching-learning strategy, with remarkable results. Throughout this experience between AQUACOL and Cinara, it has become clear that inter-community training has great potential for promoting the sustainability of investments in the rural sector, according to García (2005), because of:

- The confidence which is inspired by communities who have been able to stand out in specific areas, such as in the efficient handling of the business aspect, the management of waterways through pumping systems, the recovery of small water supply basins, the efficient use of water, etc.
- The credibility of training promoted by the communities themselves, because they teach based on their experiences, and what they teach can be proved when it is put into practice.
- The greater ease of grasping the knowledge and its authority because they spring from cultural codes and the narratives of struggle, which are basically similar.

5.3.2 RESOURCE CENTRE DEVELOPMENT PROGRAMME

Within the Program for Developing Resource Centres (Spanish acronym DCR), 2005–2007, under the auspices of the International Water Supply and

Sanitation Centre (IRC), the Cinara Institute, as one of these centers[4], took on the mission of improving the flow of information as well as the communication channels within the water and sanitation sector, especially at the national level. They also looked into transmitting the wealth of knowledge and experience which they had acquired while implementing projects in different countries, using strategies of participatory action research.

One initiative of AQUACOL and the Cinara Institute, within the framework of Phase Two of the Resource Center Development programme, was to strengthen four organizations affiliated to AQUACOL. Due to their high level of development, these four organizations could facilitate and train other communities. They were designated as Community Learning Centres for Water and Sanitation, CLCs.

5.3.3   COMMUNITY LEARNING CENTRES

Community Learning Centres for Water and Sanitation (CLCs), are affiliated to AQUACOL. They have been organized by rural and peri-urban communities to share information and knowledge created by those communities. The centres are placed in communities that have succeeded significantly in one or more facets of community management of water and sanitation services. The idea is to take advantage of the experiences of these communities for training, orienting and offering information to other communities that are also managing their own water systems.

Cinara facilitated the strengthening of the community learning centres. They did this through a series of workshops, whose topics were chosen by the communities themselves, taking into account their weaknesses. The workshops included leadership training, development of creativity, preparing teaching materials, and developing their skills as facilitators of community development processes, among others. Each workshop considered the knowledge base of the participants, thus reaffirming their capacity, before advancing to new information.

In the process of creating the community learning centres, it was the community leaders themselves who defined the mission, vision and activities of the centres.

To expand training activities, the meeting places were rotated between the communities. This allowed for integration and mutual recognition, as well as learning beyond that proposed in the workshops. In this way, people could become familiar with different systems of water treatment, and with the physical and cultural realities of each community.

---

[4] A resource center is defined as "an organization or network of institutions that supports and provides services to the water and sanitation sector, ensuring that knowledge and information are accessible to the different actors within the sector, according to specific demands, and in such a way that all of them can use it" (IRC, 2004).

Figure 5.1 Mission: "Centres which lead, facilitate and promote continuous improvement of locally provided public services ... through holding knowledge in common and putting into practice values such as: solidarity, equity, responsibility and honesty; and seeking to maintain a harmonious interaction between human beings and their environment." (Workshop IV).

The expanded process, besides reinforcing knowledge, strengthened teamwork.

We have been taught that to be a leader you should be a person who rises to the occasion and takes on responsibility, and does everything. In these sessions I have learned that to be a leader is to learn how to share so that the load is not too heavy, because otherwise you have to do it all, play lead and back-up at the same time, so the better the team, the easier it is[5], Elvia Álvares, member of the board of directors of La Sirena Waterworks between 2004 and 2007.

5.3.3.1   Functions of the centres

- To share information and knowledge with interested communities.
- To maintain contact with the communities who visit the centres.

---

[5] Comment extracted from one of the documents from the training workshops of the community learning centres.

- To plan support visits to communities that ask for them.
- To promote participatory action within each of the user communities, thus increasing their sense of belonging.
- To channel community requests for information and pass them on to the appropriate offices.
- To generate conditions for access to information and to processes for updating knowledge.
- To continue networking with other community centres and institutions in the area.
- To give feedback to AQUACOL based on their experiences.

5.3.3.2 Results of the centres – Dialogues with government agencies at the national level

The Public Service Superintendent (Spanish acronym SSPD) is responsible for the surveillance and control of water utilities. The Unified Information System (Spanish acronym SUI) is one of the tools created to implement this task.

The SUI is an Internet-based system for collecting, unifying, classifying and storing the information that the state requests from water and sanitation service providers. The office of the SSPD is responsible for establishing, administering, maintaining and operating the SUI. The information reported to this system provides the source for policies which regulate, plan and control, and allocate resources. The quality of the data is the responsibility, first of all, of public service providers who confirm reported information.

Although this system is fundamental to the state system, it represents a problem for community organizations because it has been structured for the large water utilities. The forms for reporting use complex terminology and ask difficult questions. Moreover, the advanced level of accounting and financial

Figure 5.2 Visit from the Superintendent to the Community Learning Centres, Cinara Institute, 2007.

information requested by the system is not what is kept by the rural water systems.

Furthermore, members of rural water organizations do not know how to use a computer, and in some cases do not have computers at all. The state system was designed to submit information on-line, without taking into account the fact that in rural areas the electricity supply is intermittent and that the technological platforms in the countryside are inadequate. As a consequence, many organizations either remain isolated from these processes or have to pay the high cost of having others do the work for them.

In 2007 there were meetings between the SSPD and several community-based organizations from Caldas, Cauca and Valle del Cauca (Colombia) promoted by AQUACOL with Cinara's support. There were also four SUI needs-assessment workshops promoted by SSPD in different parts of the country. These activities revealed the weaknesses of the SUI, especially considering the precarious technological conditions in the rural areas.

The efforts of AQUACOL and its CLCs to show the problems faced by rural communities when using the SUI have yielded important results. AQUACOL has been admitted as a representative of small rural service providers on the team made up of the SSPD, the Cinara Institute, the Systems School of the Universidad del Valle, UNICEF – Colombia and the government of the region of Caldas, to create an exclusive module for small rural service providers within the SUI. This module will ask for information in an easy manner. The fields in the module will be completed using information that is consistent with the reality of small water supply providers in rural areas. The small community companies will now also have the ability to complete the forms, store their information on a USB memory stick, and upload it from any Internet café.

The uniqueness of this process is rooted in teamwork between national government institutions, such as the SSPD, and community-based organizations like AQUACOL's members, seeking alternatives to make better use of information and communication technologies (ICTs).

Figure 5.3 Superintendents' visit to community-based organizations, Mondomo Water Supply, Cauca, 2007.

5.4 KNOWLEDGE MANAGEMENT AT THE COMMUNITY LEVEL

Power-related and political decisions are made by means of the construction of human thought through communication. In our type of society, the mass media play a decisive role in the creation of public opinion which shapes decision-making. Politics is, above all, media politics, and this has important consequences, as it makes politics more personal, a politics of scandal. However, mass communication is being transformed by the spread of the Internet and of Web 2.0, as well as by wireless communication ... allowing greater citizen involvement. This is helpful to social movements and to alternative politics. Hence, contradictory social tendencies are expressed both in the mass media and in the new communication media. This means that, more and more, power is being decided within a multi-modal communication space. In our society, power is the power to communicate (Castells, 2008).

As Castells (2008) suggested, communication is the process of constructing human thought, but it is also a process of social construction; that is to say, it is one of the main foundations of society, and is therefore a process which is natural to the human condition.

The experiences of AQUACOL and CLCs can be seen as a successful communication strategy. Learning among equals, knowledge from a 'horizontal' point of view, is the product of dialogue. The methodology used begins by reaffirming what both communities already know; using this, new information is constructed, and it is strengthened by tools which facilitate its transmission to other communities. Up until now, knowledge exchange has been basically oral, a direct exchange which took place during the visits of leaders to the communities, or when visitors were received in the offices of the community centres. Still, oral communication and visits have their limitations and their costs, and these are beginning to be overcome by combining these strategies with published written documents which collect the experiences of the organizations. Even so, the reach and the dissemination of these experiences can be improved by using the media and the new information and communication technologies.

5.4.1 THE MEDIA AND INFORMATION AND COMMUNICATION TECHNOLOGIES IN COMMUNITY MANAGEMENT

Communication should not be a vertical process, and it will never be a transparent process. This means that all communication should be two-way: an exchange. This exchange is affected by its context and by the reference points which lead us to an interpretation. No communicated message is understood in the same way by all members of the audience.

For communities, having access to current information from different sources – as well as to information pathways, to institutions, and to people – is a fundamental tool for developing *empowerment*. However, communities themselves can share their knowledge with institutions, individuals or other communities. A community that has effectively appropriated knowledge for itself, and has become empowered, represents a great forward move for the sector. But a community with a successful process, which starts to share its experience on a large scale using media or information and communication technologies, and, more importantly, which uses the same language as the communities themselves, takes a truly giant step towards knowledge management within the sector.

Creating mass communication items, such as newspapers and videos, and using information technologies, such as website creation or multimedia presentations to disseminate the experiences, allows AQUACOL to: give greater visibility to community problems, recognize their achievements and pass them on to other communities and, finally, to energize processes of participation and lobbying for the improvement of policies within the sector.

As Castells (2008) would say, *"power is the power to communicate"*, which, in this context, implies that offering communication tools to public service associations not only continues the process which has been achieved so far, but also reaffirms the work of the CLCs at a political level, because *"the mass media are decisive in the creation of public opinion which shapes political decision-making"*.

In the same fashion, using communication media allows for other possibilities, such as opening discussion spaces in which all levels – local, regional and national – can participate to support campaigns promoting health practices, efficient water use and the integrated management of water resources.

5.4.2 INFORMATION AND COMMUNICATION TECHNOLOGIES AND MULTIMEDIA TOOLS

Multimedia systems use more than one communication medium at the same time to present information: text, images, animation, video and sound. This concept is as old as human communication, as when we express ourselves in a regular conversation we speak (sound), we write (text), we observe our conversation partner (video), and we make facial expressions and gestures (animation).

5.4.2.1 Uses of multimedia tools

Multimedia tools have enormous advantages in improving training and strengthening skills. When a computer program, document or presentation combines media appropriately, there is notable improvement in attention, comprehension and learning because we are closer to the day-to-day experience of human communication. This reality is fundamental in places where there is little reading culture.

5.4.2.2 Advantages and disadvantages of these tools

We live in a society that is immersed in technological development, where the advances in new information and communication technologies are changing our way of life. Rural communities and peri-urban communities often lack sufficient resources for access to these technologies. Still, this cannot be an excuse for them to remain excluded from a process which is changing and revolutionizing the whole world.

It is important to investigate the real possibilities for the use of these types of communication in the work of community management. In many cases, there are other different technologies that have become widespread, such as the cell phone. For AQUACOL, the cell phone is the main tool for communication between affiliates, so it would be interesting to seek possible large-scale uses for this tool, such as text messages containing key information.

According to Cinara's experience with community work, another very prevalent technology is the use of DVDs. In this respect, resources such as video or digital photography can also have a wide reach in these areas.

5.4.3 KNOWLEDGE MANAGEMENT AT THE COMMUNITY LEVEL – STRATEGIES

The following descriptions represent two communication strategies that have made community management stronger by using communication media to promote knowledge management in the sector. Both strategies received support from Cinara and were developed through participatory processes with AQUACOL through the CLCs.

5.4.3.1 The AQUACOL bulletin

A few years ago AQUACOL had the idea of producing a bulletin which would circulate periodically among its members and Colombian institutions working in the water sector. The bulletin would contain information that was not easily accessible, but which was needed for effective operation of this public service.

During 2008, with the financial support of the Water Supply and Sanitation Collaborative Council (WSSCC), Cinara trained a team of community leaders who belonged to AQUACOL to create and publish a news bulletin that would appear regularly over the long term.

The first issue of the bulletin is now ready. It was written and designed by the community leaders themselves, using simple language which respects and communicates with the cultural identities of the communities it represents.

The methodology for this process was developed in phases:

Phase I – Needs assessment: a survey was sent to the 33 AQUACOL members in order to learn about their needs and their information channels. Research on and analysis of other news bulletins in the water and sanitation sector, particularly in Latin America, and two workshops with leaders of AQUACOL were conducted to understand their previous experiences in this field and their expectations about this new experience.

Phase II – Selection and evaluation of the initial group: in participatory workshops, group members developed their writing abilities and learned the skills of writing and researching news articles.

Phase III – Creation and strengthening of the editorial committee: once the group was selected and trained, they designed strategies for distributing and financing the product while continuing their individual learning in the fields of writing and research.

Phase IV – Standardization of the production process: the process is analysed and key tasks which would need to be repeated for each edition are identified (this phase is a continuous process).

Phase V – Delivery of the final product – first issue of the bulletin.

Editorial committee: One of the main achievements of the process was creating the editorial policies that would guide each issue of the bulletin.

After a close look at existing models for editorial committees and their hierarchical structures, a different model was chosen – one in which all decisions are discussed and decided on within the group, using a horizontal system.

The general committee is made up of the ten participants in the process. They sorted themselves into three groups, all of which are at the same level of importance, so they could make decisions relating to different areas. In general, they all belong to the writing committee, as all are responsible for having enough information and articles for each edition. A financial committee

is in charge of ensuring that there are financial resources for each issue; and, an editorial committee reviews the material and maintains the quality of the articles.

Producing a periodical news bulletin involves a formal communication process which differs from personal, spoken communication. Initially, this generated difficulties for participants in the project.

Developing writing and research skills for news articles involved a change in cultural and communication practices, which are deeply rooted in the communities benefiting from the process. Knowledge management involves formalizing and organizing processes which are otherwise often natural and spontaneous: informing others, recounting experiences, and sharing ideas through anecdotes and jokes. Because of the need for greater formality, although many trained leaders have experience and strong speaking skills, they are not successful in expressing their ideas in writing.

From now on, this product will allow a more constant flow of information from CLCs and other AQUACOL leaders to the rest of the affiliates. Also, because it gives national visibility to the difficulties of community organization providers of water services, it will strengthen lobbying for political decisions.

On the other hand, all community leaders should be aware of their rights and duties with regard to managing water and sanitation systems. With this information, many companies can learn about the legal requirements and the technological, administrative and financial alternatives. They can also become familiar with the experiences of other communities and the lessons

they learned. This will allow them to take the necessary steps to function better and to take on greater challenges, such as those related to sanitation systems and health.

5.4.3.2   La Sirena multimedia

Within the framework of Phase Two of the DCR programme, one of the proposed tools for improving information sharing from the CLCs to other communities was the development of a multimedia programme that would pull together the history of one of those centres: La Sirena.

The multimedia project contains these modules:

1. Information on La Sirena
   1.1.   Location (animation)
   1.2.   Brief historical summary
   1.3.   Current (2007) situation
      1.3.1.   Population
      1.3.2.   Education
      1.3.3.   Economic activities
      1.3.4.   Transportation

2. La Sirena and its water system
   2.1.   History of the water system
      2.1.1.   Creating the water system
      2.1.2.   Previous boards of directors (members, achievements)
   2.2.   The water system today
      2.2.1.   The current board of directors; description of the association; organizational chart, duties of each director, number of users, suggestions for effective operation

2.2.2.   Bylaws
2.2.3.   Brief explanation of the water treatment system being used

3. Magic
3.1.   Magical tales about mermaids (*sirenas* in Spanish)
3.2.   Magical tales told in the community about local rivers

For six months community-based meetings, interviews and information surveys were conducted so that this project could be created. The final result had over 600 photographs going back thirty years, and seven videos with interviews and texts giving the history of the community and of the water supply.

This product is oriented towards raising awareness about the community management processes. People from La Sirena are actually using this product to share their experiences with different local institutions such as schools, service users and healthcare centres. According to Anyela Gutierrez[6]: "The multimedia is a really useful material for visitors. It is also used in our user meetings. Its design is beautiful, didactic and easy to work with."

Cinara staff are using this multimedia presentation to show that community management of the water supply can be very successful and sustainable. It is worth noting that many communities are eager to end the learning process quickly, as soon as they begin to see improvement in their water supply systems, but they fail to understand that results cannot be obtained overnight.

5.5   CONCLUSIONS

It is hoped that these products, developed using the participatory approach with AQUACOL and the CLCs, will be tools that harmonize with their mission and will help them accomplish their objectives. Having mass-communication products which originated in the communities circulating in the water and sanitation sectors also gives a voice to social actors who previously remained silent. Relationships that have developed between communities and with national and international institutions can now be solidified.

As it is now, information and communication technologies allow us to process, store, retrieve and communicate information in any of its forms: spoken, written or visual; independent of distance, time and quantity. Similarly, these technologies have allowed the production of knowledge to add a feedback phase: this is because of direct interaction between information producers and their audience, or due to a process in which a person can be simultaneously a creator and user of information. However, all this increased information is no guarantee of sound information management; it

[6]Secretary of the community-based organization in charge of the water supply system.

must be linked to solid learning processes. For this reason, information and communication technologies framed by a process, such as the one used by AQUACOL and the CLCs, have enormous possibilities.

It is becoming more and more necessary to apply communication resources and information and communication technologies to the water and sanitation sector; furthermore, this presents an opportunity for all social actors within the sector to express themselves and be informed, and to let others know about all those circumstances in which water, sanitation and health play an influential role.

As information producers, communities reaffirm their roles, as well as the horizontal viewpoint, within which effective access to information allows for a knowledge dialogue in which the goal is shared: to improve and accelerate the process of reaching the Millennium Development Goals.

REFERENCES

Castells, M. 2008, Comunicación, poder y contrapoder en la sociedad red (II). Los nuevos espacios de la comunicación. *TELOS 75.* http://www.telos.es/home. asp?idRevistaAnt=75&rev=80

García, M. 2005, Asociación entre comunidades. Asociación de organizaciones comunitarias prestadoras de servicios de agua y saneamiento en el suroeste de Colombia. Cinara Institute. http://www2.irc.nl/manage/cinara/bibliovirtual/ articasociacion.pdf

García, M. and Bastidas, S. 2000, La gestión comunitaria en proyectos de abastecimiento de agua y saneamiento como base de sostenibilidad y de construcción de tejido Social. Cinara Institute. http://www2.irc.nl/manage/cinara/bibliovirtual/ ponentejidosocial.pdf

García, M., Jimenéz, C. and Gómez, C. 2000, La gestión comunitaria como una alternativa en la prestación de servicios de abastecimiento de agua y saneamiento. Discussion Paper. Cinara Institute. http://www2.irc.nl/manage/cinara/bibliovirtual/ articgestion.pdf

IRC, 2004, Making WATSAN Knowledge Sharing and Development work! The conceptual framework of the IRC Resource Centre Development 18 Countries Programme (RCD18). http://www.irc.nl/page/16507

Pérez, M. 2001, Balance de la gestión en empresas de servicios de acueducto y alcantarillado de pequeña escala en Colombia. This document is a part of the project, The Role of the Community in Managing Rural Water Supply Systems in Developing Countries.

Chapter 6

DEVELOPING CAPACITY FOR CONFLICT RESOLUTION APPLIED TO WATER ISSUES

Léna Salamé
UNESCO PCCP, France

Larry Swatuk
Programme for Environment and International Development, Faculty of Environment, University of Waterloo, Canada

Pieter van der Zaag
UNESCO-IHE Institute for Water Education, The Netherlands

ABSTRACT

The world's water resources are under great pressure from increasing human demands. Given water's key role in development, it is inevitable that conflicts over this resource are on the rise. However, conflict is often a necessary first step towards cooperation. To ensure positive-sum outcomes, it is essential that stakeholders have appropriate knowledge and capacity to collectively move away from potentially destructive actions and behaviours. Litigation is the most widely understood form of formal conflict resolution. However, decisions rendered in court often leave lingering feelings of resentment, resolving the issue but leaving the underlying conditions of the conflict untouched. This chapter reviews the current context for water resources management and describes various approaches to conflict resolution. Alternative Dispute Resolution (ADR) is regarded as an important method for achieving mutually beneficial results on conflicted situations. Drawing on their extensive experience with conflict resolution training programs, the authors reflect on these methods and describe effective approaches to capacity building for ADR. While the authors are confident in ADR approaches to conflict management, they argue that it is not a fixed template: among other things, context matters, as does the willingness of all parties to participate in resolving the dispute.

6.1 INTRODUCTION

Conflict may be defined as a disagreement through which the parties involved perceive a threat to their needs, interests or concerns. Conflict is an unavoidable aspect of human social systems. Indeed, many argue that conflict is a necessary fact of life, for it is only through struggle that lasting and meaningful change can be brought about.

Given the central importance of water resources to all human communities, it is natural that conflicts arise with regard to their access, allocation, development and management. It is equally clear, however, that necessity is not only the mother of invention, but also the basis for extensive cooperative activities concerning the management of water resources. Thus, both conflicted and cooperative behaviours – across time and space, and at all levels of human social organization – constitute the norm where water resources are concerned.

Water resources of all types are under increasing pressure from a number of actors, forces and factors manifest in the early 21st-century world (UN, 2006). Of particular concern is the impact of population growth; by 2050 the world's population will have increased by 50 per cent. Also of concern is the way in which sovereign states will deal with increasing (seasonal, absolute, natural, man-made) scarcities in transboundary river basins. Geography is thought to play a special role, with location in the basin (upstream/downstream) and in the environment (arid/semi-arid ecosystems) regarded as key factors in future water conflicts. Global warming is also thought to pose particular challenges to water-stressed societies and communities that must develop adaptation and mitigation mechanisms in order to survive. At the national level, important questions have arisen concerning the optimal use of limited resources.

Debates and disputes are now popping up between and among a wide variety of users (e.g. urban/rural; industry/agriculture; humans/the environment, rich/poor people) within and across watersheds, ecosystems, basins, political jurisdictions and increasingly crowded cities. Given the diversity of needs and interests that surround water, disputes over the resource are normal. That is to say, they are to be expected. Not all lead to conflict, however; and not all conflicts turn violent. Some fester perpetually beneath the surface and, as with limited access to potable water in many parts of urban areas, are part of settled social relations. However, a change in the setting – such as an unexpected drought or flood, or a change in government policy – can bring long suppressed grievances to the surface. At the same time, other longer-term changes, such as population growth, urbanization, land use, and climate variability, can create new grievances or worsen already existing ones in a slow and creeping manner.

What is to be done about such events and eventualities? Clearly, we must be prepared to anticipate, prevent and address water conflicts as and when they arise. The intention of this paper is to reflect on the many and varied efforts to move water resources management away from its inherent conflict potential toward cooperative behaviours. Within the field of conflict resolution and negotiation for water resources management, there has been consistent emphasis on training actors in Alternative Dispute Resolution (ADR) skills, in particular, principled negotiation: an approach that seeks to embed outcomes and processes that will serve sustainable, equitable and efficient long-term social needs.

This paper reflects on the following questions:

– What training tools are available to water users/managers interested in conflict resolution?
– How do training modules for water resources-specific conflict resolution capacity building vary? What are their strengths and weaknesses?
– What tools have proven most effective in controlled training settings?
– Given the tendency to borrow widely from generic conflict resolution training modules, are standard conflict resolution training methods suitable to water resources settings or should we be thinking "outside the box"?

There are several important questions for which we do not yet have answers, despite quite a few years of training. For example:

– Has the emphasis on application of an Integrated Water Resources Management (IWRM) framework served to enhance the ability of water managers to resolve conflicts? Or are we ill-preparing our water managers for the real-world setting?
– Is there hard evidence that *any* of our training efforts have borne fruit?

This paper, therefore, makes suggestions regarding necessary information gathering and research, if the future of training and capacity building for the anticipation, prevention and resolution of water-related conflicts are to be relevant in real-world settings.

6.2 A WORLD WATER CRISIS

Water is central to human development. The ability to harness water resources for human use has enabled the rise of complex civilizations. Globally, aggregate national water use varies directly with both the Gross National Income and Human Development Index values. Water is both a common and precious commodity. It exists in abundance but is not always located where or when we humans need it. Of course, we have not helped matters. For most of human history, we have had limited impact on the resources around us. With rapid technological and social change throughout the last 500 years, however, our environmental footprint has grown such that we face the greatest challenge yet to human civilization in the form of global warming. Where water resources are concerned,

[f]rom a situation of limited, low-impact and largely riparian uses of water, we have now reached a point where, in many parts of the world,

cumulative uses of river resources have not just local but basin-wide and regional impacts. The result is that water resources in many river basins are fully or almost fully committed to a variety of purposes, both in-stream and remote; water quality is degraded; river-dependent ecosystems are threatened; and still-expanding demand is leading to intense competition and, at time, to strife. (Svendsen et al., 2004)

Thus, today we face a world water crisis, some characteristics of which are as follows:

- Water resources are increasingly under pressure from population growth, economic activity and intensifying competition among users;
- Water withdrawals have increased more than twice as fast as population growth, and currently one third of the world's population live in countries that experience medium to high water stress;
- Groundwater withdrawals frequently exceed natural recharge levels, leading to the lowering of water tables and depletion of aquifers;
- Pollution is further increasing water scarcity by reducing water usability at the source and downstream;
- Shortcomings in the management of water, a focus on developing new sources rather than managing existing ones better, and top-down sector approaches to water management result in uncoordinated development and management of the resource;
- More and more development means greater impact on the environment;
- Current concerns about climate variability and climate change demand improved management of water resources to cope with more intense floods and droughts.

Access to water is fundamental to human survival, health and productivity. But, there are many challenges related to ensuring the sustainability of people's access to water for various purposes. Many development projects have not viewed water within the environment as being an exhaustible supply and the approach has been mostly sectoral and non-integrated, causing many pressures on the limited resource. The results of this approach, together with external factors (most notably population increase and climate change), have produced situations in which the water source has either run out or is severely stressed. These situations include disasters such as pollution, the overexploitation of aquifers, the drying-up of springs, floods, and funds being wasted on many inappropriate projects.

6.3   A CRISIS OF GOVERNANCE

While an understanding of water resources, their dynamics and limitations on abstraction is considered to be essential to permit the development of sustainable water management strategies, the problems of today and tomorrow are as much a consequence of poor governance as they are of absolute scarcity (see UN, 2006, Chapter 2 for details).

Governance is both an outcome and a process, involving a variety of legitimate and authoritative actors. As an outcome it reflects settled social relations. If it is good, it suggests widespread – if not universal – social approval of its practices. Good governance can never reach an end point; as a process it depends on the reiteration of activities that deepen trust.

With regard to water governance, a commonly accepted definition comes from the Global Water Partnership: "Water governance refers to the range of political, social, economic and administrative systems that are in place to develop and manage water resources, and the delivery of water services at different levels" (Rogers and Hall, 2003). According to the authors of the UN *World Water Development Report 2* (UN, 2006: 46, 47), water governance has four dimensions: a social dimension concerned with "equitable use"; an economic dimension concerned with "efficient use"; an environmental dimension concerned with "sustainable use"; and, a political dimension concerned with "equal democratic opportunities". Each of these dimensions is "anchored in governance systems across three levels: government, civil society and the private sector". To realize "effective governance", the UN Report proposes a checklist that includes the following:

- participation;
- transparency;
- equity;
- effectiveness and efficiency;
- rule of law;
- accountability;
- coherency;
- responsiveness;
- integration;
- ethical considerations.

The absence of some or all of these practices is supposed to result in "bad" or "poor" governance, a simple definition of which is the inability and/or unwillingness to alter patterns of resource allocation, use and management despite clear evidence of resource degradation, uneconomic behaviour, and abiding poverty and social inequality (UN, 2006: 49).

6.4   TRANSBOUNDARY WATER GOVERNANCE

Complicating the issue further is the fact that most of the planet's people live within one of the estimated 276 river basins shared by two or more states. These basins cover more than 45 per cent of the earth's surface, and "of the

145 states occupying international river basins, almost two-thirds (92) have at least half of their national territory lying in an international basin, and more than one-third (50) have 80 per cent or more of [their] national territory in an international basin" (Conca, 2006). Given that sovereign states arrogate to themselves the right to develop resources located within their territory, and given that water is a fugitive (i.e. without respect for international political boundaries), as demands for water increase across communities, states and sectors, so the likelihood of conflicts over water will increase.

6.5 INTEGRATED WATER RESOURCES MANAGEMENT

Avoiding or minimizing the negative effects of physical and human-induced resource scarcity "will require institutional innovations that allow focusing simultaneously on the goals and trade-offs in food security, poverty reduction, and environmental sustainability" (Molden et al., 2007: 62). Such a perspective has now crystallized in the concept Integrated Water Resources Management (IWRM). According to the UN (2006: 17), "Humanity has embarked on a huge global ecological engineering project, with little or no preconception, or indeed full present knowledge, of the consequences .... In the water sector, securing reliable and safe water supplies for health and food, the needs of industrial and energy production processes, and the development of rights markets for both land and water have hugely changed the natural order of many rivers worldwide". We are now coming to grips with the enormity of the problems we have created for ourselves through the unselfconscious manipulation of nature for particular ends, and hence the need for change. With change comes challenge, and with challenge comes opportunity. For many scholars, policy-makers and activists, IWRM provides a solid framework for thinking systematically about a future in which water use is ecologically sustainable, socially equitable, and economically efficient. Many approaches to conflict resolution and negotiation regard IWRM as a primary goal, assuming that, through IWRM, mutual benefits will be realized. However, the evidence on the ground suggests that the pursuit of IWRM may itself be a contributor to disputes and conflicts within basins and across multiple users (Merrey, 2008). Nevertheless, the primary challenge is to turn the inevitable conflicts that will arise into productive, win-win, mutually beneficial outcomes, but sometimes choices have to be made.

6.6 FACTS ABOUT WATER AND CONFLICT/ COOPERATION

To date, there are numerous global agreements/statements/conventions that are in place and also in the making, to address the issues of the prevailing or expected conflicts, one being the 1997 UN convention on the non-navigational uses of international water courses. Despite such cooperative models, what happens in reality is that more powerful actors unilaterally determine the ways in which the resource is allocated, used and managed. However, many opportunities for cooperation instead of conflict present themselves to those willing to look for them. For example:

- Equitable sharing of water from a common source,
- Sharing of data and expertise for flood forecasting,
- Watershed management, and soil and water conservation to the benefit of the entire watershed,
- New opportunities that may be created by institutionalising linkages and dependencies between individuals and groups within catchments and watersheds,
- Hydropower generation,
- Cooperation in flood management,
- Cooperation in navigation systems,
- Establishment of rules for environmental flows that not only benefit nature but also people living off riverine ecosystems,
- Leakage control to reduce water losses so that we may "do more with less",
- Cross-border pollution management,
- Cooperation in river training works.

Indeed, the evidence shows that while there are many conflicts, there is much more cooperation on the use of surface waters of all kinds.

6.6.1 WATER WARS

Animating much of the research conducted on transboundary waters over the last decade or so is the persistent sense that water will be "the oil of the future" and that future wars will be about water. The available data, however, suggest a more complex relationship between transboundary water and social stability. Gleick (2000) shows that water has been involved in conflict as:

- a political or military tool,
- a military target,
- an object of terrorism,
- part of a development dispute,
- an object of control.

In the first three cases, one cannot talk of a water war or a water conflict, since water is not the objective of the event but only a tool in it, or a victim of it. On the other hand, when water is part of a development dispute, or part of a larger dispute, and/or when it is an object of control, one can say that the resource is the object of the event and the event can consequently be considered a water dispute or conflict.

Most of Gleick's cases involved inter-state activity, although intra-state conflicts were sometimes reported. In no case was water the principal cause of two states going to war. Given the large proportion of the world's population that resides in shared international river basins, this is not an insignificant finding: the opportunities for violent conflict are abundant, yet such instances are extremely rare. According to Wolf et al. (2005: 84), "[N]o states have gone to war specifically over water resources since the city-states of Lagash and Umma fought each other in the Tigris-Euphrates Basin in 2500 BC. Instead, according to the UN Food and Agricultural Organization, more than 3,600 water treaties were signed from AD 805 to 1984".

In a recent summary of work conducted at Oregon State University, Wolf et al. (2005: 84–85) highlight four key findings:

1   "the incidence of acute conflict over international water resources is overwhelmed by the rate of cooperation";
2   "despite the fiery rhetoric of politicians ... most actions taken over water are mild";
3   "there are more examples of cooperation than of conflict";
4   "despite the lack of violence, water acts as both an irritant and a unifier".

The authors state, "The historical record proves that international water disputes do get resolved, even among enemies, and even as conflicts erupt over other issues. Some of the world's most vociferous enemies have negotiated water agreements or are in the process of doing so, and the institutions they have created often prove to be resilient, even when relations are strained" (Wolf et al., 2005: 85). The authors put great stock in institutional capacity, arguing that it is the key to cooperation in situations of increasing scarcity. Allan (2003) argues that water has not been an object of conflict in the Middle East partly because the region's chronic water deficit is compensated for through the importation of food – in essence, the region imports "virtual water" contained in the manufacture of foodstuffs, allowing people in Middle Eastern countries to enjoy a standard of living beyond their natural water barriers. So, there may be many other factors at play in either the onset or the absence of interstate violent conflict. Intuition (i.e. increasing populations + finite water supply = conflict) is no substitute for empirical analysis.

Alongside such analyses of inter-state behaviour over shared water resources are a number of studies and approaches that aim, through conceptual innovation, to assist decision-makers in reaching better informed decisions. Given that most international law has been negotiated about the quality and quantity of visible "blue" freshwater resources, such as lakes, rivers, streams and wetlands, the world's water experts have taken great pains to alter this narrow understanding of what water is, what its values are, and how it interrelates with other aspects of the ecosystems in which it is found (e.g. Falkenmark and Rockström, 2004; UN, 2006). In addition, there is a growing effort among scholars to explore (and possibly exploit) the cooperative potential of shared water resources. Practical programmes and hands-on resources on conflict resolution and transboundary water management are increasingly on offer (see below).

6.7 APPROACHES TO CONFLICT MANAGEMENT

While conflict may be difficult, it is by no means only a destructive process. Conflicts often have positive functions, and may be key drivers of constructive change (Coser, 1956). For example, prior to the formation of environmental protest groups, the dumping of raw sewerage into the world's rivers was regarded as normal practice. During the 1960s and 1970s, many environmental groups opposed to this practice were regarded by governments and mainstream society as extremists (Carter, 2007). Yet, their persistence through confrontational tactics eventually led to a change in the dominant value system. It is now nearly inconceivable that anyone living along a river regards treating it as a sewer an acceptable practice. Conflict also helps to define boundaries, clarify who and what belongs where, and helps to establish procedures for managing resource access, allocation, use and management (albeit not always in an equitable or sustainable fashion). Conflict often brings with it creative potential that helps social groups, organizations, communities and entire states to (re)define themselves, to change and adapt, and to innovate and create.

However, conflict has a positive role to play if only we have the necessary skills to create the synergy for the well-being of all contending parties. There are many techniques, both formal and informal, to manage conflicts. The following are the most commonly known methods of conflict resolution. However, it must be said that conflict resolution is more art than science.

6.7.1 LITIGATION

The formal and ultimate mechanism for conflict resolution is taking recourse through the legal system of the country. In a legal proceeding, the parties to a dispute are heard by a court of law, which decides upon the case on the basis of existing laws in force in the country. In many instances, this is the only way to resolve a conflict, but in many other cases, it may not be so. For example:

a Many conflicts involve the use of a common resource over which no party has a clearly superior and exclusive legal claim;

b Legal rules sometimes prevent parties from bringing an action to court if they do not have some right that has been directly infringed upon;

c Legal rules may also prevent a party with a grievance from having access to the courts even to have its case heard;

d Narrow procedural and legal issues take precedence over policy issues, thereby failing to resolve the real differences between the contending parties;

e Parties represented by their attorneys usually focus on the past to place or deny blame, thereby damaging the relationship between the parties;

f Legal rules present judge and/or jury with a limited range of remedies. In most cases, the outcome is equivalent to a win-lose equation;

g While decisions are enforceable, the compliance rate is often low;

h Litigation processes are usually costly, complex and relatively long;

i Evidence that is required to support a case brought before the courts (e.g. a pollution event, an illegal abstraction of water) is often cumbersome and costly to collect, and, if available, its validity is frequently questioned on scientific grounds due to the inevitable uncertainties of the measurements;

j Many conflicts are either transnational where international law is of limited utility, or are so local they are not anticipated by national law.

6.7.2   ALTERNATIVE DISPUTE RESOLUTION (ADR)

ADR developed in the West in the 1970s as an acceptable alternative to the dominant approach of litigation, with its focus on confrontation and "winner takes all". We continue to use the term ADR while recognizing that it, too, has become rather "mainstream" over the last three or more decades. Today, truly alternative methods of conflict resolution are inspired by indigenous practices, such as forgiveness rituals, strategies for face-saving, reconciliation, re-establishing harmony – being forward-looking (i.e. how, from now on, we can find a way of collaborating without fighting) rather than looking in the past ("who was the culprit, who is to blame?"). Experiences outside the water world are relevant, such as the Truth and Reconciliation Commissions of Uganda and South Africa, including the latter's emphasis on the importance of *Ubuntu* ("humans are humans through/because of other humans"); the role of traditional tribunals called *gacaca* in Rwanda after the genocide; and, the role of spirituality in social organizations across Africa (Ellis and Ter Haar, 1998). Perhaps a key challenge for knowledge and capacity building, then, is for those of us trained in Western modes of ADR to develop capacities to operate with local vernaculars and approaches that may or may not be compatible with our own – largely secular – approaches to dispute resolution and conflict management.

ADR covers a broad spectrum of approaches, from party-to-party engagement in negotiations as the most direct way to reach a mutually accepted resolution, to arbitration and adjudication at the other end, where an external party imposes a solution. ADR allows parties to speak for themselves, is future-oriented and avoids making judgements. It usually also clarifies or mends relationships, preserving them rather than breaking them (which is often the result of litigation). Parties to ADR shape their own settlement, directly or indirectly, and they have access to the broadest range of remedies. In most cases, the outcome is equivalent to a win-win equation, and although the decisions are not enforceable, the compliance rate can be quite high. This is because the parties to the decision feel ownership for the decision they made. An ADR process is usually less costly, less complex and much shorter than litigation. We shall have a quick review of those techniques.

6.7.2.1 Negotiation

Negotiation is a process in which the parties to the dispute meet to reach a mutually acceptable solution. There is no facilitation or mediation by a third party: each party represents its own interest. Representatives of interested parties are invited to participate in negotiations to agree on new rules governing issues such as industrial safety standards and environmental pollution from waste sites.

6.7.2.2 Facilitation

Facilitation is a process by which an impartial individual participates in the design and conduct of problem-solving meetings to help the parties jointly diagnose, create and implement jointly-owned solutions. This process is often used in situations involving multiple parties, issues and stakeholders, and where issues are unclear. Facilitators create the conditions in which everybody is able to speak freely, but they are not expected to volunteer their own ideas or participate actively in moving the parties towards agreement. Facilitation may be the first step in identifying a dispute resolution process.

6.7.2.3 Mediation

Mediation is a process of settling conflict in which an outside party oversees the negotiation between the two disputing parties. The parties choose an acceptable mediator to guide them in designing a process and reaching an agreement on mutually acceptable solutions. The mediator tries to create a safe environment for parties to share information, address underlying problems and vent emotions. It is more formal than facilitation and parties often share the costs of mediation. It is especially useful when the parties have reached an impasse.

6.7.2.4 Arbitration

Arbitration is usually used as a less formal alternative to litigation. It is a process in which a neutral outside party or panel meets with the parties in a dispute, hears presentations from each side and makes an award. Such a decision may be binding or not according to agreements reached between the parties prior to the formal commencement of hearings. The parties choose the arbitrator through consensus and may set the rules that govern the process. Arbitration is often used in the business world and in cases where parties desire a quick solution to their problems.

6.7.3 FROM POTENTIAL CONFLICT TO COOPERATION POTENTIAL: CONSENSUS BUILDING/STAKEHOLDER APPROACH

Stakeholder participation is key to sustainable resource use and management. Conflict resolution techniques are generally employed once a dispute has already arisen. However, anticipating and preventing the forms of future conflict is an important element of conflict resolution itself. In the context of a river basin, where disputes arise from time to time, it is useful to give a home to these issues through the creation of a setting where stakeholders can regularly meet and communicate with each other regarding interests, needs and positions. While there are no uniform methodologies for undertaking the process, the important thing is to create an enabling environment whereby the stakeholders are able to actively participate in the policy dialogues and subsequent planning and design processes. Among others, these may include the following steps:

- Defining the problem rather than proposing solutions,
- Focusing on interests,
- Identifying various alternatives,
- Separating the generation of alternatives from their evaluation,
- Agreeing on principles or criteria to evaluate alternatives,
- Documenting agreements to reduce the risk of later misunderstandings,
- Agreeing on the process by which agreements can be revised and the process by which other types of disagreements might be solved,
- Using the process to create agreement,
- Creating a commitment to implementation by allowing the stakeholders specific roles in the execution of the agreed-upon action/programme,
- Reaching agreement over the content and procedures for joint monitoring of water quantity and water quality (which parameters, to be measured where, when, and how; who will have access to these data, etc.).

6.7.4 REQUIREMENTS FOR A SUCCESSFUL RESOLUTION TO CONFLICT

The techniques discussed above need to fulfill certain conditions for successful outcomes. Some of these are:

6.7.4.1 Willingness to participate

The participants must be free to decide when to participate and when to withdraw from a conflict resolution process, should that be necessary. They should set the agenda and decide on the method to be followed in the process. It is sometimes not possible even to agree to discuss a problem if either of the parties hold a deeply entrenched position or system of values.

6.7.4.2 Opportunity for mutual gain

Linked to the above condition is the requirement of opportunity for mutual gain. The key to success of conflict resolution is the probability that the contending parties will be better off through cooperative action. If one or both believe that they can achieve a better outcome through unilateral action, they will not be willing to participate in the process.

6.7.4.3 Opportunity for participation

For successful conflict resolution, all interested parties must have the opportunity to participate in the process. Exclusion of an interested party is not only unfair but also risky because that party may obstruct the implementation of the outcome by legal or extra-legal means.

6.7.4.4 Identification of interests

It is important, in working toward consensus, to identify interests rather than positions. Conflicting parties often engage in positional bargaining without listening to or appreciating the underlying interests of the other parties. This is the classic "dam: no dam" oppositional position. This creates confrontation and a barrier to consensus. A strategy of mutual gain is often impossible to identify, if we limit ourselves to consider only water in only this river. If we include resources other than water, and/or we consider also the water resources in other (adjacent) rivers or basins, then it is easier to identify options of mutual gain (cf. Sadoff and Grey, 2002).

6.7.4.5 Developing options

An important part of a conflict resolution process is the neutral development of possible solutions and options. An impartial third party can be a great asset to the process as it can put forward ideas and suggestions from a neutral perspective.

6.7.4.6 Carrying out an agreement

Not only must the issue be capable of resolution through the participatory process, but the parties themselves must also be capable of entering into and carrying out an agreement, i.e. enacting the decisions they have mutually agreed upon.

6.8 RESOURCES AVAILABLE FOR IN-BUILDING COOPERATION

Presently, there are several training networks operating in regional or global settings, most being linked to one another. For example, Cap-Net (capacity building network for IWRM) has launched a global network of actors interested in IWRM. Many of these regional networks have been developing their own capacity to train stakeholders in conflict resolution and negotiation for IWRM. UNESCO has also created within its PCCP (from potential conflict to cooperation potential) initiative, a web-centred conflict resolution database and management system. Cap-Net has relied a great deal on the materials developed by the UNESCO PCCP programme. Many universities and think-tanks have also established IWRM and/or conflict resolution training programmes.

As resources are freely available on the Internet, most of those programmes borrow from each other, openly acknowledging their debts of gratitude along the way. Understandably, the various training programmes are similar in form and content, varying mainly in the application of the local case study to the programme.

6.9 WHAT WORKS IN PRACTICE

6.9.1 PARTICIPATORY METHODOLOGY

In our experience, training-of-trainers sessions are most successful when those around the table to be trained are actively involved, in a sense being mobilized to train themselves. Often, actors have much experience and a wealth of information – what they usually lack is a structure through which to channel their knowledge. How to think about an issue, and how to put those thoughts into action should be the centre of any training exercise. A complementary factor is deliberately soliciting participation across disciplines, sectors, ages, sexes and genders. In our experience, mutual learning and appreciation of each participant's unique perspective is both a key learning point and a trust-building exercise.

6.9.2 KNOWLEDGE DISSEMINATION AND INFORMATION SHARING

At the heart of water conflicts is knowledge and information. The two are not the same thing. With regard to "knowledge", people have been trained to think – or not think – about water in many different ways: for example, as the central driver of an ecosystem, or as the central driver of an industrial economy. Such very different ways of knowing water make it very difficult to find common ground when and where disputes arise. A related but different issue concerns information and the parties' willingness or unwillingness to share information. Successful training, therefore, involves joining the sharing of relevant information about water resources that is drawn from across scientific disciplines and the water users. New ways of understanding the resource itself (e.g. "virtual water", "green water") can have revolutionary impacts on how water managers think about the availability of the resource. New data sources and technologies, such as information derived from satellite images that are in principle accessible to any actor, also provide new opportunities for trust building. As withholding information has become a less effective strategy, openly showing a willingness to share data builds trust.

6.9.3 SIMULATIONS/MODELLING

There are numerous decision-support tools (DSTs) available for mapping and modelling water and related resources. Visual representations of the state of a resource and its many possible future states, depending on decisions made, help dispel commonly held myths, prejudices or assumptions about either the resource, its future, or its users.

6.9.4 GAMES/ROLE PLAYING

Assisting stakeholders to "walk a mile in each other's shoes" is a cliché, but also a central strategy in ADR. There are several simple games that have been devised so that actors can inhabit each other's geographic, socio-economic, political, or hydrological roles and experiences. Such activities are assumed to assist decision-makers beyond their staked-out positions toward the articulation of possibly mutually agreed-upon interests.

Other games have also been designed in a way that allows one party to watch another playing its role. Trainees from a country or river basin being discussed in a role play learn a lot while observing how trainees from another country (not concerned with the basin at stake) discuss issues concerning their own basin. This arrangement is used to establish a kind of detachment and creates situations in which participants can witness how outsiders may view the management of their basin objectively without having any cultural, religious, political or historical prejudice.

6.9.5 SKITS

Negotiations and/or conflicts over water are beset by numerous difficulties: social, cultural, political and economic. Short, participatory dramatizations (i.e. 'skits') illustrating these various barriers to successful negotiation are both fun and effective. They need not be water-specific, but can be drawn from any real-life setting where negotiation is hampered by, for example, an inability to communicate effectively, or where geography and/or political power enables one actor to withhold information, or to foot-drag their way toward dispute resolution.

6.9.6 LOCAL CASE STUDIES

An extremely valuable and effective element of conflict resolution training is the local case study. Indeed, involving the various stakeholders in the case study during the overall training is also useful, although it is rare to get a full commitment by all stakeholders to participate openly over the entire course of the workshop.

6.10 TRANSLATING WORKSHOP EXPERIENCE INTO REAL-WORLD SETTINGS

6.10.1 WHO IS BEING TRAINED?

Short, focused training activities are legion across the "water world". Three- and five-day or two-week training sessions are available through a wide array of international actors, be they universities, governments, (I)NGOs, or multi-actor networks such as GWP. Frequently, where governments are concerned, those being trained are junior civil servants with little decision-making power. One wonders if these many training sessions are actually reaching those with ultimate authority. In their absence, we, as trainers, often comfort ourselves with the idea that, over time, we are creating a culture of IWRM- and ADR-trained actors who will inevitably influence the way decisions will be made in future.

6.10.2 WILLINGNESS TO COOPERATE

Training for conflict management is anticipatory: it draws on past experience to help prepare stakeholders for situations that are likely to arise in the future. Rarely do such activities become part of resource management's standard operating procedures. So, when a conflict or dispute does flare up, it is not clear that parties have internalized all or any of their training. What is evident, however, is that where there is clear willingness to cooperate, ADR is an effective tool for arriving at mutually beneficial outcomes.

6.10.3 THE ROLE OF (INTERNATIONAL) LAW: ADR VERSUS MEDIATION/ARBITRATION VERSUS POWER POLITICS

Much ADR activity assumes a set of circumstances that rarely inhere in the 21st century real world. ADR is most easily applied on a limited geographic scale, where users are limited, uses are not complex and there is a common culture in place. So, cooperation along streams and (ephemeral) rivers in rural, low consumption world settings are most amenable to ADR methods. However, most of the world's water resources lie in highly complex settings with many actors, many competing uses, and different languages and cultures. Power is also unequally distributed throughout these settings. At the national level in mature capitalist, democratic societies, dispute resolution relies heavily on law and legal precedent. ADR is not commonly used; rather, disputes are usually settled by lawyers in or out of court through processes of either mediation or arbitration. These outcomes, far from being mutually beneficial, are primarily winner-take-all. Mutual respect for the law is the glue that holds the system together.

In Third World settings, the "law" is often arbitrarily devised and imposed. There are limited avenues for popular participation in dispute resolution, especially where socio-economic inequalities are pronounced. Dispute resolution is largely limited to inter-elite dialogue that tends more toward power politics than even respect for (international) law. Given water's central role in economic development and the creation of socio-economic opportunity, where cooperation does occur is often among elites in service of their mutually identified interests (e.g. through infrastructure development). Does ADR have a place here? It is hard to say. We continue to train actors in the art of ADR in the hope that a commonly held culture of cooperation can be fostered: a culture in which parties stand side by side to address a difficulty and solve a problem, rather than opposite each other across the table. The evidence, to the contrary, is that most river basins are treated as the site of power politics, and leaders will use ADR if and only if they have no recourse to more winner-take-all scenarios. Are we therefore training people correctly? Should we not try to focus on squeezing mutual benefit out of power political settings? Or is this simply not possible?

6.10.4   KNOWLEDGE OF OPTIONS AND THE PRESENCE OF POLITICAL WILL

Despite what we have just said in 6.10.3 above, based on our experience, it is clearly important that decision-makers, or their advisers (if the trainers manage to reach the right level of civil servants), know that ADR techniques exist. Even though they would not be able to apply them after a two-day course, they would know that such techniques exist and that they can call for resource people to support them. This is the role of PCCP: to train the civil servants and let them know these techniques exist, and give them access to a network of resource people who can support them should the need arise.

A related point is that if there is a political will to solve a problem, it gets solved, irrespective of the setting (e.g. complex, large scale or small scale with a limited number of users). Thus, what really makes the difference is the political will to solve the conflict, or the lack of that will. When there are ADR techniques available, a positive-sum outcome is more likely.

6.11 CONCLUSION

In this chapter we have reviewed the state of thinking about water and conflict/cooperation. We have highlighted the various approaches to conflict management/resolution and have articulated what we believe to be effective teaching tools for water managers who face or will face conflict now and in the future. We have also highlighted several caveats regarding the recurrent focus on ADR as the chosen method of water conflict management. This should perhaps be regarded as the ideal-typical form of conflict management, and we should investigate further the ways and means of resolving conflicts and/or unsustainable and inequitable resource-use practices in settings in which stakeholders are more inclined toward litigation (at the national level) and/or power politics (especially at the international level). Our feeling is that creating and maintaining open channels for dialogue (such as between North and South Korea on the Han River) in politically fraught settings is as important as actively promoting the de-escalation of negative, winner-take-all behaviours.

It is also our feeling that in addition to the promotion of ADR techniques, those engaged in conflict management should also learn from their specific setting. As highlighted above, there may be truly alternative and relevant approaches to conflict management specific to a particular cultural and socio-political setting unknown to outside facilitators.

Lastly, it is our belief that cooperation on water may have a positive spinoff to the governance of other sectors of a society. This is because water is widely considered to be "special" and acknowledged as vital in all societies and cultures (Savenije, 2002).

This specialness is documented as far back as the Roman Empire (1000 BC – AD 500), where fresh water was classified as *res omnium communes* – things belonging to each and every one – in terms of the *ius gentium* (law of all people) and *ius naturale* (natural law). No living creature (including all non-human forms of life) could be deprived of its right to use water, just as was the case with the sea and the air. In accordance with the principles of justice and equity of natural law, no one who was in need of water could be deterred from utilising it, as long as such use did not limit the equivalent right-of-use of others. Such an approach remains embedded today in communal law throughout much of the global south (Uys, 1996).

REFERENCES

Allan, J.A. 2003, 'Virtual Water – the Water, Food, and Trade Nexus. Useful Concept or Misleading Metaphor?' *Water International*, 28(1): 4–11.
Carter, N. 2007, *The Politics of the Environment: Ideas, Activism, Policy*, 2nd ed. Cambridge: Cambridge University Press.
Conca, K. 2006, *Governing Water: Contentious Transnational Politics and Global Institution Building*. Cambridge, Mass: MIT Press.

Coser, L. 1956, *The Functions of Social Conflict*. New York: The Free Press.

Ellis, S. and Ter Haar, G. 1998, 'Religion and politics in sub-Saharan Africa'. *Journal of Modern African Studies*. 36(2): 175–201.

Falkenmark, M. and Rockström, J. 2004, *Balancing Water for Humans and Nature*. London: Earthscan.

Gleick, P.H. 2000, *The World's Water 2000–2001*. California: Island Press.

Merrey, D. 2008, 'Is Normative IWRM Implementable? Charting a practical course with lessons from Southern Africa'. *Physics and Chemistry of the Earth*, 33: 899–905.

Molden, D., Frenken, K., Barker, R., de Fraiture, C., Mati, B., Svendsen, M., Sadoff, C. and Finlayson, C.M. 2007, 'Trends in water and agricultural development'. In: D. Molden, ed., *Water for Food, Water for Life: A Comprehensive Assessment of Water Management in Agriculture*. London: Earthscan and Colombo: IWMI.

Rogers, P. and Hall, A.W. 2003, *Effective Water Governance*. TEC Background Papers 7. Stockholm: Global Water Partnership.

Sadoff, C.W. and Grey, D. 2002, 'Beyond the River: the benefits of cooperation on international rivers'. *Water Policy*, 4: 389–403.

Savenije, H.H.G. 2002, 'Why water is not an ordinary good, or why the girl is special'. *Physics and Chemistry of the Earth*, 27: 741–744.

Svendsen, M., Wester P. and Molle, F. 2004, 'Managing River Basins: an Institutional Perspective'. In: M. Svendsen, ed., *Irrigation and River Basin Management: options for governance and institutions*. Cambridge, Mass: CABI.

UN. 2006, *Water: A Shared Responsibility. World Water Development Report 2*. New York and Geneva: UNESCO and Berghahn Books.

Uys, M. 1996, A structural analysis of the water allocation mechanism of the Water Act 54 of 1956 in the light of the requirements of competing water user sectors. Pretoria: Water Research Commission.

Wolf, A.T., Kramer, A., Carius A. and Dabelko, G.D. 2005, 'Managing Water Conflict and Cooperation'. In: *Worldwatch Institute, State of the World 2005: Global Security*. London: Earthscan.

Chapter 7

CAPACITY CHANGE AND PERFORMANCE:
Insights and implications for development cooperation

The European Centre for Development Policy Management,
The Netherlands (ECDPM)*

ABSTRACT

The European Centre for Development Policy Management (ECDPM) recently published the final report of a five-year research programme on capacity, change and performance.[1] This research provides fresh perspectives on the topic of capacity and its development. It does so by highlighting endogenous perspectives: how capacity develops from within, rather than focusing on what outsiders do to induce it. The research also embraces ideas on capacity development drawn from the literature outside the context of development cooperation. Although the research draws implications for international development cooperation, it does not specifically examine donor agency experiences in capacity development, or related issues of aid management and effectiveness.

The final report, which this chapter is based on, provides a comprehensive analysis of the findings and conclusions of the research programme. In total, 16 case studies[2] were prepared that embrace a wide spectrum of capacity situations covering different sectors, objectives, geographic locations and organisational histories, from churches in Papua New Guinea to a tax office in

* The European Centre for Development Policy Management (ECDPM) works to improve relations between Europe and its partners in Africa, the Caribbean and the Pacific. The brief from which this chapter was taken was prepared with inputs from: Tony Land, Niels Keijzer, Anje Kruiter, Volker Hauck, Heather Baser and Peter Morgan. This chapter has been reproduced from an ECDPM Brief with permission and is based on an extensive study on capacity development called Capacity, Change and Performance (www.ecdpm.org/capacitystudy).

Rwanda to nationwide networks in Brazil. The case studies are complemented by seven thematic papers and five workshop reports.[3]

The final report is written for people interested and involved in capacity development work. It offers insights as much for managers and staff of public sector and civil society organisations as it does for external agencies, either those providing capacity development services to local organisations or donors that finance capacity development work. This chapter highlights key findings and conclusions of the final report and presents implications for external agencies engaged in capacity development in the context of international development cooperation. It contains a bibliography listing the publications produced in the context of this study.

7.1   CONTEXT AND CONTRIBUTION

There has been an upsurge of interest in capacity development over recent years. The 2005 Paris Declaration on Aid Effectiveness recognised the centrality of core-state capabilities to the effective management of domestic and international resources for development. The Accra Agenda for Action (AAA) endorsed in Accra at the HLF in September 2008[4] has further raised the profile of capacity development as a fundamental ingredient of development effectiveness. An increasing number of studies and papers have been produced on the topic, including the DAC's 2006 "The Challenge of Capacity Development – Working towards Good Practice".[5] This document distinguishes capacity as an outcome, capacity development as a process, and support for capacity development as the contribution that external actors can make to country processes.

ECDPM's research offers insights on this three-way distinction. It takes the position that if the creation of capable country systems, organisations and individuals is a fundamental objective of development cooperation, then donors and their colleagues in developing countries need to better understand how capacity emerges and how it is sustained. This is all the more important given the mixed record of achievement of development cooperation in supporting the development of sustainable capacity.

7.2   COMPLEXITY AND UNCERTAINTY:
THE CONTRIBUTION OF SYSTEMS THINKING

The balance of issues in development cooperation is shifting against predictability and control towards complexity and uncertainty. Capacity development itself has shifted from a focus on implementing discrete projects aimed at skills enhancement or organisational strengthening, to addressing much broader societal and systemic challenges of building modern states in sometimes highly contested environments characterised by uncertainty and insecurity.

Formal planning models and technocratic approaches in such circumstances are not necessarily appropriate. More experimental and incremental approaches are required. Against this background, the concept of *complex adaptive systems thinking* can be helpful.[6]

While by no means a panacea, this can help to see the deeper patterns of behaviour and relationships that lie beneath individual events and actions. Because it puts less faith in planning and intentionality, it implies looking differently at causation, attribution and result chains. It also encourages people to think more creatively about disorder, uncertainty and unpredictability, and about the processes through which capacity develops.

7.3   CAPACITY DEVELOPMENT: A FUNDAMENTAL DEVELOPMENT CHALLENGE

The study contends that finding ways to develop and sustain capacity is a fundamental development challenge to which country partners and external agencies need to give greater attention. Effective systems, such as institutions and organisations, need to be seen as crucial elements of the development challenge. This is because they house the collective ingenuity and skills that countries need to survive and prosper.

Recognising capacity as more than a means to an end, but as a legitimate end in itself, is thus fundamental to any serious effort to improve the understanding and practice of capacity development. Such a perspective brings into question the deeper purposes of development cooperation. Are substantive gains such as those in health, education, agriculture, environmental protection the only true results of development? Or is the ability of a country to choose and implement its own development path – its basic capacity – also a development result?

7.4   INSIGHTS ON CAPACITY

Poor performance is often attributed to a lack of capacity. Probing to understand which capacity is lacking often leads to identification of resource shortfalls such as too few staff or the wrong skills, lack of equipment and infrastructure, out-dated systems, and inappropriate incentives.

The ECDPM study proposes a complementary lens for exploring organisational or system capacity. It encourages stakeholders to look beyond the formal capacities to deliver development results (such as technical and managerial competencies) and to identify other factors that drive organisational and system behaviour.

It identifies five core capabilities (see Box 7.1) which enable an organisation or system to perform and survive. All are necessary, yet none is sufficient by itself. A key challenge, therefore, is for an organisation to balance and integrate these five core capabilities.

Implications: Exploring capacity through this lens can help stakeholders diagnose capacity strengths and weaknesses, monitor capacity change over time and thus contribute to organisational learning. It can also be used to gauge the contribution of external support.

7.5   INSIGHTS INTO CAPACITY DEVELOPMENT

The final report highlights the many ways that organisations and systems go about developing capacity. It concludes that there are no blueprints for capacity development and that the process tends to be more complex, nuanced and unpredictable than is sometimes assumed. On the basis of the case studies, it identifies some generic characteristics of capacity development processes, which carry implications for the way external agencies go about supporting capacity development.

Box 7.1 Five core capabilities[7].

1   To commit and engage: *volition, empowerment, motivation, attitude, confidence.*

2   To carry out technical, service delivery & logistical tasks: *core functions directed at the implementation of mandated goals.*

3   To relate and attract resources & support: *manage relationships, mobilise resources, network, build legitimacy, protect space.*

4   To adapt and self-renew: *learning, strategising, adapting, repositioning, and managing change.*

5   To balance coherence and diversity: *encourage innovation and stability, control fragmentation, manage complexity, balance capability* mix.

7.5.1   VOLUNTARY COLLECTIVE ACTION

Crucial to capacity development is the collective energy, motivation and commitment of stakeholders to engage in a process of change. Voluntary collective action arises from leadership as well as from the ability of groups to be motivated and driven by leaders.

There is, however, a diversity of leadership styles that can influence collective action, the appropriateness of which varies according to the organisational context. Whilst in some situations leadership may be associated with a heroic, dominating and charismatic individual, in others it may be associated with a more facilitating and distributive style of leading.

Implication: External interveners can only facilitate capacity development indirectly by providing access to new resources, ideas, connections and opportunities. While they can support and nurture different styles of local leadership, they cannot substitute for it, nor drive the process.

7.5.2   OWNERSHIP – AN ILLUSIVE PROPERTY

It follows that ownership is key to building and sustaining capacity. It is a function of both the willingness and ability of stakeholders to engage in and lead change. But ownership can be elusive, ebbing and flowing over the life of any intervention. Ownership can exist at the highest levels of an organisation (where negotiations and planning takes place) but may be absent lower down, and vice-versa.

Interests can change and supporters at the outset may become detractors later on. Those who have the ability to exercise their ownership may not share the same interests and objectives as other stakeholders with less voice (see Box 7.2). In politically unstable environments, ownership can quickly shift as alliances and allegiances form and reform.

Box 7.2 Commitment and ownership: The SISDUK programme, Indonesia.

The SISDUK programme in Indonesia, which sought to empower communities to plan and implement their own projects, contended with a variety of commitments during the course of the programme. Policy makers and bureaucrats in Jakarta wanted fast-disbursing and widely spread programmes that would generate visible benefits in the short term. Provincial technocrats wanted to manage the programme using tested planning and budgeting techniques. Village heads wanted direct control over programme budgets. Community participants and programme fieldworkers who understood the system best ended up having the least power.

Implication: External partners need to be aware of the formal and informal processes that can shape and modify patterns of ownership over time. This implies having a good understanding of the local context and of stakeholder interests and influence, and staying engaged.

The 'aid relationship' has a built-in tendency to undermine ownership. Imbalances in resources, power and knowledge can give a feeling of mastery to the helper and dependence to the helped. It can confer 'expert' status on the helper that may be justified in terms of technical knowledge but is usually unwarranted in terms of process skills or country knowledge. It is likely to focus attention on gaps and weaknesses that can further add to the feelings of dependence and disempowerment of country actors. External initiatives quickly become "owned" by development agencies.

Implication: The way development initiatives are identified and formulated is critical. Creating space and opportunity for local priorities to be expressed, and ensuring country leadership in conceptualisation and design, is fundamental. So, too, is ensuring that the roles and responsibilities of local partners throughout implementation is made explicit.

7.5.3 MANY TACIT MODELS – FEW EXPLICIT STRATEGIES

Everyone, be they analysts or practitioners, has some sort of *tacit mental model* of capacity and how it develops. These are based on certain principles and assumptions about what makes people and systems be effective, or what capacity issues matter more than others.

Few, however, operate on the basis of an explicit strategy of change which is then tracked systematically. Too often, a shared vision is assumed and remains implicit. This makes it difficult to align approaches among stakeholders and can easily lead to difficulties and disagreements during implementation.

Implication: External partners need to have an open discussion with their country partners about capacity development in order to understand underlying assumptions, and as a basis for crafting a shared strategy for change.

7.5.4 DIAGNOSIS AND ENTRY POINTS

Assessing capacity through some kind of diagnostic process can help participants arrive at a shared understanding of their capacity challenge, agree on aspects of capacity that need attention and take account of factors that may promote or inhibit change. Such insights provide a basis upon which an intervention strategy can be conceived, including the identification of appropriate entry points.

In practice, a combination of entry points may be needed. These might include: organisational development work, adjusting internal and external incentives, promoting knowledge and understanding, tackling underlying organisational values and meaning, and adapting formal and informal structures and systems (see Box 7.3).

Implication: The five capabilities discussed earlier offer a lens to explore organisation and system capacity, to diagnose strengths and weaknesses and to monitor change over time. It can help broaden perspectives on the relationship

---

Box 7.3 The influence of the informal.

In some instances, there is an informal structure or 'shadow' system that has its own pattern of relationships, access to power and information flows. This informal structure can be the main repository of capacity, with the formal being in place for symbolic rather than operational reasons. In many cases, informal structures, both inside and outside the formal system, are intertwined with the formal system in ways that both support and hinder capacity development. An example is the influence of Wontoks – clan-based affiliations – on modern politics and governance in Papua New Guinea.

between capacity and performance by highlighting some of the informal and intangible aspects of capacity that can influence behaviour and motivation.

7.5.5 TO PLAN OR NOT TO PLAN?

Good design does not imply having to have a fully worked out implementation strategy. Indeed in some situations, this can be counter-productive. The research distinguishes three forms of strategy:

Planned approaches: Planned approaches tend to rely on prediction, goal setting, hierarchical structures and a top-down strategy. These can have a genuine advantage in situations where tasks are clear, in terms of ends and means, and in situations that respond to a disciplined, systematic approach. Such approaches also help to reduce confusion in the early stages, so that participants feel more comfortable. They also allow for coordinated action.

Incremental approaches: These emphasize learning and adjustment and tend to work better in situations of greater uncertainty and rapid change, where stakeholders are less able to predict their capacity and performance needs, or when the constraints or the degree of commitment are not fully understood.

Under these circumstances, strategies can still include objectives and milestones, but they function more as guidelines than as fixed targets. Using adjustments and small interventions, stakeholders are able to seek out opportunities, try different changes, move as the context allows, and try to learn what might work under different conditions.

Emergent approaches: These work where the driving forces for change are relationships, interactions and the system's energy. Capacity emerges out of multiple interdependencies and causal connections operating within the system. Capacity is partly about functional expertise, but also about system cohesion and energy. It is frequently a messy process and works best in complex situations. It needs space and freedom to explore the best way forward (see Box 7.4).

Box 7.4 The emergence of capacity: Two networks in Brazil.

The process of emergence can be seen at work in the COEP and Observatório cases in Brazil. They were first energised by the pursuit of key values to do with democratisation and social justice. They grew organically through informal connections and relationships. They refused to set clear objectives at the outset. A direction and an identity emerged over time. Facilitation, connection and stimulation worked better than the traditional directive management. There was no attempt to develop formal hierarchies at the outset. They experimented throughout the network with small projects and interventions. There was a constant exchange of experiences, information and knowledge. They spun off many working groups, informal communities and associations. Collective networking capabilities emerged through linking and connecting capabilities at the individual and organisational levels.

Implication: Few capacity development strategies work well in all cases. There is no 'code' or recipe for effective capacity development. Partners need to think carefully about the appropriateness of different strategic approaches as stakeholders become more aware of the nature of their capacity challenge, the demands of stakeholders, and the dynamics of their own organisation or system. Often customised approaches work best – in particular for highly specific jobs.

7.5.6   BEYOND 'MACHINE BUILDING'

Capacity development is often thought of in terms of 'machine building' – the bolting on of different parts to form a whole. While elements of capacity can be supplied in this way, other less tangible elements such as ownership, identity, legitimacy and values cannot (see Boxes 7.5 and 7.6). Because capacity development has to take account of politics and power relations, the process is also as much about negotiation and accommodation as it is about the supply of resources and tangible assets.

Box 7.5 Legitimacy – an element of capacity?

What are the connections between legitimacy and capacity?
- The legitimacy issue suggests that capacity is as much conferred from an organisation's stakeholders as it is developed internally.
- Stakeholders – clients, peers and oversight organisations – develop their views on the legitimacy of an organisation based on how well it performs in terms of its core capabilities, especially its commitment or motivation; its ability to carry out tasks, particularly delivery of services; its relationships; and, its adaptability and, hence, its ability to survive.
- The need for legitimacy encourages actors to earn support and approval from other groups in society.

Box 7.6 The importance of identity and confidence.

Capacity frequently emerges out of a technocratic combination of functional skills, assets and resources, and mandate. But in many cases, intangibles such as identity and confidence assume major importance. In ESDU, ENACT, IUCN in Asia, the Rwanda Revenue Authority, the COEP and Observatório networks in Brazil, the PSRP in Tanzania and CTPL Moscow, the participants worked, both directly and indirectly, to foster a collective identity that could be recognised both internally and externally. Coupled with this sense of identity was the growth in confidence and mastery, which led participants to develop a belief in their ability to make a special contribution to those with whom they worked. This belief, in turn, generated feelings of loyalty and pride that deepened the emotional and psychological relationships underlying the capacity, and expanded the range of activities that people thought they could attempt.

Implications: External partners seeking to support endogenous capacity development processes need to expand their tool box so as to be able to tackle more political and less tangible aspects of change. They need to be able to identify the factors that can stimulate or inhibit capacity development, which will differ from one context to another and which will evolve over time.

7.5.7 BALANCING OPERATING SPACE AND ACCOUNTABILITY

The term 'operating space' refers to a protected area within which organisational actors can make decisions, experiment, and establish an identity. Such a space can be physical, organisational, financial, institutional, intellectual, psychological or political. Operating space can be critical for capacity development for two reasons:

- it creates the conditions that allow a psychological sense of ownership to take hold. Without the freedom to move and decide, actors soon lose motivation and engagement.
- it allows the key processes of capacity development to evolve, especially at the middle and lower levels of the system.

The notion of operating space suggests that capacity is more likely to develop where organisational actors are given sufficient space to shape their own destiny. Thus, some degree of detachment can, in fact, benefit capacity development.

Space can be obtained in different ways. An organisation or a programme could be positioned outside the main political battleground in order not to attract attention or predatory behaviour. Its existence could be protected by law, custom or legitimacy (see Box 7.7). History could endow a system with a sense of identity and independent purpose. Powerful protectors, including development agencies, could buffer it from intrusion.

Nimble and politically-astute actors could benefit from chaotic contexts because these can contain niches, spaces and possible relationships that were impossible to find and exploit in older, more formal and ordered systems, such as those in many high-income countries.

This perspective contrasts with the view that sees capacity development contingent on the relationships that exist between an organisation/system supplying goods and services and its constituency(ies) that place(s) demands on it to perform. Ideally, the relationship is one in which a self-reinforcing cycle is created in which demand and supply react in a positive way to create a virtuous cycle of increasing capacity. Over time, a pattern of reinforcing demand and supply ratchet the relationship and the organisation up to new levels of performance and legitimacy.

Creating and maintaining spaces, therefore, requires a complex and delicate balance. Too little space can lead to the withering of innovation,

---

Box 7.7 Space and the Rwanda Revenue Authority (RRA).

Although the RRA is a public organisation created by an act of parliament, its agency status has freed it from some of the normal constraints imposed by public service rules and regulations and enabled it to manage its own affairs at arms length from political authority. Agency status offered a 'space' within which the RRA could chart its own growth and establish its identity, while at the same time shielding it from an otherwise harsh external environment. The importance of this protected space cannot be underestimated.

- It offered the RRA a new lease on life, making a clean break with the past and allowing it to cultivate a new organisational culture based on values of integrity, accountability and performance, and to generate internal ownership and identity.
- It allowed the organisation to put in place a human resource management system that has attracted, developed and retained capable and committed personnel.
- It helped shield it from political interference in management, personnel and resource allocation decisions, while helping to sever any pre-existing patronage relations.
- It allowed it to craft its own change strategy.

At the same time, it did not sever the organisation's accountability to its political masters. In this sense, the space enjoyed had to be earned. Safeguarding space, therefore, represented a further incentive to perform.

---

energy and commitment. Too much space can be equally damaging. People and organisations lose a sense of accountability and responsiveness. They become isolated and cut off from other sources of energy and collaboration.

Implication: External development partners need to read situations carefully to ensure that any support provided contributes appropriately to creating spaces for experimentation, learning and protection while also ensuring that mechanisms of local accountability are reinforced.

7.5.8   THE POTENTIAL OF SMALL INTERVENTIONS –
BIG IS NOT ALWAYS BETTER

An emphasis on small interventions may seem counter-intuitive at a time when large-scale impacts are wanted and needed and where emphasis is increasingly placed on comprehensive, holistic, integrated approaches.[8]

Lessons from the research point to the fact that smaller, more manageable interventions can have a better chance of success in the short term, and can even lead to bigger capacity gains in the medium and long term. Small interventions can be appropriate when absorptive capacity is weak and demand uncertain. In particular, small interventions can deal directly with what is perhaps the biggest constraint on capacity development – the implementation gap.

Many country organisations simply do not have the capacity to take on complex programmes, even if they have the drive and the commitment. Implementing smaller interventions allows them to build skills and craft their own capacity development strategy.

Small interventions can also more easily target pockets of country commitment. The in-close, high-involvement, high-energy process nature of small interventions makes it easier for country participants to build confidence and awareness.

Implication: Development and country partners need to strike the right balance in combining large-scale comprehensive approaches and smaller-scale niche interventions. A key challenge is to identify complementarities between the two approaches and ensure that field experiences support national processes, and vice versa.

7.5.9   STAY THE COURSE – BUT INVEST IN QUICK WINS

At the heart of the 'time' issue lies one of the most difficult of capacity challenges, that of combining short-term responsiveness, usually in the form of some sort of change in performance or technical capabilities, with the ability to focus over the long term on the development of more complex capabilities such as slow, incremental, collective learning.

The long-term approach, no matter how appropriate in terms of the evolution of capacity development, has to face the hard reality of donor impatience and loss of internal legitimacy.

In an era of 'demanding' and 'proving' results, most long-term efforts have little chance of surviving without facing the need to demonstrate just what has been gained for the money expended. The research thus shows the opposite of the long term, i.e. the need for speed and urgency in *the short term*. Windows of opportunity open briefly and create the space for capacity entrepreneurs to act. The challenge, particularly for organisations in the public sector, is to sustain both processes over a long period in a context of shifting political trends and bureaucratic dynamics.

Implication: A balance needs to be struck between seeking opportunities for quick wins and keeping an eye on the long-term. Quick wins can be an effective way of convincing stakeholders that investing in capacity development is worth their while.

7.5.10   TAILOR M&E FOR CAPACITY DEVELOPMENT

Much attention is being given to the need to improve ways of monitoring and evaluating (M&E) capacity and its development. While many existing M&E approaches that are linked to results-based management (RBM) work well where tangible outputs and outcomes are being measured, they are less suited to making judgements about less tangible and predictable aspects of organisational or system change.

Both organisational stakeholders – for whom the monitoring of capacity can contribute to organisational learning – and external partners – who are interested in gauging the effectiveness of their support – need tools and methods that can explore the different dimensions of capacity and change.

Participatory approaches, including forms of self-assessment, such as the five capabilities framework discussed earlier, can contribute to organisational learning. Methodologies, such as Outcome Mapping (that focus on the impact of change processes on individual and organisational behaviour, actions and relationships), or the Most Significant Change technique (which involves the collection of stories about significant changes in capacity as perceived by stakeholders), may be appropriate in this context.

Yet, in other situations, it may be appropriate to conduct a more formal and independent assessment that is peer-based and benchmarked against recognised standards (e.g. for central bank or auditor general operations).

Usually, a combination of qualitative and quantitative indicators is needed to capture the different dimensions of capacity, including factors in the environment that facilitate or inhibit change. Indicators need to be accepted and understood by all stakeholders concerned, so that the process of analysis and conclusions reached are fully owned and can feed into learning and action processes.

Implication: Development partners need to think carefully about the appropriateness of any M&E system for capacity development. Consideration must be given to both organisational learning and external accountability requirements. M&E Systems should be able to capture both tangible and intangible aspects of capacity. They also have to take account of process as well as product outcomes. Yet, in the process, systems should to be kept simple in order to avoid burdening organisations with complex and time-consuming demands that may lose support.

7.6   EFFECTIVE SUPPORT FOR CAPACITY DEVELOPMENT

Building on the implications that have been listed throughout this chapter, this final section suggests some *additional actions* that development partners can take to improve the effectiveness of their capacity development support.

- Pay more attention to unleashing the potential for capacity development. This potential is present in all situations in all countries. But participants need to focus more on finding, inducing, igniting, and unleashing endogenous human energy and commitment. This means paying less attention to gaps, and more on strengths.

- According to a strength-based theory of action, the deeper capacity of human systems comes not from fixing things and solving problems, but from affirmation, from tapping into sources of commitment and imagination.
- Encourage effective leadership to help groups to work together. At the core of effective capacity development is endogenous energy, motivation, commitment and persistence. These add up to more than a vague notion of country ownership and they imply more than conventional 'leadership'. They require a process of encouraging and stimulating individuals to act either alone or, more likely, together. The leadership involved can take many forms from the individual heroic to the collective.
- Emphasise learning and adaptation. At the core of capacity development is the practice, in some form, of learning and adapting. In the majority of cases, the process needs to be shaped by adaptation, experimentation, learning and adjusting. That has implications for 'design', management, evaluation and all the other conventional aid functions. It also implies maintaining a broad range of types of interventions to match different conditions.
- Be more wide-ranging and creative about capacity development. External actors need to think about the potential of using more indirect approaches to intervening, including buffering and protecting, the provision of information, providing tangible and intangible resources, networking and connecting, working to shift contextual factors and encouraging learning.
- Put more emphasis on understanding country context, identifying appropriate partners and building relationships. The analysis from this study suggests that capacity development is a challenging process, and that an understanding of country conditions is crucial. This means having an appropriate level of staffing in-country with an appropriate skill mix. It also means finding the right balance between coordination work and engaging in policy dialogue and the acquisition of knowledge through interpersonal contacts and field experience.
- Develop the capabilities required to address capacity issues. Addressing the implications of capacity development outlined in this chapter will require increased investments in the issue by outside interveners. It will need to be seen as a specialty requiring dedicated resources. It will require more incremental planning processes and more organisational incentives to encourage staff to develop in-depth cultural understanding of partner countries. Monitoring and evaluation will have to put more emphasis on intangible aspects of development such as legitimacy and self-empowerment, as well as on the tangible outcomes.

NOTES

1 The research programme arose from a request from the UK's Department for International Development (DFID) to build on earlier work by UNDP on technical cooperation and capacity development. It was subsequently included in the work plan of the Network on Governance and Capacity Development (Govnet) of the OECD's Development Assistance Committee (DAC).
2 The case studies adopted an appreciative perspective, focusing on what has worked well and why.
3 See bibliography, below. Also consult www.ecdpm.org/capacitystudy to access published and unpublished documents linked to this research programme.
4 See www.accrahlf.net
5 See DAC (2006).
6 Although not part of the study's original conceptual framework, many of the findings of the research are informed by complex adaptive systems thinking. For further information, see (Morgan, 2005).
7 In an earlier phase of this study, the capabilities were referred to as (1) the capability to act, (2) the capability to generate development results, (3) the capability to relate, (4) the capability to adapt and self-renew, and (5) the capability to achieve coherence (see Morgan, 2006; Engel et al., 2007).
8 As reflected in international agreements such as the 2005 Paris Declaration on Aid effectiveness and the recently adopted Accra Agenda for Action (see www.accrahlf.net). Pressures placed on development agencies to increase spending and disbursement rates also encourage investment in larger programmes.

ACKNOWLEDGEMENTS

Funding the study were: Australian Agency for International Development (AusAID); African Capacity Building Foundation (ACBF); Canadian International Development Agency (CIDA); Committee of Entities in the Struggle against Hunger and for a Full Life (COEP), Brazil; Department for International Development (DFID), UK; Directorate General for Development Cooperation (DGIS), the Netherlands; IUCN in Asia, World Conservation Union; Japanese International Cooperation Agency (JICA); L'Organisation Internationale de la Francophonie (OIF); Swedish International Development Agency (Sida); SNV Netherlands Development Organisation; St Mary's Hospital Lacor, Uganda; United Nations Development Programme (UNDP).

REFERENCES

DAC 2006 "The Challenge of Capacity Development – Working Towards Good Practice" http://www.oecd.org/dataoecd/4/36/36326495.pdf
Engel, P., Keijzer, N. and Land, T. 2007. *A Balanced Approach to Monitoring and Evaluating Capacity and Performance: A Proposal for a Framework.* (Discussion Paper 58E). Maastricht: ECDPM. www.ecdpm.org/dp58e
Morgan, P. 2005. *The Idea and Practice of Systems Thinking and their Relevance for Capacity Development,* Mimeo.

Morgan, P. 2006. *National Action Committee Western Cape (NACWC), South Africa*. Mimeo.

BIOGRAPHY

Banerjee, N. 2006. *A Note on Capabilities that Contribute to the Success of NGOs.* (Discussion Paper 57P). Maastricht: ECDPM. www.ecdpm.org/dp57p

Baser, H. and Morgan, P. 2008. *Capacity, Change and Performance. Study Report.* (Discussion Paper, 59B). Maastricht: ECDPM. www.ecdpm.org/dp59b

Brinkerhoff, D. 2006. *Organisational Legitimacy, Capacity and Capacity Development.* (Discussion Paper 58A). Maastricht: ECDPM. www.ecdpm.org/dp58a

Brinkerhoff, D. 2007. *Capacity Development in Fragile States.* (Discussion Paper 58D). Maastricht: ECDPM. www.ecdpm.org/dp58d

Morgan, P. 2003. *Bibliography on Study on Capacity, Change and Performance.* Mimeo.

Morgan, P. 2006. *The Concept of Capacity.* Mimeo.

Morgan, P., Land, T. and Baser, H. 2005. *Study on Capacity, Change and Performance. Interim report.* (Discussion Paper, 59A). Maastricht: ECDPM. www.ecdpm. org/dp59a

Taschereau, S. and Bolger, J. 2005. *Networks and Capacity.* (Discussion Paper 58C). Maastricht: ECDPM. www.ecdpm.org/dp58c

Watson, D. 2006. *Monitoring and Evaluation of Capacity and Capacity Development.* (Discussion Paper 58B). Maastricht: ECDPM. www.ecdpm.org/dp58b

CASE STUDIES WITH SUMMARIES

Hauck, A. 2005. *Resilience and High Performance amidst Conflict, Epidemics and Extreme Poverty: The Lacor Hospital, northern Uganda.* (Discussion Paper 57A). Maastricht: ECDPM. www.ecdpm.org/dp57a

Lacor Hospital in Gulu District, northern Uganda, formerly an isolated missionary hospital, is now fully integrated into the Ugandan health system. The case study describes how the hospital has grown into a centre of excellence, setting an example for the rest of the health system and helping to build health care capacity for the whole country. It is an extraordinary example of capacity development, adaptation and performance in a region characterised by civil war, extreme poverty and outbreaks of virulent epidemics.

Land, A. 2005 *Developing Capacity for Participatory Development in the Context of Decentralisation: Takalar District, South Sulawesi Province, Indonesia.* (Discussion Paper 57B). Maastricht: ECDPM. www.ecdpm.org/dp57b

This case study examines how Takalar district in the Indonesian province of South Sulawesi took up the challenge of tackling rural poverty through the use of participatory development and community empowerment methodologies. The study looks at the capacity that was required of various local stakeholders, traces the processes through which the district, in partnership with JICA, undertook to develop the necessary capacity, and discusses the challenges encountered in sustaining interest in and the capacity for participatory development.

Saxby, C. 2005. *COEP – Comitê de Entidades no Combate à Fome e pela Vida – Mobilising against Hunger and for Life: An Analysis of Capacity and Change in a Brazilian Network.* (Discussion Paper 57C). Maastricht: ECDPM. www.ecdpm.org/dp57c

This case examines a Brazilian social solidarity network, COEP (the Committee of Entities in the Struggle against Hunger and a Full Life) through the lens of organisational and social capacity and change. COEP is committed to building a just and inclusive society, one without hunger and poverty. Its members include government agencies, parastatals and organisations from the private sector and civil society. The case examines how the network has evolved and identifies the capabilities that have enabled it to become a thriving and dynamic network active throughout the country.

Land, A. 2005. *Developing Capacity for Tax Administration: The Rwanda Revenue Authority.* (Discussion Paper 57D). Maastricht: ECDPM. www.ecdpm.org/dp57d

This case study explores the process through which Rwanda's revenue collection capability was transformed from that of a moribund government department into to respected and performing organisation in just six years. The Rwanda Revenue Authority was established in 1997 as a semi-autonomous executive agency. With substantial financial and technical support from the Department for International Development (DFID) and driven by high level political commitment to change from Rwanda's leadership, the authority has helped raise revenue collection from 9.5% of GDP to 13% of GDP.

Hauck, V., A. Mandie-Filer and J. Bolger. 2005. *Ringing the Church Bell: The Role of Churches in Governance and Public Performance in Papua New Guinea.* (Discussion Paper 57E). Maastricht: ECDPM. www.ecdpm.org/dp57e

This case examines the role of Christian churches as institutional actors within Papua New Guinea's governance and service delivery landscape. It discusses their existing capabilities to engage in advocacy and policy-related work, as well as to function as a partner of government in the delivery of social services. In so doing, it looks at the interplay of endogenous change processes and the development of capabilities to see how this has translated into the performance of various church-based institutions and the capacity of the church sector as a whole.

Bolger, J., A. Mandie-Filer and V. Hauck. 2005. *Papua New Guinea's Health Sector: A Review of Capacity, Change and Performance issues.* (Discussion Paper 57F). Maastricht: ECDPM. www.ecdpm.org/dp57f

This case study looks at the recent reforms in Papua New Guinea's health sector from a capacity development perspective. It addresses a number of factors influencing capacity development, change and performance including issues internal to the National Department of Health, capacity issues at sub-national levels, the institutional "rules of the game" that guide attitudes, behaviour and relationships in the PNG context and in the emerging health sector SWAP, and broader contextual factors.

Watson, D and A.Q. Khan. 2005. *Capacity Building for Decentralised Education Service Delivery in Pakistan.* (Discussion Paper 57G). Maastricht: ECDPM. www.ecdpm.org/dp57g

The Pakistan case is one of two that examine capacity issues in relation to decentralised education service delivery. The study explores the institutional environment and broader governance context within which institutional reform and capacity development has taken place. The study identifies factors that have either facilitated or constrained capacity development across the sector from the classroom level to the policy-making level within central government.

Watson, D and L. Yohannes. 2005. *Capacity Building for Decentralised Education Service Delivery in Ethiopia.* (Discussion Paper 57H). Maastricht: ECDPM. www.ecdpm.org/dp57h

The Ethiopia case is one of two that examine capacity issues in relation to decentralised education service delivery. The study explores the institutional environment and broader governance context within which institutional reform and capacity development has taken place. The study identifies factors that have either facilitated or constrained capacity development across the sector from the classroom level to the policy-making level within central government.

Watson, D. 2005. *Capacity Building for Decentralised Education Service Delivery in Ethiopia and Pakistan: A comparative analysis.* (Discussion Paper 57I). Maastricht: ECDPM. www.ecdpm.org/dp57i

This comparative analysis of the two cases on decentralised education service delivery in Ethiopia and Pakistan does three things. It summarises the main features of the two cases in terms of their contexts and features of the capacity-building experience the two countries have had; compares the main features of contexts and capacity-building experience with a view to drawing conclusions about the apparent significance of various aspects of this experience; distils conclusions about what factors appear to matter most in the relationship of change, capacity and performance, and as determinants of the feasibility of building effective capacity for devolved education service delivery over time. The paper is not a substitute for reading the full texts of the two cases.

Morgan, P. 2005. *Organising for Large-scale System Change: The Environmental Action (ENACT) programme, Jamaica.* (Discussion Paper 57J). Maastricht: ECDPM. www.ecdpm.org/dp57j

This case looks at the Environmental Action (ENACT) Programme, a collaboration between Jamaica's National Conservation Resources Agency and the Canadian International Development Agency. ENACT's mandate was to work with Jamaican public, private and non-profit organisations to improve their capabilities to identify and solve national environmental problems. Programme design began in 1990 but field activities only got under way in 1994. It took until 1999 to put in place all the pieces to make ENACT a high-performing support unit.

Morgan, P. 2005. *Building Capabilities for Performance: The Environment and Sustainable Development Unit (ESDU) of the Organisation of Eastern Caribbean States (OECS)*. (Discussion Paper 57K). Maastricht: ECDPM. www.ecdpm.org/dp57k

This case looks at the experience of the Environment and Sustainable Development Unit (ESDU) of the Organisation of Eastern Caribbean States (OECS) located in St. Lucia. The unit, originally conceived as the regional implementing arm for projects funded by the Deutsche Gesellschaft für Technische Zusammenarbeit (GTZ) at the beginning of the 1990s, has since become a facilitating and bridging organisation responding to the needs of the Member States of the OECS. The study explains ESDU's effectiveness in enhancing its organisational capabilities for performance over the period 1996–2003.

de Campos, F. and V. Hauck. 2005. *Networking Collaboratively: The Brazilian Observatório on Human Resources in Health*. (Discussion Paper 57L). Maastricht: ECDPM. www.ecdpm.org/dp57l

This case looks at a network of university institutes, research centres and one federal office dealing with human resources questions in the health sector of Brazil. The study explores how informal networking developed into a formal network delivering outputs and outcomes and potentially impacting the well-being of society. The network is seen as a unique and successful case of state–non-state interaction in health, and has raised interest among regional and international observers from the health sector, from network specialists, and from development agencies dealing with institutional development.

Rademacher, A. 2005. *The Growth of Capacity in IUCN in Asia*. (Discussion Paper 57M). Maastricht: ECDPM. www.ecdpm.org/dp57m

This study explores the growth of capacity in IUCN in Asia over the period from its inception in 1995 to early 2005, and assesses how capacity was built, maintained, and strengthened. In the first decades of its existence, the management of the global IUCN programme was highly centralised. The effort to create an Asia Regional Programme followed a global directive to decentralize in the mid-1990s. A regionalized IUCN was expected to be more responsive to its membership, more financially sound and sustainable, and more likely to realise IUCN's overarching goals.

Agriteam Canada. 2006. *The Philippines–Canada Local Government Support Program: A Case Study of Local Government Capacity Development in the Philippines.* (Discussion Paper 57N). Maastricht: ECDPM. www.ecdpm.org/dp57n

This study explores capacity development and related performance improvement within the context of local governance in the Philippines. The paper describes the local government units and the enabling and regulatory environment as a system that is evolving and becoming stronger at the same time as the individual local government units are developing. Given the political and social context conducive to devolution and democratisation, after the fall of the Marcos regime, and the enabling policy environment created by the Local Government Code, endogenous local government capacity development began to occur. Certain external interventions were very effective in enhancing this endogenous local government capacity-development process because of the specific approaches and methodologies that characterised these external interventions.

Rourke, P. 2006. *Strategic Positioning and Trade-related Capacity Development: The case of CTPL and Russia.* (Discussion Paper 57O). Maastricht: ECDPM. www.ecdpm.org/dp57o

This case explores the evolution and transformation of a trade-related capacity-development initiative aimed at supporting Russia's accession to the WTO. At the outset, the programme focused on training provisional and short-term technical assistants and worked mainly with government authorities. Lack of impact led to a rethinking of strategy and an overhaul of the programme. The new programme aimed at developing a sustainable institutional capability across the government and private sector divide that could respond to emerging needs and serve itself as a provider of capacity development support and a catalyser of change. This was achieved through a partnership arrangement between the Canadian Centre for Trade Policy and Law and a newly established policy institute led by Russian trade experts.

Morgan, P and H. Baser. 2006. *Building the Capacity for Managing Public Service Reform: The Tanzania Experience.* (Discussion Paper 57Q). Maastricht: ECDPM. www.ecdpm.org/dp57q

This case study is about how Tanzania built its capacity to manage a complex process of institutional and organisational change. It explores why the support for reform and change has been more pronounced in Tanzania than in other African states and how this allowed the government to position its programme in the mainstream of global public sector reform. The heart of the strategy was based on the "new public management" concept and has been largely top-down and supply-driven, with many simultaneous institutional and organisational reforms. It is likely, however, that more pressure from the demand side will be needed to keep the reform programme energised over the medium and long term.

Morgan, P. 2006. *National Action Committee Western Cape (NACWC), South Africa.* Mimeo. Summary at www.ecdpm.org/capacitystudy

The National Action Committee Western Cape (NACWC) case was unique in a number of respects. It came at a time of rapid institutional and organisational change in South Africa as the country moved into the post-Apartheid period. It was an effort by groups in the non-profit sector to introduce a 'new institutional form' that would be aimed at reforming the public sector from the outside. And despite many advantages, the experiment ended up with outcomes that none could have foreseen at the outset.

Part 3

Areas of application

Chapter 8

CAPACITY CHALLENGES IN THE INDONESIAN WATER RESOURCES SECTOR

Klaas Schwartz
UNESCO-IHE Institute for Water Education

Iwan Nursyirwan
Ministry of Public Works, Republic of Indonesia

Aart van Nes
DHV-Water Consultants, Amersfoort, The Netherlands

Jan Luijendijk
UNESCO-IHE Institute for Water Education, Delft, The Netherlands

ABSTRACT

The Indonesian water resources sector has been subject to reforms in recent years. The most significant of these reforms is the new water law adopted in 2004. This law reflects the principles of IWRM and represents a shift from previous development oriented towards more water management-oriented practices. Implementation of this law is quite challenging, as the capacity of organizations to undertake the required tasks is often missing. The capacity needs facing the water sector in Indonesia are two-fold. On the one hand, new skills are needed to address the challenges brought on by the reforms, which emphasizes a shift from development of new facilities and infrastructure to management of existing infrastructure, and from specific users to management of the whole cycle from source to users. On the other hand, the sector is facing enormous challenges resulting from the consequences of a decade-long zero-hiring policy. Whilst capacity needs are severe, the (institutional) mechanism in Indonesia to address these needs does not yet appear to be in place. For these needs to be addressed, action has to be taken at both the sector level (such as changes in university curricula and increasing coordination in the sector) and at the organizational level (such as developing better human resources management practices).

8.1 INTRODUCTION

Since the International Conference on Water and the Environment in Dublin, in 1992, Integrated Water Resources Management (IWRM) has emerged as a driving concept behind the management of water resources. The Global Water Partnership (2000) has defined IWRM as "a process, which promotes the coordinated development and management of water, land and related resources, in order to maximize the resultant economic and social welfare in an equitable manner without compromising the sustainability of vital ecosystems". Also in Indonesia, the concept of IWRM has been adopted as the guiding concept on how to manage water resources. The Law of Water Resources (UU SDA), passed by Indonesian Parliament in February 2004, introduces basic principles for water resources management consistent with IWRM.

For most of the 1990s, the discussions regarding IWRM as the guiding concept behind water resources management have been rather conceptual in nature. In recent years, however, the discussion has increasingly shifted towards actual implementation of the widely discussed principles and concepts. With the shift in focus on implementation of IWRM, capacity of organizations in the water sector to actually implement IWRM has also gained increasing attention. Capacity building closely supports and helps to guide the required institutional strengthening as well as the development of reform programmes that are needed to make effective integrated water resources management operational.

In this chapter, the capacity challenges in the Indonesian water sector regarding the implementation of IWRM are analyzed. The next section of the chapter presents the historical context of the Indonesian water sector. In the third section, the capacity needs and challenges are estimated based on an extrapolation of three case studies. The fourth section highlights the ability of Indonesian organizations to address the capacity challenges highlighted in the previous section. The chapter finishes with a conclusion concerning the existing capacity challenges and how these should be addressed.

8.2 THE INDONESIAN WATER SECTOR: A HISTORICAL PERSPECTIVE

Over the past fifteen years, the Indonesian water sector has been subject to considerable changes. Below, an overview is provided of these changes and what the impact of these changes has been on water resources management. The section is divided into three main parts. The first part is the period prior to 1999. The second part begins with the decentralization policies of 1999 and ends with the introduction of the Law on Water Resources in 2004. The third part starts with the introduction of the Law on Water Resources.

8.2.1 WATER MANAGEMENT PRIOR TO 1999

8.2.1.1 General policies

Until 1999, Indonesia's governance system was strongly based on the concept of the 'Unitary State'. One of the main characteristics of this 'Unitary State' is that the Government of Indonesia largely resisted any real decentralization of authority. A number of reasons for this resistance can be pinpointed (Devas, 1997):

- First, there was a widespread belief that local governments lacked the required knowledge and skills to exercise this authority.
- Secondly, the maintenance of authority at the central level was viewed as a way to ensure the proper use of public funds.
- Third, according to Devas' more cynical view, a reluctance existed in those at the center to decentralize, as this would mean sharing income-earning opportunities more widely and, as such, foregoing some of the rent-seeking opportunities derived from control systems.
- Fourth, central control allowed more possibilities for ensuring that benefits of development be more equally spread over the country. In this view, local autonomy was seen as an obstacle to inter-regional distribution, which could result in increasing inequalities between different regions within Indonesia.
- Fifth, and possibly the most important of the reasons why the central government resisted decentralizing its authority, is that decentralization was seen as threatening national unity. Indonesia has a large socio-ethnic diversity and maintaining national unity was seen as requiring central control.

Although this "Centralized Public Administration System" (Ramu, 2007) applies to the governance system in Indonesia in general, it also specifically applied to the water sector prior to 1999. Speaking of Indonesia's water sector in 1991, Cervero (1992:4) found that "In Indonesia, decentralization has not meant the transfer of political power and financial resources to lower levels of government, but rather has involved cultivating a local capacity to plan, manage, implement, and co-finance development projects under the auspices of the central government. As a unitary state, all levels of government in Indonesia are, in effect, branches of the central government". In this scheme, engineering work was done by the central level, with the assistance of projects in the provinces, which operated under the responsibility of the central agency, and with little or no coordination with the regional government or water users. Infrastructure would subsequently be transferred to the provinces and districts. The problem, however, was that in most cases the transfer of infrastructure, which was constructed without involvement of the regional governments, was 'assumed' to have taken place, meaning that it took place without matching budgets for operation and maintenance. The result was that regional governments, which frequently would not consider these transferred works to be of priority, would often decide not to spend funds on the operation and maintenance of these structures.

8.2.1.2   Focus of water management

Development of the water sector from the 1960s until the 1980s was strongly focused on development of new infrastructure, mainly for irrigation (Houterman et al., 2004). An important goal of this development focus was to achieve a situation of food security. In this context, the lack of (technical) capacity at the local level was an important consideration in maintaining a centralized administrative system. The focus on development of irrigation systems had a strong impact on food security, and by 1984 Indonesia had achieved self-sufficiency in its rice production.

By the late 1980s and early 1990s it became apparent that the supply-side driven approach had become unsustainable; it resulted in the neglect of operation and maintenance, and in financially unsustainable schemes. "Some of the schemes rehabilitated in the early seventies were in urgent need of re-rehabilitation by the mid 80s" (Herman, 2007). In response, the Government of Indonesia adopted the Irrigation and Maintenance Policy (IOMP) in 1987. The policy incorporated five major initiatives (Herman, 2005):

1   An efficient operations and maintenance funding program;
2   Direct cost recovery for O&M through property tax land classification and an Irrigation Service Fee Program;
3   Institutional development for water-user associations (WUAs) to facilitate full transfer of management control for schemes smaller than 500 ha to WUAs;
4   Needs-based programming, budgeting and monitoring for O&M;
5   Special maintenance programs for small schemes to be transferred to WUAs.

The adoption of the IOMP marked a policy shift on paper as it changed "the balance between central and provincial government and created a role for farmers as decision-makers in limited areas" (Herman, 2005:9). Houterman et al. (2004) also consider the IOMP to be the first systematic effort "to shift from development to management of water resources systems". The actual implementation of the policy proved difficult in practice. Houterman et al. (2004) conclude that "despite the good ideas and intentions, little of the IOMP survived the test of time".

The village-level WUAs proved ineffective as the Government was reluctant to provide incentives for WUA empowerment that conferred governance authority beyond the village level, and the WUAs were too small to act as an effective management institution. The small scheme management turnover program achieved only about one-third of its original targets. Irrigation O&M funding proved unsustainable as externally-aided rehabilitation was preferred over routine maintenance by provincial governments, and the Irrigation Service Fee managed to raise only 2% of O&M expenditures at its peak in 1994–1995.

The result of emphasizing the supply-side approach and the failure of the IOMP was that the period of rice self-sufficiency lasted only for a period of 10 years, and by 1993 Indonesia became an importer of rice again. Moreover, a growing population and economic development led to a stronger demand from domestic and industrial water users, frequently resulting in conflicts over use of limited water resources. Therefore, in 1994, new policies were adopted, shifting focus from supplying end-users to management at the source. River basin territories (*Wilayah Sungai*) were identified based on hydrological boundaries and comprising one or several river basins. Water Management Committees were established to coordinate between government agencies at provincial (PTPA) and river territory levels (PPTPA). The provinces of Java established Water Resources Management Centers (Balai PSDA) at most river territories, later followed by several other provinces outside Java, most notably North Sumatra.

8.2.2 1999–2004: DECENTRALIZATION … AND RECENTRALIZATION?

8.2.2.1 Decentralization laws of 1999

Although the development focus of the 1960s, 1970s and early 1980s had become unsustainable by the early 1990s (as illustrated by the introduction of the IOMP in 1987), it proved difficult to change the actual policy of the central government in practice. This changed in 1998 when the 'New Order Government' collapsed and led to radical changes in the country's governance, with profound consequences for the water sector. The government adopted the Law on Regional Government (Law 22/1999) and the Law on Fiscal Balance between the Center and the Region (Law 25/1999). Under these laws the prior central government's functions and responsibilities (except for defense, foreign affairs, the judiciary, and national finances) were decentralized to the regional governments.

Also for the water sector, this had significant consequences. "As a result of administrative and fiscal decentralization, all sector management tasks and responsibilities were given to district governments except where a river or canal crosses a district boundary. The same applied if the river, lake, or canal crosses a provincial boundary, in which case, the national government may assume control" (Herman, 2005:13–14). With responsibilities for water resources decentralized to provincial and district administrations, the focal point for water resources management would become the Provincial and District Water Resources Services (PSDA), Balai PSDAs, and irrigation committees.

8.2.2.2 A move back to the center?

The radical shift from a centralized administrative system to a decentralized administrative system appears to have been driven with the purpose of easing mounting political pressures and alleviating some of the administrative problems that contributed to the erosion of the central government's legitimacy and control. As such, the decentralization process was done without "serious assessment and discussion across the regions and provinces as to what decentralization would mean for them and the appropriate level of decentralization" (Ramu, 2004:14).

The decentralization policies quickly revealed constraints relating to the provincial and district governments' capacity and financing ability to implement the tasks they had become responsible for. In addition, the decentralization policies 'balkanized' water resources and irrigation activities with inter-government jockeying for authority, revenue and budgets. Realizing that the decentralization policies of 1999 may have gone too far, the Government of Indonesia issued revised laws in September of 2004. These revisions seemed to lead to a more 'Mixed Administration System' than the 'Decentralized Administration System' envisaged in Laws 22/1999 and 25/1999 (Ramu, 2004).

8.2.3   2004: INTEGRATED WATER RESOURCES MANAGEMENT

8.2.3.1   National water resources policy

In December of 2001, the National Water Resources Policy (NWRP) was issued. The NWRP served as "the basis for formulation of a new water management resources law and stipulates an integrated and sustainable approach to water resources management, and recognizes the economic value of water" (Herman, 2005:15). The NWRP's goals are to:

- Achieve synergy and prevent conflict between regions, between sectors, and between generations;
- Encourage the process of integrated water resources management between sectors at the central, provincial, and district/town levels, and in river basins;
- Balance conservation efforts and the efficient utilization of water resources with a view to ensuring that water is sustainably used;
- Balance the social functions and the economic functions of water to ensure fulfillment of the basic water requirements and, to achieve efficiency in the use of water as an economic resource that takes into account the costs of conservation and asset maintenance;
- Implement the regulation of water resources management in an integrated and balanced manner;
- Improve and develop a financing system for water resources management so that water resources management can be carried out effectively, efficiently, equitably, and sustainably;
- Take into account the principle of cost recovery, but consider community social and economic conditions; and,

- Develop a water resource management institutional system based on separation of regulatory and operational functions (Herman, 2005:15).

As these goals of the NWRP illustrate, a considerable shift in thinking about managing water resources occurred in comparison with the early 1980s, in which the focus was still strongly on the development of infrastructure.

8.2.3.2   The law on water resources 7/2004

On February 19, 2004, the Government of Indonesia adopted the Law of Water Resources (7/2004). The Law of Water Resources is a complete revision of the previous water law dating back to 1974 (11/1974). In a recent speech, the Director General[1] for Water Resources explained the changes resulting from Law 7/2004 as follows: "In this law the focus changed from new development of facilities to management of existing facilities, and from specific users, mainly farmers through irrigation, to management of the whole cycle, from source to users. As such, the focus changed from specific output (dams, structures, canals) to outcome (better services/performance) [...]. The government is changing from the provider to the facilitator".

Law 7/2004, though not entirely uncontroversial[2], appears to be designed in such a way that it is consistent with and would promote IWRM. A number of principles underlie Law 7/2004. First, the law promotes the need to find a balance between social, economic and environmental values. Second, the law states that the management of water resources should consider the principles of:

- Provision of effective and efficient public benefits;
- Balanced development meeting principles of integration and harmony in balancing different interests;
- Sustainability, justice, autonomy; and,
- The principles of transparency and accountability that imply development and management are regarded as an open and publicly accountable process (Herman, 2005:16).

8.2.3.3   River Basin Organizations (RBO)

The management structure and responsibility of RBOs is clearly described in Water Law 7/2004 and further detailed in the Government Regulation on WRM PP 42/2008. River basins (*DAS*) in Indonesia are combined in river territories (*Wilayah Sungai (WS)*), comprising one or more river basins,

---

[1] The speech was read by the Director for Planning and Programming, Mr. Hartoyo, on behalf of the Director General on February 25, 2008 in Jakarta.
[2] The controversy surrounding the adoption of Law 7/2004 is centered on the issue of how much room the Law leaves for 'privatization' of water services (see Al'Afghani 2006).

under one RBO. The boundary of this river territory and the corresponding responsible government layer (national, provincial and district level) will be set by a new presidential degree. Permen11A/2006 from the Minister of Public Works proposes a total of 133 river territories and their related government layers. These territories and responsible government levels are shown in Table 8.1.

Permen11A/2006 revised the earlier division of 90 river basin territories of the early 1990s. The earlier established provincial RBOs (*Balai PSDA*) were meant to manage the river basin territories under authority of the provincial government. The *Balai PSDA* also operate in river territories which fall under the national government. The existence of multiple RBOs (national and provincial) in a given river territory has created some confusion about the exact tasks and responsibilities of the *Balai PSDA* and the *Balai WS*. In these cases, the *Balai WS* are seen as having strategic role to play, whilst the provincial Balai should concentrate on operational functions. The Ministry of Public Works created a total of 30 River Basin Organizations for River Basin Territories under national government authority (*Balai Wilayah Sungai (BWS)*) by putting together existing infrastructure development projects and creating new organizations. The BWS are to function directly under supervision of the DGWR and are the technical executing units for water resources management. Permen12/2006 arranges the establishment of six larger organizations called *Balai Besar Wilayah Sungai* for large river basins and Permen13/2006 allows for the establishment of 24smaller *Balai Wilayah Sungai* for the other centrally managed basins (Ramu, 2007). Later ministerial regulations raised 5 BWS to the level of Balai Besar (BBWS), Table 8.2 provides an overview of the existing Balai.

8.2.3.4　Community involvement and stakeholder participation

Law 7/2004 also gives a strong role to the community and stakeholders in the management of water resources. In Chapter 11 of the Law, the rights, obligations and roles of the community are elaborated upon. The community, which is obliged to consider the public interest in the utilization of water, is entitled to:

Table 8.1 Overview of river territories and related government layers.

Responsible government level	Number of RBOs	Types
National government	69	International river territories (5) Cross-provincial river territories (27) River territories of strategic importance (37)
Provincial government	51	Cross-district river territories
District government	13	District river territories

Table 8.2 Existing *Balai Besar Wilayah Sungai* and *Balai Wilayah Sungai.*

Province/Region	BBWS	BWS	Total
Sumatra	1	8	9
Java/Bali	9	1	10
NTB/NTT	–	2	2
Kalimantan	–	3	3
Sulawesi	1	3	4
Maluku	–	1	1
Papua	–	1	1
Total	11	19	30

- Get information related to water resources management;
- Get reasonable compensation for the loss suffered as the result of the implementation of water resources management;
- Express objections to a water resources management plan which has been announced;
- Submit a report and/or complaint to the authorized official for the loss suffered in connection with the implementation of water resources management; and,
- Submit legal claims on various problems of water resources management causing loss to the community.

The community is furthermore determined to have an equal opportunity to play a role in the planning, implementation and supervision of water resources management.

8.2.3.5 Water resources councils

Law 7/2004stipulates the need for water resource councils whose main tasks are to prepare and formulate water resources management policies and strategies. These councils are to operate at national, basin, provincial and district/municipal levels and have a strong coordinating role. These coordinating institutions are to have members that consist of government elements and non-government elements in a balanced number. Non-government representatives would include experts in the field of water resources, water-user community organizations, self-supporting water resources community organizations, and professional associations in the field of water resources. Details for the selection of such representatives are to be issued under a Government Regulation.

8.2.3.6 Basin water resources management plans

Each river basin should have a Basin Policy or Strategy (*Pola*) reflecting the development and management views of all basin stakeholders through a public consultation process. The government responsible for river basin management in given basins must determine a Basin Water Resources Management Plan (BWRMP) for the river basins under their responsibility. The BWRMP must be based on the water resources policy and should express all basic principles of Law 7/2004. A BWRMP for every river basin is to be broken down into development and management programs related to water resources management by government agencies, the private sector, and the community. Preparation of a BWRMP is carried out in a coordinated way by the authorized agency in accordance with their task by involving parties connected to water resources. The community has the right to express their objections to the draft of a BWRMP and the draft should be revised to take into account community and stakeholder objections.

8.3 CAPACITY ASSESSMENT FOR THE WATER RESOURCES SECTOR IN INDONESIA

In this section the capacity needs for the Indonesian water sector are assessed. First, this assessment analyzes the types of new skills and knowledge required as a result of the paradigm shift in the sector as well as tries to quantify this capacity gap by linking it to a number of people who would have to be trained. Secondly, the assessment focuses on the staff shortages resulting from a decade-long zero-hiring policy.

It should be noted that the focus of the analysis is mainly on the Ministry of Public Works, and provincial and district water agencies. The reality of IWRM is that more organizations (such as, the Ministries of Agriculture, Forestry, Environment, etc.) at different levels play a role in the management of water resources. Their capacity needs are not included in this assessment and, as such, creates a limitation on the capacity assessment presented.

8.3.1 THE TYPES OF NEW SKILLS AND KNOWLEDGE REQUIRED AS A RESULT OF THE PARADIGM SHIFT

Within the DGWR of the Ministry of Public Works, a general consensus exists that Law 7/2004 does indeed represent a shift from the traditional development focus to a focus on existing facilities. There is also the realization that this shift will require new capacities. Illustrative in this sense is the recent speech by one of the directors of the DGWR[3] in which it was stated: "It is clear that this implies a change in paradigm and requires a thorough adjustment of the personnel involved in Water Resources Management, both at central and at local level". In highlighting the new 'facilitative role' of the national government, the director continued by stating that "a facilitator

---

[3] Presentation of Imam Anshori on 25 March 2008 in Jakarta.

needs awareness and active involvement of the community to be facilitated, and this requires again more skills not included earlier".

Box 8.1 The consequences of the reforms for the directorate of planning and programmes

The Director of the Directorate of Planning and Programmes of the DGWR explained the main changes in the water sector as follows:

- Increased participation
- Increased decentralization
- Increased transparency
- Freedom to express aspirations
- Democratization

In explaining what this means for the water sector and his Directorate in particular, Mr. Hartoyo replied that it means people are to be provided with information about the outcomes and outputs of water management activities and that the process of water management from planning to implementation has to be "more open to people". Before the reforms everything was centrally driven. Involvement of stakeholders and local government was minimal. Planning for water resources management was relatively simple and easier to implement. Now the process of planning and programming has changed. His staff need to manage many aspirations of stakeholders, address conflicts between stakeholders, and manage reconciliation efforts between stakeholders in this process.

In terms of capacity building Mr. Hartoyo mentioned that the skills/capacities needed at the moment are the following:

- Communication skills
- Leadership
- Negotiation skills
- Conflict management
- Public Policy and Policy Analysis regarding water resources management

Source: Interview with Mr. Hartoyo on November 21, 2007, in Jakarta

---

The change described above is also highlighted in a capacity-building study for the six 'Ci-basins' in West Java. In the study it is argued that the central change that needs to be made is one from "design, construction, operation and maintenance of water projects in a single sector/sub-sector to responsibility for managing the water of the river basin as a resource which must be protected, conserved and allocated" (Taylor, 2007). The study continues to identify the type of change in capacity that this would require. The required capacity concerns:

- Awareness of integrated water resources management and its challenges;
- An inter-sectoral outlook and collaborative working mode;
- New capabilities in planning and coordinating the activities of various sectors within the basin;
- Capacity to develop policies and guidelines to address a range of issues, or participate with those who have expertise;
- Appropriate technical skills in analysis, modeling and evaluation of alternatives (from Taylor, 2007:20).

8.3.2   QUANTIFYING THE NEW SKILLS AND KNOWLEDGE REQUIRED AS A RESULT OF THE PARADIGM SHIFT

Having identified the types of capacities required in the current context, the next step is to try to quantify the capacity-building needs resulting from the reforms in the Indonesian water sector. This involved identifying the types of training activities required at different government levels to support the reforms and to estimate how many of these trainings would be required. Table 8.3 provides an overview of the capacity needs of the Head Office of the Directorate General for Water Resources (DGWR) in Jakarta.

In order to establish capacity needs at the level of river basin organizations, the provincial level and the district level, three case studies were undertaken in December 2007 and January 2008. These case studies concerned the Jratunseluna Basin, the Musi Basin and the Citarum Basin[4]. On the basis of the case studies, a rough estimate was made concerning the capacity needs at the level of the Balai, at the provincial level and at the district level. The estimate was based on an extrapolation of the findings concerning capacity needs in the case studies. The case studies selected represent relatively well-developed areas within Indonesia, for which the organizations with water management tasks and responsibilities are relatively well-capacitated. In this sense, the estimate of capacity needs is likely to be relatively conservative (meaning that it is more likely to underestimate rather than overestimate the required training activities).

Tables 8.3 and 8.4 show that an estimated 3,500 people need to be trained within the next few years to address the capacity needs that result due to reforms introduced by Law 7/2004. This number will increase considerably when new members of the water councils require additional training.

---

[4] The three case studies were undertaken using a standardized analytical framework which allowed for a comparison of them. In these case studies, organizations involved in water management were assessed in terms of their tasks/responsibilities and the existing capacity of these organizations in terms of their expertise/knowledge and staffing levels.

Table 8.3 Estimate of new skills and knowledge required for staff at DGWR and *Balai WS* (number of persons).

Capacity need	Head Office in Jakarta	Balai (Besar) WS
Formulation of Policies, Regulations and Guidelines	10	–
Public Awareness, Communication and Networking	100	300
Asset Management	30	60
Hydrological Modeling and Simulation	10	120
Formulation of Strategic River Basin Plans (Pola) and Formulation of Master Plans (BWRMP)	20	60
Formulation of DIMP	10	
Design Process (Survey and Investigation, System Planning, Feasibility Study, Detailed Design, Tender Documents)		60
Construction Supervision		60
Financial Management, Project Administration		60
Environmental Monitoring and Management		60
Total	180	780

Table 8.4 Capacity needs at the Balai, province and district levels (number of persons who require training).

Capacity needs	Balai	Province	District
Formulation of Policies, Regulations And Guidelines		66	88
Public Awareness, Communication and Networking	300	330	440
Design Process (Survey and Investigation, System Planning, Feasibility Study, Detailed Design, Tender Documents)	60	66	
Construction Supervision	60	66	88
Planning of O&M, Asset Management	60	66	88
Financial Management, Project Administration	60	66	
Hydrological Modeling and Simulation	120	132	
Formulation of Strategic River Basin Plans (Pola) and Master Plans (BWRMP)	60	66	
Formulation of DIMP		66	88
Environmental Monitoring and Management	60	66	88
Licensing of Water Utilization by Private Sector		66	
Total	780	1,056	880

8.3.3   STAFF SHORTAGES DUE TO THE ZERO-HIRING POLICY

In addition to the capacity requirements resulting from the reforms, another urgent capacity constraint at the moment, and especially in the near future, can be linked to the zero-hiring policy which has been in effect for over a decade. Whereas the capacity gaps highlighted under Section 8.3.2 are the result of changes in the way of managing water resources, the zero-hiring policy has greatly impacted the overall number of people working in the water resources sector. The zero-hiring policy has been in place for almost 14 years, beginning in the early 1990s, and in practice meant that the organization was not able to recruit new employees. The consequence of this policy has been that the age structure of many organizations in the water sector is 'top heavy' with relatively few staff being younger than 46 years of age.

In the Directorate of Water Resources in the Ministry of Public Works, all sub-directors and chiefs of staff are over 50. Illustrative is also the age structure of the sub-Directorate of Coasts and Swamps. As Table 8.5 illustrates, less than 40% of the staff is younger than 46. Considering that the retirement age in Indonesia is 56, it means that within 10 years 60% of the Directorate will reach the age for retirement. Moreover, 38% will reach retirement age within the next five years.

For DGWR as a whole the picture is similar to that of the Directorate of Coasts and Swamps. Similar findings have been found at other government organizations within the water sector which have been subject to the zero-hiring policy. Figure 8.1 gives an overview of the age structure for nine river basin organizations at the provincial level (*Balai PSDA*).

Figure 8.1 shows that the employees in the *Balai PSDAs* are mainly to be found in the 46–56 age category. With 56 being the retirement age, it means that for most of these organizations approximately 70% of staff will retire within the next 5–10 years.

Based on an extrapolation of the in-depth case studies of three river basins, a rough estimate can be made of current staff shortages at the level of the *Balai*, at the provincial and district levels (Table 8.6). It should be noted that the rough estimates are rather conservative, as the main impact of the zero-hiring policy will be felt in the coming 5–10 years (i.e. an extrapolation done in a few years is likely to paint a much bleaker picture than our estimates do).

Table 8.5 Age Structure for the Sub-Directorate of Coasts and Swamps (number of persons).

Age group	Technical Staff	Non-technical staff
≤35	8	1
36–40	10	4
41–45	13	3
46–50	19	2
51–55	30	–
56≤	10	1

Source: Djohan, 2007

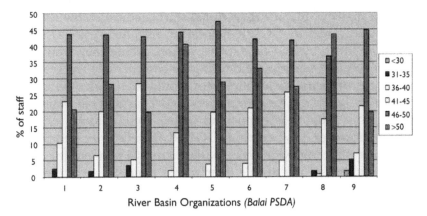

Figure 8.1 Overview of the age structure at nine *Balai PSDAs*.

Source: Presentation by Imam Anshori on February 25, 2008, in Jakarta

Table 8.6 Conservative rough estimate of present staff shortages in the Indonesian water sector (number of persons).

Education level	Balai (Besar) WS	Province (Balai PSDA)	District
Masters-level (technical)	50	30	50
Masters-level (non-technical)	10	5	50
Bachelor-level (technical)	250	60	1,000
Bachelor-level (non-technical)	200	200	1,000
Vocational level	500	200	5,000
Total	1,000	500	7,100

In itself, retirement is not a problem as long as adequate replacements can be recruited to replace the departing staff. However, the new recruits who are coming into the water sector are not able to compensate (both in quantity and in quality) for the wave of retirements that will occur in the coming 5–10 years. Quantitatively, the number of new staff recruited is insufficient to replace the staff who are retiring. For every 100 staff that retire, only five new recruits are recruited. The reason for the low level of recruitment is that recruitment of staff by water organizations in the public sector has to be approved by the central or provincial governments and these agencies only allow the water organizations to recruit a small percentage of what they actually request. As such, public sector recruitment policies appear to hamper the replacement of retiring staff.

In addition, the water sector does not appear to be very appealing to young recruits, as a large percentage of new recruits resigns within a few months. In 2007, for example, out of a total of 227 new recruits recruited by the Ministry of Public Works, more than 10% had resigned within 3 months.

Also qualitatively, concerns exist about new recruits. First of all, the work experience of many of the newly recruited staff is considerably less than that of employees recruited by DGWR before the zero-hiring policy. Previously, employees would first spend time in the region, where they would work on different projects and schemes (and thus gain considerable work experience). After a number of years, these employees would then make the move to the Head Office in Jakarta. These employees had considerable knowledge and experience about what was happening in the region. Many of the recent new recruits, however, have come to DGWR directly from university. These employees lack the knowledge and experience that is required for their work. Secondly, a number of the people interviewed expressed concern about the quality of the university education received by many new recruits, a quality they saw as decreasing. Some water-focused programs in Indonesia have gone down in quality, according to these respondents. Moreover, the universities have not yet altered the curriculum of their water-related courses to reflect the paradigm shift that was introduced by Law 7/2004 (see Section 4). The curricula offered at the moment, and which have been followed by the new recruits, continue to be strongly infrastructure-development focused. The result is that the new recruits are not able to deliver the work that is required of them.

8.3.3.1   The consequences of the staff shortages and the role of the private sector

The limited replacement of retiring staff will have a profound impact on the way that water organizations undertake their tasks. If in the future the current trend continues, it is likely that the water organizations, which currently incorporate a large number of specialists, will increasingly become organizations of 'generalists' who will out-contract specialist activities to the private sector, as the organizations themselves will no longer have the capacity to undertake these tasks. The question is, however, whether the private sector is, at present, able to fulfill such a role. A scenario in which the public sector organizations out-contract more and more activities to the private sector may require considerable capacity building of the private sector. Moreover, it will also require the public sector to increasingly develop capacities such as contract management and performance auditing. At current, it seems that sector professionals hold differing opinions about the capacity of the private sector to fulfill this role. Some argue that the private sector does not have this capacity and that considerable (private sector) capacity would need to be built before the private sector is able to play this role[5]. Others argue that such capacity already largely exists and would require only a limited capacity building effort[6].

[5] Comments from Jos Houterman during a conference in March 2008.
[6] Comments from Guy Alaerts, received in January 2009.

8.3.4   SUMMARY OF CAPACITY NEEDS

The capacity needs facing the water sector in Indonesia are two-fold. On the one hand, new skills are needed to address the challenges brought on by the paradigm shift which emphasizes a shift from development of new facilities and infrastructure to management of existing infrastructure, and from specific users (mainly farmers) to management of the whole cycle from source to users. In addition, this paradigm shift strongly emphasizes increasing participation in the management of water resources. Addressing these needs requires an estimated 3,000 training activities in the coming years.

On the other hand, the sector is facing enormous challenges resulting from the consequences of the decade-long zero-hiring policy. Based on an extrapolation of the current situation at the Head Office of DGWR, as well as the three in-depth case studies, it is possible to develop a rough estimate of the staff shortages for 2018 at different educational levels. This rough estimate is presented in Table 8.7. It is emphasized that the extrapolation below is a very conservative estimate, as the true impact of the zero-hiring policy is only likely to be felt in 5–10 years time.

Without diminishing the seriousness of the capacity challenges facing the Indonesian water sector, the limitations of the capacity needs assessment presented above should be noted. First of all, the sector is still trying to come to grips with the recent reforms and the exact roles and responsibilities of the different organizations have yet to be clearly defined. This requires a flexible approach to capacity building as well as periodic review of capacity needs to ensure that the capacity needs assessment accurately reflects existing needs. Secondly, conflicting legislation has lead to uncertainty of roles and responsibilities of different organizations. Harmonization of legislation is required before the roles and responsibilities can be clearly defined. Third, the assessment above mainly concerns the Ministry of Public Works, provincial and district water agencies. The reality of IWRM is that more organizations (such as the Ministries of Agriculture, Forestry, Environment, etc.), at different levels, play a role in the management of water resources. Their capacity needs are not included in this assessment.

Table 8.7 Estimated staff shortages in 2018.

Level	Shortages (Extrapolated for 2018)
Masters level	400
Bachelors level	5,000
Vocational level	12,000
Total	17,400

8.4 ADDRESSING THE CAPACITY NEEDS

Assessing the capacity needs of the Indonesian water sector highlights the demand-side of capacity needs in that sector. As important as the demand-side is, the supply-side of capacity building is no less important, i.e. the education and training activities that are available for addressing the capacity needs that have been identified.

8.4.1 EDUCATIONAL PROGRAMS

Educational programs offered through the Ministry of Public Works are an important component of capacity development in the water sector. This is particularly true because the public sector recruitment policies limit the number of recruits and because the universities, at present, still have not adapted their curricula to reflect the reforms in the Indonesian water sector. This means that the educational programs offered through the Ministry of Public Works are one of the few ways in which their capacity needs can be addressed. At present, a number of educational courses are being offered which are relevant to the water resources sector. The majority of programs offered, however, still have a strong infrastructure development-oriented focus rather than one focusing on the management of existing infrastructure. In this sense the courses offered through the Ministry of Public Works show the same development-oriented bias that the 'regular' courses of the universities have. Exceptions appear to be the asset management courses at the State Polytechnic in Bandung and the ITS in Surabaya. Also, the Water Resources Management course at the University of Brawijaya in Malang appears to have a more management-oriented focus. These are the exceptions, however, and the majority of programs offered to sector professionals in the Indonesian water sector do not reflect the reforms initiated under Law 7/2004 (and the 'new' capacities required as a result).

Apart from the types of programs being offered through the Ministry of Public Works, another problem relates to the number of people who are educated through these programs. The current level of educational programs provided are insufficient to address the current gap in capacity needs (which, as highlighted in the previous section, requires education for about 1740 people per year; ten times more than the current number of graduates from programs provided by this ministry). Moreover, if we take into account that the estimates are rather conservative, it means that, currently, not enough graduates are being produced through the Ministry of Public Works courses to address the capacity gaps in the Indonesian water sector.

8.4.2 THE MISMATCH BETWEEN SUPPLY AND DEMAND

If the demand for capacity building is compared to the supply, there seems to be a mismatch between the demand for education and training activities by

Table 8.8 The mismatch between demand and supply.

Demand-side	Supply-side
Large amount of training and education activities required	Limited number of graduates per year
Diverse capacity needs that are to be addressed	'Traditional' development-oriented curricula
Changing capacity needs as a result of reforms	Difficulty in establishing what the demand is

the water sector institutions and the supply of education and training activities by capacity-building organizations (Table 8.8).

One of the consequences of this mismatch is that the Indonesian water sector will remain dependent on a large number of international and a few national capacity-building organizations which have the ability to address the capacity needs that currently exist in the sector. At the same time, it is clear that the magnitude of the capacity problems facing the Indonesian water sector is such that the international and capable national universities are unable to deliver sufficient educational programs and training activities to address the existing needs, let alone the capacity needs for the coming 5–10 years.

In order to ensure that capacity needs in the sector can be addressed in a sustainable manner, the capacity of national, and especially regional, capacity-building organizations in Indonesia need to be increased. In other words, if the capacity needs in the Indonesian water sector are to be addressed in a sustainable manner, it will require building the capacity of the capacity-building organizations.

8.5   CONCLUSION AND RECOMMENDATIONS

Addressing the capacity gaps in the Indonesian water sector will involve action to be undertaken at different levels within the water sector.

8.5.1   NEED FOR COORDINATION AT THE SECTOR LEVEL

The mismatch between demand and supply highlights the need for a coordinated mechanism in the sector. This coordinated mechanism should address a number of issues. First of all, the sector as a whole would greatly benefit from the development of a strategy on capacity building in the sector. Capacity development requires actions at three levels: the sector, the organization, and training and educational activities at individual levels. Specifically at the sector level, a capacity-development strategy is required. This strategy should be based on a thorough assessment of capacity needs which should be updated periodically. Secondly, the strategy should incorporate a detailed analysis of the existing 'regular' university courses in the fields of water management and water engineering as well as the educational and training programs offered through the Ministry of Public Works. In case the capacity needs as identified and the educational and training programs offered do not match (as is currently the case), the strategy should highlight how the supply of capacity-building activities can be adapted to suit the present needs. In addition, the strategy should identify, on the one hand, sources of funding to address the capacity needs, and, on the other hand, provide opportunities for Indonesian training and educational organizations to adapt their curricula to address the existing (and future) capacity gaps.

Coordination should not only focus on government agencies, but also on the private sector. The coming wave of retirements suggests that more services and activities are going to be outsourced as the (public) organizations have increasingly inadequate capacity to undertake these tasks and activities. As a result, a coherent effort is required for capacity building of the private sector, based on the new requirements stemming from the recent reforms.

8.5.2 CHALLENGES FOR CAPACITY DEVELOPMENT AT THE ORGANIZATIONAL LEVEL

In light of the constraints in the environment in which many water organizations operate (such as public sector recruitment policies and limited training and educational programs which reflect the reforms), the challenges facing water organizations are enormous. Within the present context, these organizations have to develop their human resources management and development programs. These Human Resources Development Programs would have to focus on the following issues:

1 Obtaining new promising staff:
   • Increase the appeal of the organization – As mentioned earlier, almost 10% of the staff that entered the Ministry of Public Works in 2007 had resigned within 3 months. Considering the difficulties of recruiting new staff, it becomes paramount to increase the appeal of the organization to attract qualified staff and at the same time, ensure that new recruits stay with the organization. At present this is the aim of a system of mentors assisting the new recruits during their beginning years.

2 Developing and utilizing existing human resources:
   • Introduce career development plans (identifying potential leaders) – The wave of retirements that will be experienced in the

coming 5 to 10 years means that many young employees will likely have to follow an accelerated development through the organization. This accelerated career development requires career development plans to optimally prepare employees for future positions that will be vacated by retiring staff;

- Formulate training and education needs – Capacity development should focus on addressing the problems that are being faced. This requires an accurate and up-to-date assessment of these problems and the required capacity to address them. This means formulating clear training and education needs on a periodic basis.
- Utilize trained staff – Once staff has received training or education, it becomes paramount to provide the opportunity to the trained staff to actually utilize their 'new' capacities in their daily work.

3 Retaining valuable human resources:
- Make use of retiring employees who have valuable knowledge and experience – A danger lurks: with retirement of staff, a wealth of experience and knowledge is lost. Organizations have to develop policies to maintain the services of retiring staff who are able to provide added value to the organization.

All the issues that should be addressed in a human resources development program require a pro-active HRD policy. It appears, however, that only a few organizations seem to be implementing such a management policy. Part of the lack of the pro-active policy can be explained by the magnitude of the challenges that are being faced, which causes many personnel departments to concentrate on plugging leaks rather than on developing and implementing the required pro-active HRD policy.

REFERENCES

Al'Afghani, M. (2006), 'Constitutional Court's Review and the Future of Water Law in Indonesia', 2/1 Law, Environment and Development Journal, http://www.lead-journal.org/content/06001.pdf

Cervero, R. (1992), 'Regional Distribution of Development Grants in Indonesia's Water Supply Sector', Review Of Urban & Regional Development Studies, (4)1, pp. 3–16.

Devas, N. (1997), 'Indonesia: what do we mean by decentralization?', Public Administration and Development, 17(3), pp. 351–367.

Djohan, R. (2007), Presentation September 2007 by the 'Direktur Rawa dan Pantai', Jakarta: Ministry of Public Works.

Global Water Partnership (2000), Integrated Water Resources Management, TAC Background Papers, no 4, Stockholm: GWP.

Herman, T. (2005), Implementation of Integrated Water Resources Management (IWRM) in Indonesia, Unpublished paper.

Houterman, J., Djoeachir, M., Susanto, R. and Steenbergen, F. (2004), *Water Resources Management during Transition and Reform in Indonesia*, Agriculture and Rural Development Working paper 14, Washington D.C.: World Bank.

Ministry of Public Works (2007), *Strengthening Operation and Maintenance for PJT1, Main Report*, Jakarta: Ministry of Public Works.

Ramu, K. (2004), *Brantas River Basin Case Study Indonesia*, Background Paper, http://siteresources.worldbank.org/INTSAREGTOPWATRES/Resources/Indonesia_BrantasBasinFINAL.pdf, Washington D.C.: World Bank.

Ramu, K. (2007), *Institutional Framework for Managing Water Resources in National River Basins in Indonesia*, Draft paper prepared for the Sub – Program "Aligning the Water Resources Sector with National Development Goals".

Taylor, P. (2007), *Draft Report on Action Plan Phase 2 Capacity Building for the Balai Besar Wilayah Sungai of the Six 'Ci' River Basins*, Jakarta: Ministry of Public Works.

Chapter 9

INSTITUTIONAL CAPACITY DEVELOPMENT OF WATER RESOURCES MANAGEMENT IN IRAN

Mojtaba Nikravesh
Water Resources Management Company, Iran

Reza Ardakanian
UNW-DPC

Sayed H. Alemohammad
University of Technology, Iran

ABSTRACT

According to the geographic and climatic situation of Iran, water has always had a strategic importance. Due to this importance, our ancestors in different times and by considering temporal and spatial needs have approved different rules and established organizations for appropriate and correct management. Water needs due to an increasing population, important changes in economic and social situations, and exciting new technologies, have resulted in the compiling and gradual improvement of the "Institutional Framework" for Water Management. This recent trend can be divided into four periods in Iran: Preliminary, Establishment, Development and Alliance. As water management in Iran is experiencing a unification period, the following chapter will first provide an overview of the characteristics and history of water management in Iran, then the Integrated Water Resources Management (IWRM) Model is explained briefly, based on the international definitions. Next, a description of the executed works, especially the establishment of the needed "Legal Capacities" and "Organizational Reforms" is presented, followed by the final organizational situation. Finally, recommendations are presented for the effective and principal rehabilitation of the Integrated Water Resources Management, as well as ways to prepare the appropriate opportunities for decision-makers, decision-making and other affected procedures related to water.

9.1 INTRODUCTION TO THE WATER SECTOR OF IRAN

Iran, with an area of about 1,648 million square kilometers, has a population of around 70 million. It is located in the north moderately dry region and in the middle latitudes on Earth. Iran's neighbors are Afghanistan, Pakistan, Iraq, Azerbaijan, Turkmenistan, Armenia, Turkey and the Arab countries of the Persian Gulf region.

The vastness of Iran and the existence of different natural features, such as high mountain ranges in the northern and western parts, individual mountains in the central and eastern areas, and widespread plains like the central desert inside the plateau, plus the neighboring Caspian Sea, Persian Gulf and Oman Sea, have made Iran a country with varied climates. Different climates such as arid and semi-arid, mountain-influenced and temperate can be found across Iran, but Iran is truly a dry country with very low precipitation (Faragi, 1987).

As Iran is located in the arid belt, water shortage has been one of its main challenges. The average precipitation is only 250 mm per year, which is one-third the yearly average of the world (860 mm). The spatial distribution of precipitation is such that only 4% of the total area of the country has precipitation of more than 500 mm per year. This area receives about 27% of the total precipitation of the country. The other 73% falls on the other 96% of the country, which has a yearly average of less than 250 mm (Figure 9.1).

One of the other problems of water in Iran is the "inappropriate" spatial and temporal distribution of precipitation. Most of the precipitation falls on high mountains in the form of snow, and the central and eastern parts receive little precipitation. Also, a small percentage of rain falls during the "appropriate" seasons (spring and early summer). In addition, there is a high level of evaporation and transpiration in all parts of the country, especially in the central and eastern desert, which covers more than 50% of the country (Ministry of Energy, 2001).

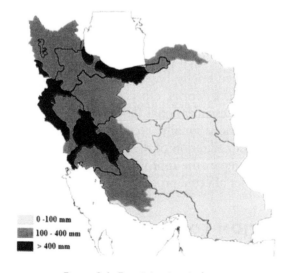

0 -100 mm
100 - 400 mm
> 400 mm

Figure 9.1 Precipitation in Iran.

9.1.1   CHARACTERISTICS OF WATER RESOURCES

9.1.1.1   Six main water basins

Research and reports show that Iran can be divided into six main water basins; namely, the Caspian Sea, Orumieh Lake, Ghareghum, the Eastern Border, the Central Plateau, and the Persian Gulf and Oman Sea (Ministry of Energy, 2001). These six main water basins are divided into 30 sub-basins (Figure 9.2).

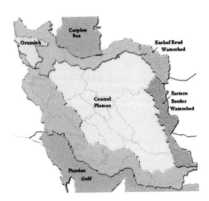

Figure 9.2 Iran's main basins.

9.1.1.2   Water resources and infrastructure

Table 9.1 Main characteristics of Iran's water resources. (Ministry of Energy, 2001, 2008).

	Year	1961	2001	2021
Available renewable water	Yearly available renewable water (m³/capita)	6850	2000	1300
	Sector	1961	2001	2021
Water use (%)	Domestic	1.3	6.0	9.5
	Agriculture	98.0	92.8	87.0
	Industry	0.7	1.2	3.5

	Situation	No.	Total volume (MCM)	Regulating volume (MCM)	Installed hydropower capacity (MW)
Storage dams	Under operation	567	37,414	29,557	7912
	Under construction	119	26,322	9,566	6917
	Under study	320	52,685	16,806	8271

(Continued)

Table 9.1 (Continued)

	Situation	Total area	Networks (ha)		
			Constructed	Under construction	Under study
	Downstream of operation dams	1,941,876	1,668,000	139,230	434,977
Modern irrigation and drainage networks	Downstream of dams under construction	815,558	34,260	94,565	686,733
	Downstream of dams under study	858,886	–	–	858,886
	Total	3,981,731	1,702,260	233,795	1,980,596

9.1.1.3   Urban water and wastewater

Table 9.2 Urban water and wastewater characteristics (www.nww.co.ir).

	Statistics items	Unit	Amount
	Total population	Million person	48.8
Urban water	Total serviced population	Million Person	48.1
	Percentage of the serviced population	Percent	98.5
	Number of serviced cities	–	935
	Supplied water	MCM per year	5,319
	Total length of networks	Kilometer	110,600
	Number of connections	–	10,640,807
	Number of working treatment systems	–	93
Urban wastewater	Total serviced population	Million	14.4
	Percentage of the serviced sopulation	Percent	29.5
	Number of serviced cities	–	221
	Number of cities in which sewage networks are under construction	–	222
	Number of cities in which sewage networks are under review	–	261
	Total length of sewage networks	Kilometer	29,802
	Number of connections	–	2,799,081
	Number of working sewage treatment systems	–	94

9.1.1.4   Human resources

The number of employees in water resources companies was 8754, and in the water and wastewater companies the total was 24277, in 2007. Moreover, about 3500 people work in affiliated public and non-governmental companies (www.wrm.ir).

9.1.2   WATER MANAGEMENT IN THE PAST

As mentioned above, most of Iran is located in an arid and semi-arid part of the world and, due to the non-uniform spatial and temporal distribution of precipitation, our ancestors always faced water shortage problems. This resulted in struggles to prepare appropriate situations through which to exploit, in the best way possible, the limited available water resources. Thus, different ways for exploitation, storage and use were considered early on, and so it has been called "irrigation art".

Preparing and supplying water requires teamwork, and organizations of people to access the water, supply the financial needs and operate the irrigation installations. All these issues led to the establishment of rules and procedures, as well as preliminary organizations for water management. Today, however, the preliminary organizations for water management and irrigation don't exist. Their responsibilities, goals and structures have changed so much due to a variety of reasons: technology development and improvement in all industrial and agricultural processes of production; the development of water-related activities and related services; technical and engineering innovations regarding exploitation; the transfer, distribution and methods of irrigation and agriculture; population growth, and the major changes in the rural and urban population structure; and, major changes in water use that consider quantity, quality and variety (i.e. agricultural, urban and industrial). These issues and changes will be discussed in the following sections.

9.1.2.1   Inventions and methods of operation

In the history of ancient Iran, water has had a special holiness and respect. Our ancestors recognized an escort angel for water named Anahita[1], as they believed that water is the manifestation of life and innocence. This angel has been respected by all of the people and, because of that, many temples and statues of her have been built, and in many religious ceremonies water has played an important role.

Water management has been considered in different ways from those ancient years, too; most of the leaders and distinctly religious people have

---

[1] Anahita, Anahid or Nahid (Venus) – In Persian it means away from pollution – it is the name of a woman who is the god of water, rain and productivity in Iranian ancient religion.

been sensitive about it and because of this have prepared protocols, scrolls, procedures or handwritings and inscriptions in which the method of supplying, distributing and rationing water has been mentioned, based on a wet or dry year. Also, many clues have been found that show the establishment of water structures, and the dredging and repairing of them to supply water, especially to establish a thriving economy and develop villages.

Archaeologists like Pope[2] and Ghirshman[3] believe that cultivating wheat was customary near the Orumieh Lake from 5000 to 7000 years ago, back in the Neolithic period. Also, there are some clues which show that aquatic agriculture was started in Iran (Farshad, undated).

By developing more and increasing the population of ancient Iran, the usual methods became inadequate to supply the increasing demand for water. So, our ancestors made an invention: using a special method, they brought groundwater to the surface by gravitational force. This invention was unique and was called Kariz or Qanat (Mahmudi Bakhtiari, 1979).

Some of the Iranian qanats date back 5000 to 6000 years ago, the same age as ancient Iran. Albeit, thousands of years have passed from the date of that invention, yet this method is still used as an important part of rural, urban and agricultural water supply. Using this method, Iranians have been successful in the sustainable exploitation of groundwater and have withstood Iran's drought conditions. These qanats have usually been managed, constructed and exploited by the local residents and under the supervision of the leader, the patriarch, in the Department of Sacred Mosques and Rural Entities. The dredger, together with inserting kavals[4] and cleaning the canals, was the responsibility of the local users (www.icqhs.org).

During that time, Iranians spread the technology of qanats to other countries such as Oman, Egypt, Spain, Germany, Mexico, America, Japan, Chile and North Africa. Today, it has been shown that 34 countries have used this system. In consideration of the importance of qanats as a water supply system which is compatible with the environment on the one hand, and the use of new techniques to dig deep wells which threaten the use of qanats on the other, and in accordance with the suggestion that Iran was a city born of qanats, the International Centre on Qanats and Historic Hydraulic Structures was established in Yazd, Iran, in 2005, as a Category II centre under the auspices of UNESCO.

In addition to qanats, our ancestors used other methods, such as big dams (some of which are still in operation), long irrigation canals, wells, water wheels, Ab-Anbars[5] and water distributors to successfully use their limited water resources (Farshad, 1996).

---

[2] Arthur Upham Pope (1881–1969) famous American historian.
[3] Roman Ghirshman (1895–1979) famous French archaeologist.
[4] Kaval: An elliptical-shaped ring made of clay or cement to cover the wells or canals.
[5] Ab-Anbar: A traditional tank to preserve water under the ground.

9.1.2.2 *Major rules*

From the date of the establishment of the government in Iran, different rules for the economical use of water were adopted. The Ammurapi[6] Act was the first among those. In this Act the possession of land and water, construction of water installations, and the management and exploitation of them were considered.

In the Achaemenid Empire[7], the role of the government in water affairs and water transferring was harmonized. In the Islamic period, according to the importance of land improvements and agricultural and economic developments, and based on the statement by the prophet Mohammad that people are partners in three things: water, plants and fire, some rules for the exploitation of water resources were adopted, some of which are the basis of rules today.

The first rule of recent times came around 1910 and was named the "Iranian Qanats Act". After that, the most important acts are: the "Water Act and the Way to be Nationalized" (approved in 1968), and the "Equitable Water Distribution Act" (approved in 1982). In total, 458 rules and 206 articles related to water have been approved over the past 70 years, and these have led to the development of the Comprehensive Water Act in the Third Development Program of the country (Memari, undated).

9.1.3 TODAY'S WATER MANAGEMENT

The hierarchical order of water management at present can be divided into the four following categories:

A   Policy-Making Structure
  - High Water Council – This council, organized under the supervision of the highest executive office (the President) and with representatives of the stakeholders present, is a mechanism to dialogue and to harmonize the policies and general plans of water resources management;
  - Parliament – It is one of the three main bodies of the political structure of the country and the highest authority of legislation in various sections.
B   Executive – Administrative Structure (Ministry of Energy)
  - Deputy Minister for Water and Wastewater Affairs
  - Holding Company of Iran Water Resources Management and related subset companies (Water Sector)
  - National Water and Wastewater Engineering Company and related subset companies (Drinking Water and Wastewater Sector)

---

[6] King of Babil (from 1750 to 1795).
[7] 550 BC – 330 BC.

C   Technical and Engineering Structure (Private Sector)
  • 126 consulting entities
  • 216 contractors
D   Other Stakeholders
  • Ministry of Jihad – Agriculture, Ministry of Mines and Industries, Ministry of the Interior, Environmental Protection Organization, Deputy President for Planning and Strategic Supervision, technical and engineering societies, etc.

9.2   INTERNATIONAL CONTEXT OF IWRM

The increasing importance and sensitivity of water issues in the world, together with the forecasts and alarms about the world's water crisis by outstanding world experts and related international organizations, on the one hand, and the lack of a scientific model which is approved and capable of being extended to water resources management because of inherent complexities and the necessity of achieving an appropriate international structure and model for water management, on the other, have attracted the world's attention to water issues since the late 1970s.

To achieve mutual understandings and attract international cooperation, different conferences and political meetings in the field of water, water management and sustainable development have been held. Some of those meetings were the UN Conference on Water, Mar del Plata (1977), International Conference on Water and the Environment, Dublin (1992), UN Conference on Environment and Development (UNCED Earth Summit), Rio de Janeiro (1992), 1st World Water Forum (1997, Morocco), 2nd World Water Forum (2000, The Hague), World Summit on Sustainable Development, Rio+10, Johannesburg (2002, Johannesburg), Bonn Fresh Water Conference (2002), 3rd World Water Forum (2003, Japan), and 4th World Water Forum (2006, Mexico). This shows how special the water management issue has been.

The main outcome of these events was about water resources management and the ways to solve those challenges. Also, it has been recommended that all the "solutions" should be based on the Integrated Water Resources Management (IWRM) System.[8]

9.2.1   WRM IN BRIEF

Until the early 1990s, different aspects of water resources management (e.g. water quality, groundwater, water exploitation, water sanitation, irrigation, hydro powers, etc.) were managed separately by different entities. To solve this problem, advanced water resources scientists suggested a method

[8] Political Declaration of the Second World Water Forum, The Hague, 2000.

for water resources management which would enable supplying the maximum benefits possible to different stakeholders. This multi-sector, coordinated, multi-field, participatory, flexible and transparent method is called "Integrated Water Resources Management" (IWRM).

Generally, the basis of integration in water resources management is the coordination between all intersectional, decision-making organizations and entities at different levels. This is illustrated in Figure 9.3.

Also, in Chapter 18 of Agenda 21, which was the declaration of the world leaders at the United Nations Conference for the Environment and Development (Earth Summit), and also one of the important outcomes of the Rio Summit, the topic of water resources management and the ways to solve this important challenge are addressed. As this document's specific goal, programs and sub-programs related to water resources management have been compiled. In other words, the commitments of the governments toward water resources management have been determined. The following are among these goals:

- Water and land should be managed in an integrated method;
- Water and land should be managed at the local appropriate level;
- Water and land should be understood better and they should be considered economic products;
- While allocating the resources, the benefits to all stakeholders should be taken into account.

Some of the efforts which have been designed to reach these goals follow:

- Improvement of legislation and an administrative system of natural resources management that protects the environment;

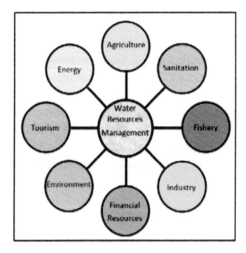

Figure 9.3 Integration in water resources management.

- Integration of different sections including agriculture, mines, industry, domestic uses and environmental needs; and,
- Participation among all users or stakeholders.

9.2.2 IWRM MAIN ITEMS

Some of the items and main axes that should be considered for laws, organizations, policies and programs to achieve Integrated Water Resources Management are: intersectional and over-sectional coordination, public participation, institutional arrangements, capacity building, other executive aspects, transboundary river and water basin management, the relationship between quality, quantity and biological aspects of water resources, and environmental sustainability (GWP-TEC, 2004a).

9.2.3 IWRM FRAMEWORK AND INSTITUTIONAL SETUP

In the Integrated Water Resources Management System, important attention has been given to an "Institutional Setup" which is simple, practical and understandable, and also capable of executing the approved policies to attain the stated goals. It has been recommended that this setup should have the ability to supply benefits to all stakeholders in different situations and over a long time, and also the ability to execute the water policies and programs correctly. This should have the following elements:

A Specific rules, regulations and procedures to control the measurement, development and use of water resources;
B A decision-making organization for the exploitation of and operational programs for water resources;
C Different levels of relationships between decision-making organizations and groups that are being affected by water programs directly, and the public.

Some of the responsibilities of an integrated institutional setup (legislation, organization and decision-making) for water management are the preparation of a qualitative and quantitative inventory of water resources; water diplomacy; work regarding water rights; programs for water use; implementation of projects for improvement; the application and conservation of water; the operation, maintenance and supervision of water works; ending the war and other conflict situations, the coordination of activities related to water resources and water resources research; and, the transfer of technology (GWP-TEC, 2004b).

9.3 INSTITUTIONAL DEVELOPMENT

The organization of water management in Iran has passed through four stages since the 1920s as follows:

I Preliminary period (1925–1963)

In this period, water management lacked many main aspects, such as organization, legislation and planning. For the first time, in 1939, by establishment of regional irrigation entities in some cities, the idea of establishing official organizations was considered. In 1943 the Independent Entity of Rivers was established under the auspices of the Ministry of Agriculture. Also, municipalities were being established and through them potable water was supplied to the cities.

II Establishment period (1964–1980)

In this period the Ministry of Water and Electricity was established, and then in 1974, by concentrating the office of Atomic Energy Activities in the Ministry of Water and Electricity, the Ministry of Energy was established. In 1978 the Deputy Minister of Water Affairs was established in this Ministry. Regional Water and Electricity companies also came into being.

III Development period (1981–1996)

In 1992 the Water Resources Management Organization was established to prepare strategies, policies and plans for exploiting water resources. Conservation, exploitation, development planning, and the execution of basic research were among its missions. Setting up water and wastewater companies, exploiting irrigation and drainage networks companies, and starting research for integrated planning were among the important activities of this period. Governmental consulting and contracting companies were established in this period, as well.

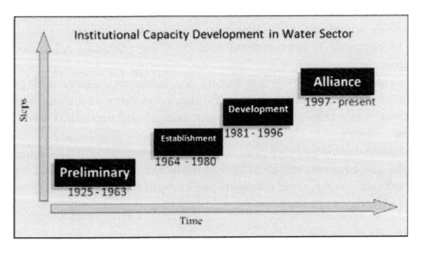

Figure 9.4 Institutional capacity development in water sector.

IV   Alliance period (1997–present)

In this period, based on the country's previous experiences and in recognition of its existing problems, and in consideration of the results of and recommendations from the technical and scientific meetings, important measures meant to harmonize and unify water management in Iran have been implemented.

Based on the importance of the implemented institutional measures and their accordance with the international entities' recommendations, these measures will be explained in details in the following paragraphs.

9.3.1   POLICY AND GOVERNANCE

As mentioned in the previous section, it is important to establish an appropriate institutional setup for the Integrated Water Resources Management System. In the following sub-section, the important measures of water management in Iran that have been designed to establish or improve the above-mentioned institutional setup are explained. These were accomplished during the Alliance Period (from 1997 to present) for water management in Iran.

9.3.1.1   General policies and laws

■ *"Fundamental" policies*

The most important fundamental policies in the I.R. Iran regarding economic safety, energy, water resources, mines, natural resources and transportation have been compiled and approved by the Islamic Republic of Iran's Expediency Council and with the cooperation of other organizations. These policies have been approved and signed by the leader of the I.R. Iran and have been communicated throughout the government[9]. The main elements of this water resources policy are as follows:

- Establish an integrated management system in the water cycle according to sustainable development and land use planning rules for water basins;
- Improve the efficiency while noting the economic, safety and political value of water in exploiting, distributing, conserving and using water;
- Increase water exploitation and lower natural and unnatural water losses in the country;
- Compile a comprehensive plan to implement dam projects, manage water basins and aquifers, set up irrigation networks, equip and level land, conserve water quality, repel droughts and prevent floods, use unconventional

---

[9]Document No. 1.76230, dated 22/1/2001 from the Leader of I.R. Iran to the President of I.R. Iran.

water resources, promote knowledge and technology, and improve people's roles in exploitation and operation; and,
- Consider the importance and feasibility of using shared waters.

■ *"Long-term development strategies for Iran's water resources"*
These strategies, which were approved by the Government of the I.R. Iran in 2002, are an appropriate guide for compiling short-term and mid-term plans in water management. They will also cause an effective exploitation of water resources due to the links which are created in interdepartmental management. In this document, the long-term strategies have been compiled based on the general view of the different sectors and by considering collaborations between those sectors. They have been introduced in 18 articles under the following topics: Macro-management, Water Resources Management, Consumption Management, Economic Value, Quality Control, Water Supply Costs, Water Exchange, Land-Use Planning, Inter-Basin Water Transfer, Management and Structure, Water Basin Compositions, Risk Management, Urban Water Distribution, Public Training, Shared Waters, Information Management, Preservation of Historic Hydraulic Structures and Interdepartmental Management.[10]

■ *Basic policies in the 4th development program*
After approval of the 20-year outlook plan, which highlights the development route of the country, the policies of the 4th Development Program were compiled and signed by the leader of the I.R. Iran and were shared with the relevant entities (government, Parliament and the I.R. Iran Expediency Council). In these policies, two pivotal points about water were emphasized[11]:

- To pay attention to the economic, security, political and environmental values of water in exploitation, maintenance and use; and,
- To consider the importance and feasibility of using shared waters.

■ *Recognition of water as an interdepartmental arena*
After compiling the 4th Development Program, considering the necessities of and based on the position of water as a part of the infrastructure for developing other economic and social sectors, the water sector was considered an interdepartmental section and the "National Document of Interdepartmental Water Resources Management" was developed under the auspices of the representatives of all stakeholders and received Cabinet approval in 2004.

---

[10]Approval dated 19/10/2003 of the Cabinet entitled Long-Term Development Strategies for Iran's Water Resources.
[11]Document No. 1.5885, dated 2/11/2003 from the Leader of I.R. Iran to the President of I.R. Iran.

9.3.1.2   Water resources management organizational principles

The most important principles that were considered during this period to organize the Water Resources Management Body in Iran are as follows:

- Integration and collaboration of water resources management activities with other sectors: environment, agriculture, industry, sanitation, reusing wastes, energy production and other stakeholders;
- Consideration of using the water basin as the basic model for water resources management and adopting the principle of unity for water basin management;
- Integration and collaboration between various levels of water resources management and elements of the water resources management cycle to maximize the convenience for people to respond;
- Attention to the process of the water resources management cycle (basic information and statistics, research, planning, construction, exploitation and maintenance);
- Consideration of the principle of decentralization in the organizational structure;
- Consideration of the economy-of-scale to minimize administrative and organizational costs.

9.3.2   IMPLEMENTED MACRO-REFORMS

During the Alliance Period (from 1997 to present), fundamental reforms were implemented in the water management structure to prepare for rehabilitation of the integrated water resources management theory in practice. The most important of those are:

- Joining the two ministries of Jihad and Agriculture to correct and improve the governmental structure and prepare the means for sustainable agriculture and natural resources development; (2000)
- Establishing the High Water Council under the auspices of the President of the I.R. Iran in order to harmonize policy-making activities related to supplying, transferring and distributing water in various sectors and with the presence of representatives of various sectors such as legislators, investors, producers and users; (2001)
- Establishing the "Water Research Institute" to integrate all research activities related to water in the country; (2001)
- Establishing the Regional Centre on Urban Water Management (RCUWM-Tehran) under the auspices of UNESCO, to meet the scientific, educational, awareness, research and consulting needs in the region; (2002)
- Establishing the Holding Company of Iran Water Resources Management as the General Assembly of all regional water companies and other related and affiliated companies of the water sector (2002). The achievements of Holding Company include following:

- Transferring the Iran Water and Power Development Company (the employer of hydropower projects in Iran) from the Ministry of Energy to the Holding Company of Iran Water Resources Management; (2002)
- Establishing the Water and Soil Development Company of Sistan Province (as a model, 2003);
- Approving the act regarding the establishment of regional water companies in the provinces; changing the provincial water administrations to provincial water companies, which is the basis for the formation of the provincial management of water resources along with political divisions (2003).

9.3.3 REFORMED WRM GOVERNMENTAL STRUCTURE

■ *Deputy minister for water and wastewater affairs*
To aggregate all of the governance activities for water in the urban water and wastewater management and prepare for the possibilities of rehabilitating Integrated Water Resources Management in Iran, the Deputy Minister for Water and Wastewater was established in the Ministry of Energy.

■ *Holding company of Iran water resources management*
The Iran Water Resources Company was established in 1996. This company was then reformed in 2002 with a new structure and became responsible for managing the stocks and capital of the company in affiliated companies and also for executing all activities related to the recognition, study, development, conservation and effective exploitation of water resources; installations; and hydraulic structures and hydropower capacities (excluding the affairs related to water distribution, collection, transfer and treatment of wastes). The affiliated companies are:

- Regional water companies (30 companies)
- Iran Water and Power Development Company
- Water and Soil Development Company of Sistan
- Farab Company
- Irrigation and drainage networks exploiting companies (20 companies).

■ *Holding national water and wastewater engineering company*
The National Water and Wastewater Engineering Company was established in 1990. This company, with the mission of ordering the activities of the Ministry of Energy in Water and Wastewater including right management, supervision and assessing the results, leadership and steering, increasing efficiency and appropriate use of the facilities, was reformed in 2002. The affiliated companies now, are:

- Urban and water and wastewater companies (38 companies)
- Rural water and wastewater companies (31 companies).

Before 1990, the municipalities were responsible for the distribution of water. At that time, because of incompetence and with the goal of concentrating the activities related to water, Parliament approved an act that transferred the responsibility of water distribution to the Ministry of Energy, and established water and wastewater companies under the supervision of the National Water and Wastewater Engineering Company.

9.3.4 MOST RECENT WRM STRUCTURE

The Water Resources Management Structure in Iran was established in 2007 on the basis of three national, water basin and provincial levels and was concerned with the following key principles:

- Unity of water basin management
- Decentralization
- Convenience for people to respond
- Economy-of-scale to minimize administrative and organizational costs.

A National level
This level is the highest technical and decision-making level for water resources management and is responsible for acting as the national governance on water resources in the country on behalf of the government. Now this level of management is headed by the Deputy Minister.

B Water basin level
As mentioned before, Iran has 6 main basins and 30 sub-basins. This level of water resources management was first considered in 2005, and was ultimately established within the structure of the Iran Water Resources Management Company with one deputy and 4 independent directors. In 2007, water basin management was established and continues to work.

C Provincial level
Provincial water resources management is responsible for water resources management missions at the level of the political divisions of the country. As mentioned previously, because of decentralization and the structure of the Ministry of Energy, and because water resources management is involved with other political and executive departments, important problems developed: complicated methods of collaboration with other organizations, political and social challenges at the provincial level, problems in receiving the provincial budget for facilities and public partnership, and problems related to exploiting water installations. After necessary investigations, the proposed reform of this structure was approved by Parliament in 2004. In accordance with this proposal, a regional water company was established in each province.

9.3.5   THE MOST IMPORTANT ADVANTAGES AND CHALLENGES

The most important advantages and challenges of the present structural model are as follows:

Advantages
- Possibility of acting as a unified management unit at water basin level
- Facilitation of the implementation of national projects
- Possibility of making more harmonized decisions
- Existence of a small central authority
- Possibility for appropriate distribution of human resources
- Appropriate response to the people
- Possibility of using the whole financial resource
- Possibility of receiving provincial credits and public partnership.

Challenges
- Necessity to promote the establishment of this management structure
- Need for more technical resources
- Complexity in communication and cooperation between structures at national, water basin and provincial levels
- Increase in operating costs for human resources and administrative installations.

9.4   ORGANIZATIONAL REFORM

To improve and promote the current structural status, the following recommendations related to various levels of water management can be presented.

A   National level

The research done by the UN University shows that the water issue and its management is second in importance, after the population crisis, among important world challenges and problems of the 21st century (Jerum, 2002). Comparing the renewable water resources per capita in the world with Iran's and its region shows that, if water is second in importance in the world, with no doubt it is of the utmost importance on a regional scale in Iran.

Therefore, it is necessary to strengthen the comprehensive, multi-aspect and interdepartmental views held about water management and to generally improve water governance. As noted by international societies, the strategy for solving the existing problems is through improvement of the method of water governance, as one of the most important measures in reforming the governance structure.

Based on this, it is supposed that the successful execution of the above-mentioned policies and strategies in the decision-making structure of the country needs the water issue to weigh in more appropriately in the country's governance structure.

So, it is necessary that the national level of water resources management be improved, aiming at:

- Establishing a strong national structure which is capable of understanding the current situation and needs, and analyzing these two to better rehabilitate integrated water resources management;
- Preparing the necessary situation for decision-making and affecting other processes related to water;
- Creating necessary situations for more effective roles in regional and international water management;
- Preparing to become responsible for managing future situations in the country using criteria for water demand per capita and for forecasting the water crises.

What their implementation needs:

- The presence of the senior manager of water resources in the highest decision-making entities in the country;
- Necessary organizational and managerial orders;
- Effective collaboration with different executive departments which affect the water management process;
- Legal capacities for affecting other departments' policies (related to water).

The establishment of an independent ministry to aggregate the policy-making, planning and national management of water resources should be taken into consideration.

It is worth noting that governments usually change their structure to meet new priorities in a changing society. These new structures make possible the development of a new capacity regarding new governmental policies. Examples of these changes are the establishment of the ministries of "Women's Affairs" and "Environment", in many countries.

International experts and analysts have forecast the crisis of water for the future. They have also cautioned that water will soon be the most important challenge facing humanity. It is obvious that adopting appropriate policies and strategies and executing preventive measures, as well as making detailed plans for different sections, especially organizational and managerial re-ordering, are among the most important measures.

B   Water basin level

One of the most approved principles in water management is re-ordering the management of geographical regions according to natural water basins. As mentioned before, according to existing documents, the regional water companies were first established according to the main water basins' limits. But over time and because of a lack of local organizations in accordance with the political divisions of the country, these companies have been moved from their ideal positions.

In the last structural reform that was executed in the water sector, the six main basins' management consisted of four independent directors under the supervision of a deputy in the Iran Water Resources Management Company. As water management is now taking its first steps towards establishing Integrated Water Resources Management, the current situation is good but it should be advanced soon.

One of the recommendations is the establishment of six water basin management companies according to the main basins of the country and make them responsible for technical and legal issues. In that way, local water management units (provincial units) will be in a close relationship with their relevant basin management, and the necessary decisions related to the technical aspects of water management will be made in these companies and within the limits of their basin. After this establishment, necessary changes at the national and provincial levels will be needed.

C   Provincial level

Today, in the governmental domain of water management, 30 regional water companies (governmental) and 90 provincial (governmental), urban (semi-governmental) and rural (governmental) water and wastewater companies are working. For more effective management, it is better to have a responsible organization for all main functions at the provincial level. After doing necessary research and having the necessary regulations, activities like water distribution issues can be put to public or private sectors, like municipalities. In this way, unified water management policies can be adopted and the principle of decision- making at the lowest level will be possible.

The schematic structural model for water resources management in Iran, according to the above-mentioned description, is shown in Figure 9.5. This chart is only a perspective of the recommendations, and the exact structure of this model needs more work.

To establish the Ideal Structural Model for Iran's Water Resources Management, the following measures should be taken after considering the legal criteria in the I.R. Iran:

- Prepare a comprehensive proposal for establishing a new ministry in which water resources management is the main mandate and get it approved by Parliament;
- Investigate and prepare an appropriate proposal for categorizing and determining the responsibility and mission of each of the organizations involved in water management based on the new structure, and get it approved by Parliament;

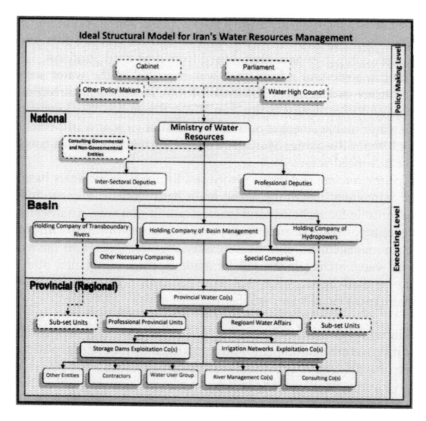

Figure 9.5 Ideal structural model for Iran's water resources management.

- Prepare a proposal for establishing coordinating committees between the decision-making organizations, groups directly affected by water plans and other stakeholders, with the goal of coordinating water resources activities and ending war and conflict situations;
- Design and plan training courses for:
  - Senior experts of water resources management, with the goal of educating them in the IWRM concepts and develop a group thinking attitude to propose plans to execute the concepts and promote working skills of the experts;
  - Employees who are directly servicing the operating companies, with the goal of increasing their efficiency and ability to solve problems and face challenges;
- Propose necessary mechanisms for inserting water resources management policies into the development plans of all related organizations (e.g. Housing and Urban Development, Industry, Agriculture, etc.)

9.5 CONCLUSIONS

Iran is a dry country and from old times people who lived in Iran tried very hard to better use the limited water resources. They established organizations and entities to effectively manage water use. These efforts have resulted in the current structure of water management in Iran. Based on the recommendations of international entities and institutions, Iran has moved toward Integrated Water Resources Management and many reforms have been made in this regard. But there is a long way to go to the establishment of an ideal structure for water resources management. In this paper some recommendations for an ideal structure were made and the necessary legal capacities were pointed out, based on the IWRM model. It is hoped that with these reforms the IWRM model can be established for Iran's water resources management.

REFERENCES

Faragi, A. 1987, *Geography of Iran*, Iran Press & Edition Co, vol. (1), pp. 28–35.
Farshad, M. undated *History of Science in Iran*, Amirkabir Press, vol. (2), p. 779.
Farshad, M. 1996, *History of Engineering in Iran*, Balkh Press.
GWP-TEC. 2004a, *Integrated Water Resources Management*, Global Water Partnership – Technical Advisory Committee, TEC Background Papers No. 4, 2004.
GWP-TEC. 2004b, *Effective Water Governance*, Global Water Partnership – Technical Advisory Committee, TEC Background Papers No 7, 2004.
Jerum, G. et al, *Future Situation – 1999* (Translators: N. Aghajani, et al.) 2002. Khazra Press, Tehran, p. 23.
Mahmudi Bakhtiari, A. 1979, *Base of Iranian Culture and Civilization, A View of the Old Time*, Ruzbeh Press, p. 295.
Memari, G. undated *A Report on the Compiling of the Comprehensive Water Act and the Problems*, Water Resources Management Press.
Ministry of Energy. 2001, *Summary of Comprehensive Water Plan of Iran, 2001*. Water Resources Management Company, Ministry of Energy, Iran.
Ministry of Energy. 2008, *Database of Iran Dams Characteristics*, Water Resources Management Company Ministry of Energy, Iran (Access Date: 2 Aug. 2008).

Chapter 10

PRACTICAL APPROACHES AND LESSONS FOR CAPACITY BUILDING:
A case of the National Water and Sewerage Corporation, Uganda

Silver Mugisha
National Water and Sewerage Corporation, Uganda

ABSTRACT

Water utility reforms aimed at enhancing the operational performance of utilities are vigorously being pursued by Africa's water utilities. In this paper, a crisp summary of key capacity-building and optimisation approaches that have been applied to turn around the ailing performance of the National Water and Sewerage Corporation, Uganda, during the period 1998–2008 is presented. The paper ends with final thoughts, alluding to the fact that a public enterprise with focused leadership can deliver good services to its citizens.

10.1 INTRODUCTION

> *"If the rhythm of the drum beat changes, the dance step must adapt."*
> *(Swahili East Africa Proverb)*

Safe and sufficient drinking water is still not a matter to be taken for granted all over the world. In developing countries, the provision of safe drinking water still remains a daunting task. And yet, the approach to solving water supply problems is not rocket science! According to Hoffer (1995), there are not many variations in water supply systems: the raw water must be abstracted, treated if necessary, distributed, and the system must be financed. Despite this apparent organisational simplicity, many water systems worldwide have shown inconsistent performance trends and have had to undergo reforms. Given the importance of water infrastructure investments and operations, improving performance of the sector is a high priority for emerging markets. An earlier study details the historical background and performance improvement initiatives that have been implemented within the National Water and Sewerage Corporation (NWSC) since 1998 (Mugisha et al., 2004a). This chapter builds on that background, describing what NWSC has done up to 2006. NWSC is a public corporation wholly owned by the government of Uganda, having been established in 1972 by Decree No. 34 (during the time of President Idi Amin Dada). The corporation's legal position was strengthened by NWSC Statute No. 7 of 1995, which was later incorporated into the NWSC Act of 2000. Under the new legal framework, the powers and structure of NWSC were revised to enable the corporation to operate on a commercial and financially viable basis.[1] Accordingly, the corporation is currently mandated to manage water and sewerage services in 19 urban areas under its jurisdiction.

The NWSC is structured in such a way that there is a head office which acts as an asset holding arm. Then, there are service providers (operators) in large towns that carry out the day-to-day operations management in those towns. The head office is responsible for large-scale investments, asset management, operational support and performance monitoring. Since 1998, there has been a progressive increase in managerial autonomy to the service providers through structured, internal incentive contracts.

According to Mugisha and Berg (2008), in 1998, NWSC was not a healthy organization. The World Bank noted in its report[2] that:

"Over the last 10 years, the GOU [Government of Uganda], in partnership with the World Bank and other donors, has made significant investments (over US$100 million) in the urban water and sewerage sector. These investments have contributed immensely to rehabilitating the existing infrastructure under the NWSC management. Unfortunately, these investments have not been matched with the necessary, efficient commercial and financial management capacity that can ensure the delivery of sustainable services in the medium- to long-term."

This conclusion, based on a thorough analysis, found that the corporation had a sound infrastructure, abundant water resources and an enabling legislative framework. However, the corporation had a large and inefficient labour force with conflicting and overlapping roles, high unaccounted-for water (more

---

[1] Before the new legislation, NWSC was operating under a decree. The powers of the corporation were constrained through cumbersome reporting requirements to the minister (government). The NWSC was not allowed to freely outsource operations management. There were a lot of overlaps in role definition between the government and the corporation. The new NWSC Act of 2000 was aimed at streamlining these inconsistencies.

[2] World Bank Aid Memoire Document (1998), Project Evaluation Report to NWSC Management.

than 50%), poor customer service, low collection efficiency (about 71%), substantial accounts receivables (a day's receivable ratio of about 420 days) and corruption within the work force, especially in the field staff. There was a running monthly deficit of about Ushs 348,000,000 (~US$300,000) despite a high average tariff of Ush 1,100/m^3 (~US$1/m^3). In other words, NWSC was nearly in a state of bankruptcy. In addition, the corporation had to contend with a number of threats, including debt-servicing obligations coming due and a value added tax law that compelled NWSC to pay taxes on any increases in bills. On the other hand, the government was willing to give support to proactive managers, and the economy was relatively stable. In addition, the government was also willing to freeze the debt (~US$100 million) for some time to give the corporation a chance to recover, if serious managerial efforts were initiated. Thus, improving operational and financial performance was essential to prevent further deterioration.

This chapter outlines the corrective actions and capacity-building processes undertaken by NWSC management and staff to turn around its performance[3], the sequencing of those steps, and the outcomes from this reform programme. The paper demonstrates the benefits of new initiatives to policy-makers, analysts, and managers of poor-performing utilities. Such initiatives are not painless, nor can they guarantee success. However, citizens and political leaders in developing countries are finding the status quo unacceptable: organizational transformation based on feasible commercial plans and team initiatives can improve performance.[4]

10.2   CAPACITY-BUILDING INITIATIVES UNDERTAKEN

"If you do not listen to good advice, you will be embarrassed in public."
(Oshiwambo Namibia Proverb)

Staying focused is crucial for reform champions (Mugisha and Berg, 2008). Sometimes, an impending crisis becomes a catalyst for change. In an effort to address managerial inefficiencies in NWSC, the government appointed a new board of directors. The new board was comprised of representatives from local governments; the business community; professional bodies; the environment; the ministries of finance, water, and health; and small-scale industries. The composition and structure of the board enabled it to exercise its governance functions properly, and it was able to shield the corporation from political interference and patronage. The new board, in turn, appointed a new

---

[3] More details on the reform measures can be found in Mugisha et al., 2004a, b and 2008.
[4] Other African nations have also instituted programmes for improving public service: Rugumyamheto (2004) outlines Tanzania's steps for reform.

managing director, Dr. William Muhairwe[5], who was given the mandate to rethink strategies for performance improvement. The appointment led to an emphasis on commercial viability, utilising "customer care" as an organising theme. The board and new management also pursued the approach of having performance contracts with the government in which roles and obligations were clearly spelled out.

Fortunately for the new team, management and staff were aware that the ship would "sink" if nothing was done to remedy the situation. Dr. Muhairwe decided to adopt the approach of working with everybody. Ultimately, all the change management programmes that are outlined hereafter were mainly implemented by incumbent staff who thankfully had acquired sufficient technical skills in the past. Changing organizational behaviour and the work culture became the focus of the new initiatives. Consequently, in response to the above challenges and with limited resources at its disposal, the NWSC devoted its efforts to enhancing its capacity and strengthening operations through numerous innovative approaches. These included effective change management, emerging managerial tools and principles, water loss management, water resource protection and stakeholder coordination, timely water production development and creative approaches.

10.2.1 CHANGE MANAGEMENT

At the tactical level, the new board and management came up with a series of change management initiatives:

* *100-Days Programme* (Feb–May, 1999) was a high-impact programme that focused on reversing operational and financial inefficiencies. It was carried out through aggressive revenue collection strategies and cost-cutting measures. A number of cost-cutting measures implemented during this programme included rationalization of the medical scheme and a reduction of travel costs (establishment costs).
* *Service and Revenue Enhancement Programme* (August 1999–August, 2000) aimed at restoring customer confidence in the ability of NWSC to deliver services. Under this programme, NWSC established customer service centres and front desks, conducted customer surveys to capture customer wants, and instituted amnesty for illegal water use.

---

[5] Dr. William T. Muhairwe is a management specialist trained in economic and business management and has been managing public companies for more than the last 15 years in Uganda and abroad. Since 1998, he has been the managing director of National Water and Sewerage Corporation, a state corporation that was almost collapsing due to mismanagement. Through his initiatives, Dr. Muhairwe has implemented innovative change management programs that have successfully turned around NWSC from a loss-making organization to a profit-making government parastatal which is now a benchmark for best practice, both nationally and internationally among water utility organizations.

- *Area and Service Performance Contracts* (2000–2003) focused on making service providers reach commercial sustainability: managers had the authority to make important decisions and were accountable for outcomes.

The corporation had to improve operating margins by reducing bureaucracy, increasing staff productivity and encouraging worker involvement. The NWSC management also collaborated with the labour union to reduce excess staff by half, from 1,800 in 1999 to 900 staff in 2001, without any industrial unrest.[6] NWSC designed these programmes to improve morale and to instil confidence in managers who were able to alter the expectations of operating staff. Two other initiatives promoted organizational change:

- *Stretch-Out Programme* (2002–2003) emphasized teamwork through work involvement and a reduction of the "boss-element" typical in bureaucratic organizations. For example, the programme promoted informal Fridays, where all staff wore NWSC "T-shirts", symbolically demonstrating that staff at all levels were making contributions to achieve a common objective.
- *One-Minute Management Programme* (2003) created procedures for promoting individual performance accountability, a problem identified when teams were the focus of organizational development. The programme promoted individual staff accountability by asking each staff person to come up with a vision, mission and goals describing his/her planned role in achieving corporate objectives. The achievement of goals was then a subject of periodic appraisals and incentive awards.

Currently, NWSC is implementing *Internally-Delegated Area Management Contracts* (IDAMCs) aimed at giving more autonomy to area managers (partners[7]), defining roles and responsibilities more clearly, and creating better incentive plans that allocate more operating risks to partners. By passing more risks to partners, the head office is able to encourage greater innovation and work commitment. In this regard, the partners are paid through increased incentives for taking on these risks. To rationalise the monitoring and evaluation activity, a "checkers" system has been introduced to strengthen the

---

[6]Kelman (2006) argues that crisis can inhibit responses by government organizations in the area of downsizing. This case study suggests that extreme pressure can improve organizational performance, but management needs to address employee issues in a participatory manner.

[7]Under the IDAMCs, the head office enters into an internal, non-legalistic contractual arrangement with partners in Areas. The partners are a team of senior managers in the Areas/Water utilities, who are bound together by a Partnership Deed detailing how business shall be conducted during the period of the internal contract. The partners are headed by a lead partner, who is the accounting officer of the water utility/Area.

IDAMC implementation process, emphasising both processes and outputs. The IDAMCs have also been recently (December, 2008) reinforced with a *Water Raving Fan* (WRF) initiative that operates on three "secrets": developing an internal vision of customer perfection, knowing what the customer wants, and delivering that plus 1%. The WRF is not a programme but a cultural change activity aimed at exceeding customer satisfaction and moving to the customer "delight" frontier. The corporation has introduced facets like serving free drinking water to visiting customers, providing a dedicated customer parking area, sweeping city streets as a way of giving back to customers, donating to the needy, among others. The WRF has the ultimate objectives of increasing customers' willingness to pay and enhance public perceptions of the services on offer.

Of course, no organization can be successful in isolation: collaboration allows managers to learn about the strengths and weaknesses of peer companies. To facilitate such exchanges, NWSC carries out regional networking through a recently established External Services unit. Through this unit, the corporation has established a mechanism for sharing experiences and rendering consultancy services to outside companies, on a cost-covering basis.

The organization's experiences in Private Sector Participation (PSP) and with other commercialization activities have provided NWSC professional staff with experience; the internal evaluation process serves as a mechanism for internal learning. When shared with peers, programmes such as those listed above provide lessons for others. Recent activities provide examples of cases that help other organizations better understand how to create value. These initiatives, which have, to a large extent, been financed by internally generated funds include:

- Reducing accounts receivable and uncollectibles: Kampala Revenue Improvement Project (KRIP): 1998–2001, under Gauff J.B.G.
- Improving financial performance further through enhanced monitoring of managerial performance: Kampala Water Supply and Sewerage Area Management Contract: 2002–2004, Ondeo Services.
- Evaluating the impacts of rate design: Tariff Review (since 2001-to date)
  - Reduced connection/reconnection fees
  - Tariff indexation against inflation
- Expanding customer base: New Connection Policy, which NWSC introduced in 2004, aimed at giving free access for pipe lengths up to 50 m (with a nominal fee of about US$30).
- Modernizing information technologies: Computerization of systems (billing, financial, procurement, payroll, voice over IP, Lotus notes, customer complaint tracking, call centre, etc). NWSC implemented major computerization initiatives in 2003, and these were funded by the German Development Cooperation through the GTZ (the German foreign aid agency).

- Improving customer services: Introduction of account balance checking system with local telephone (2003) and direct-debit (DD) system with local banks (2006). The administrative costs of these initiatives are minimal and are being met through internally generated funds.

These initiatives illustrate the range of managerial and engineering programmes introduced to promote financial sustainability and credibility with consumers and government agencies.

One factor supporting the favourable outcome to date is the performance contract between NWSC and the government of Uganda, represented by the Ministries of Water and of Finance. The targets and reporting procedures have institutionalised accountability, without introducing a separate agency to monitor the corporation. At some point, current arrangements will come under review. However, the arrangements have been beneficial to date, in contrast to patterns observed elsewhere. It may be that the second generation of performance contracts have benefited from the experiences of others.

The implementation of these change management programmes in NWSC has given rise to a number of lessons. It is important to involve and empower staff to make decisions at appropriate levels of operation. In this case, devolution of power from the centre is a key prerequisite. In addition, customer focus

Figure 10.1 The ambiance at one of NWSC Water Works created by change management initiatives.

and incorporation of "private management style" efficiency (less bureaucracy, performance-based pay, "customer-pays-for-a good service" principle, etc.) are key success factors. Furthermore, good planning and continuously challenging management teams with new performance targets keeps the momentum of performance programmes alive. Other success factors include outsourcing non-core activities, systematic use of external contracts versus short-term internal performance programmes/contracts, clear oversight and monitoring through the checkers system, and information sharing through benchmarking.

10.2.2   EMERGING MANAGERIAL TOOLS AND PRINCIPLES

Besides the above initiatives, an extensive computerization drive to network all its operating areas was undertaken in order to strengthen the communication mechanisms, which was vital for accurate decision-making and planning purposes. Further, the tariff policies were reviewed and new measures instituted, including reduction of new connection and reconnection charges. At the same time, regular tariff indexation against inflation was introduced, which enhanced financial sustainability through improved revenue generation. In all these change management programmes, one may ask: 'what type of management best enhances water availability for citizens, especially in places of abundant fresh water sources?' A number of emerging management tools (EMTs) can be distilled from the NWSC case study. These include improving governance through a series of innovative performance contracts/programmes, use of performance-based incentives and strong customer focus. Furthermore, empowerment and continuous staff development, efficient and well-targeted investments to address scarcity, and a constructive dialogue with government and donors on how to best address water scarcity are also critical ingredients. Last, but not least, utility managers MUST incorporate strong research and development to explore new ways and technologies to address water scarcity.

With respect to the principles for all the implemented programmes in NWSC, the underlying considerations were proper identification of the driving forces for key performance areas and strategy formulation with clear prioritization of the activities. In all cases, a well- and strategically-formulated monitoring and evaluation (M&E) framework was developed to ensure continuous improvement. The programmes adopted relevant PSP-like mentalities (i.e. a commercial orientation, customer care, and incentive-based pay) and encouraged ownership, collective decision-making and a balanced bottom-up and top-down management approach.

10.2.3   WATER LOSS MANAGEMENT

A substantial amount of water loss in the system was attributed to illegal use and other types of commercial losses. However, without well-established hydraulic zones and a demand management system, it was not possible to establish the percentage that illegal use contributed to such losses. In an effort to curb the rampant illegal use, dedicated and well-facilitated illegal use management units were established in all operating areas with clear outputs that formed the basis for their remuneration. The units were responsible for identifying and taking proactive action on all illegal-use cases in their respective areas. The organizational structure of the operating units was also streamlined to emphasise leakage control activity, and leakage management teams were established with delineated responsibilities and performance targets. The teams' activities were enhanced by a full-fledged call centre, which is the registration point of all reported leaks and bursts which are immediately routed to the responsible leakage management teams. The call centre database is also used to record all other actions with respect to faults that have been discovered by the teams themselves. This approach has significantly reduced the response time to leaks to within 2 hours. Apart from physical leakage management, the meter management and replacement policies were reviewed, leading to the effective definition of procedures to ensure correct levels of maintenance for customers and bulk meters. On the other hand, proactive strategic alliances with security agencies and communities played a vital role in addressing the challenges of metre vandalism in the system.

10.2.4   WATER RESOURCE PROTECTION AND STAKEHOLDER COORDINATION

NWSC experienced a number of external problems affecting service delivery that arose from climate change, and involved the quantity and quality of raw water in most areas. A continuum of approaches was adopted to respond to these challenges. Among these approaches was the enforcement of compliance with abstraction permit conditions, including, among others, the need for the utility to restrict raw water abstraction within allowable limits. In this way, emphasis was put on the 'optimisation of every drop' abstracted to ensure reliable supply, especially during times of short raw water supply. The NWSC managers also increased vigilance in the surveillance of the source and coordination with the environmental protection agencies and communities where measures were undertaken to protect the source and minimise any devastating human impact. The recent (2004–2006) drop in water levels, especially in Lake Victoria, necessitated innovative approaches that required modification of the intake system to extend the abstraction point further into the lake to guarantee reliable abstraction, in terms of both quantity and quality.

10.2.5   TIMELY WATER PRODUCTION CAPACITY DEVELOPMENT

The heavy investments carried out prior to 1998 resulted in excessive idle plant capacity for most infrastructure systems. Consequently, as in 1998, the overall plant capacity utilization for all NWSC plants was only 55%. The excessive idle plant capacity gave rise to uneconomical depreciation costs, increasing operating costs due to the oversized system. This inefficient investment activity also meant that funds were tied-up and there was not enough to carry out network expansion programmes. However, arising from the new free connection policy, expansion of the customer base and improvement in service delivery have significantly improved the utilization capacity of the system to about 75%, as in 2007. In this respect, a plant capacity utilization of 85–90% is considered a sufficient inflection/trigger point for capacity upgrading. Furthermore, in NWSC's case, the majority of heavy investments in the system have been implemented through grant financing to avoid negative tariff effects, given the limitations in implementing full cost recovery. In order to ensure long-term sustainability of the investments, all capital projects are implemented taking into full account accompanying institutional, operational and maintenance managerial implications.

10.2.6 CREATIVE APPROACHES TO ENHANCE INTERNAL REFORMS

> "Between imitation and envy, imitation is better."
> (Ekonda, Democratic Republic of Congo Proverb)

Creating Strategic Rivalry among Operating Teams: This might seem like a provocative approach to management but it worked wonders for NWSC's internal reforms. The underlying performance driver here is competition. For example, by sending an invisible signal from the supervisor telling water distribution teams, secretly, that the other team is better organised and therefore performing better creates a sense of competition. In this case, supervisors have had to be careful not to go into details of proving who is right and who is wrong, initially. The idea is to create anxiety and envy among teams so that they can streamline and improve their production technologies. Of course, the final verdict must be delivered through a transparent evaluation activity showing the teams' performance trends and target achievements. One caution that has to be borne in mind is that the teams must not take the supervisor's criticisms as 'business as usual'. The criticisms must appear like they have been well researched and should roughly predict true trends after a transparent compilation of empirical evidence. This calls for the supervisor's effective use of key invisible and genuine informants. This approach, however, requires significant situational management and readjustments to avoid demolishing teams rather than building them up. This approach has worked well, resulting in improved water leakage management in NWSC water network systems.

Flexibility in Restructuring Operating Teams: According to Chandan (1987), each organizational structure must suit the situation and be optimally useful in meeting organizational objectives. Good organizational design is a function of factors including the environment, technology, size of the company and philosophy of the central management (ibid.). Furthermore, Bennis (1956: quoted in Chandan, 1987), contrasts an organic structure with a bureaucratic structure, pointing out that the latter is more suitable under fairly stable conditions while the former is more desirable in times of dynamic and rapid technological changes. In this regard, an organic structure is more informal in nature, it de-emphasises authority and concentrates on problem solving. In water network systems management, indeed the organic model is preferred for two reasons: First, technical problems, especially in African water utilities, are complex and uncertain, requiring flexibility in the structuring of management teams. Secondly, there is significant variability and a transient character in the qualities and commitment of operating staff due to inadequate remuneration systems and subsequent coping arrangements. Such organizational flexibility has also been tried in water reforms in the NWSC and proved profoundly helpful.

10.3   NWSC BENEFITS FROM THE CAPACITY-BUILDING INITIATIVES

"Be the change you want to see"

(Mahatma Gandhi)

Implementation of the above management strategies has resulted in significant efficiency gains (see Table 10.1 below) for the corporation for the period 1998–2008. Specifically, service coverage has increased from 48% to 71% and the level of UFW has been reduced from 51% to 33% – with all towns, other than Kampala, registering UFW levels between 15–18%. The corporation registered a financial surplus after depreciation in 2008, which is a reflection of the positive impact that the changes had on financial sustainability.

Table 10.1 NWSC performance improvements.

Performance indicator	1998	2008
Service coverage	48%	71%
Total connections	50,826	205,000
New connections per year	3,317	24,000
Staff per 1000 connections	36	7
Collection efficiency	60%	93%
Unaccounted-for water	51%	33%
Proportion of metered accounts	65%	99.6%
Annual turnover	USH21.9Bn	USH84.0Bn
Profit /(Loss) after depreciation	Loss: (USH2.0Bn)	Profits: USH3.4Bn

Most importantly, the capacity-building initiatives lifted the spirits of all staff in the corporation, and enhanced performance through increased accountability, increased customer focus, prompt decision-making, increased autonomy and initiative taking. The programmes also allowed for rational allocation of operating and commercial risks through well-structured performance incentives.

10.4   FOCUSING ON REAL ISSUES: DISREGARDING RHETORIC

The successful implementation of these performance-enhancement initiatives across NWSC's operating units suggests that the conventional wisdom regarding non-performance by public companies is incorrect (Mugisha and Berg, 2008). The NWSC experience clearly shows the benefits of focusing on what works: there is no single textbook solution to the myriad of problems facing water utilities, especially in low-income countries. Most of these problems are caused by local managers, poor organizational cultures, citizen non-payments and political intrusiveness. These groups are potentially the change-agents who can address those problems. Familiarity with the problems can be an advantage when attempting to solve them, although this is not always the case.

In some cases, the person who is the source of the problem needs to be removed from his or her position of responsibility. However, the person should be given the opportunity to change – to become part of the solution. NWSC leaders believe that the one who causes the problem and lives with it is best placed to give information about it, and even provide input about how to solve it. It is essential that the internal governance system provides the right incentives and a good framework to enable such people to diagnose the situation and come up with solutions. Developing sound incentives requires several steps: strong leadership articulates the right vision for the company, guides staff in problem analysis, and motivates them to come up with strategies to address gaps. It has been established that this ought to be a continuous process. The principle "never be satisfied with the status quo" works well for any business – including water supply – where the challenges are clear.

Conventional thinking reflects the view that managerial practices under public management settings are fundamentally flawed, reflecting frustration with poor performance of public companies. Sometimes that poor performance is attributed to the non-economic objectives given to public enterprise (such as job security, employment, or regional development), so standard financial performance measures may be inappropriate. The situation has been most problematic in "basic needs" infrastructure utilities where governments say that they do not want citizens to lack access to clean water or to suffer the health impacts of contaminated water. The general record suggests that nations often become trapped in what Savedoff and Spiller (1999) label a "low-level equilibrium" involving low prices, low quality, slow network expansion, operating inefficiencies and corruption. The situation is "stable" in a sense – as managers pretend to manage, their utilities pretend to deliver services, and customers pretend to pay. However, the outcome is most unfortunate.

When the water utilities lack cash flows, those responsible for public funding recognise that using limited treasury resources to finance inefficiencies is not sensible, so promised funding disappears; the situation remains bleak. This dilemma has often led to what could be called "desperate solutions": the privatization of such companies. However, an independent consultant for NWSC remarked, "Governments should never expect to privatize their problems to international companies and think it will work" (Richards, 2003). Indeed, the private sector participation experience at NWSC suggests that international operators do not come to manage problems but to earn returns on their investments. They will exit and leave the country if the combination of their skills and the institutional environment lead to low cash flows. Those cash flows may be low because of internal problems (their lack of needed skills and misunderstanding of local conditions) or external circumstances (their reasonable expectations regarding institutional changes were unfulfilled). This fact partly explains why NWSC could not continue contracting arrangements with Gauff at one time and later with Ondeo.

Fortunately, nobody has a monopoly on approaches to improving performance. From NWSC's experience, excellent performance can be home-grown, but such an outcome requires a set of conditions. Do-it-yourself works if the implementing team has strong leadership, the right tools (legal framework), appropriate skills and a clear set of (shared) objectives. Nevertheless, one wonders why current managers have knowledge and skills, but they often lack the incentives to put forth the extra effort and make some difficult decisions. For many utilities, local managers are the starting point for improvements in performance. In the case of NWSC, Dr. Muhairwe's insistence on self-actuated internal reforms paid off in this way. Initially, some stakeholders thought that NWSC's problems needed *imported* management. The management and board insisted that they had the tools and vision that more than matched the capabilities of outsiders. Of course, weak performance would have been evidence that the new MD and his reform team were incapable of turning around the organization. That meant establishing baselines, identifying past trends and setting realistic targets. No longer were numbers proprietary (or, as has been the case for many public enterprises, "unavailable": transparency and accountability guided the process). Data reflected reality, not the wishful thinking of managers or the hopes of politicians. The results supported the team's confidence and gained stakeholder support.

In some cases, particular stakeholders (consultants, reform managers and technical advisors) might overestimate institutional problems (or underestimate the capabilities of state-owned enterprises). Such groups must be handled carefully, since their actual intentions may not be consistent with public statements (or with local values or aspirations). Some of them are representatives of stakeholders (such as national donor agencies) identifying national investment opportunities for the firms they champion. Other stakeholders, such as those representing international banks, might support the policy "flavour of the month". While both groups can provide capital through a variety of arrangements, they can prematurely damage initiatives that the company is planning and/or implementing. "Conditionality" becomes another word for "policy being dictated by those who are unfamiliar with issues" on (and under) the ground. Communicating with powerful, multilateral donor agencies and convincing their representatives that the local team can succeed requires a substantial investment of time. However, that investment is necessary: in some cases, the dialogue leads to better plans; in others, the local team is able to persuade these important stakeholders that the highest payoff will come through local talent, insulated from volatile political forces. As NWSC entered the process, managers realised that they needed to identify their roadmap, accept constructive advice, respect differences of opinion, and ensure that company values and objectives were at the centre of everything. International consultants have a role in the process, but the ultimate decisions cannot be delegated to professionals who lack a deep understanding of national institutional constraints and unique local opportunities.

10.5 MAKING THE CASE FOR PUBLIC ENTERPRISES

According to Muhairwe (2008), managers of public enterprises seem to take their usefulness or their roles in the economy and society for granted. They seem to assume that they deserve to exist simply because they were established to perform preordained statutory mandates. Otherwise, why would they have been established in the first instance? Managers see their corporations as victims of government interference, meanness and under-capitalization, and of donor conditionalities and technical and capacity constraints. This self-pity, mourning and "blame-storming" can only lead to a dead end. Public enterprises have to earn a living in accordance with the rationale for which they were established in the first place. They have to justify their existence and continuity through their performance and achievements. They have to make their case to win over the shareholders, development partners and other stakeholders through results, not words.

If a public enterprise performs well, other stakeholders will notice and back it. Should public enterprise managers present credible diagnoses and prescriptions, other stakeholders – including the government and donors – will listen, or at least give such propositions the benefit of the doubt. Since 1998, the capacity-building initiatives at NWSC have demonstrated that even sceptics and doubting Thomases can be converted into friends, allies and partners by incremental performance improvements. Take the case of the government, the sole shareholder of NWSC. By 1998, the government was bent on privatising the corporation, preferably to "international operators" with lots of resources and expertise. Accordingly, the privatization of public utilities was supposed to resolve their management and performance problems at a stroke. However, the government was stuck with NWSC because there were no serious international bidders. Public utilities like NWSC are much more than ordinary commercial enterprises whose paramount purposes are to make money for the shareholders. Over and above commercial considerations, public utilities are vested with "social-mission" objectives, but few international operators are willing to carry such a burden. In any case, at the moment, our water market is too small to attract foreign investment.

When NWSC's home-grown internal reforms strategy began to turn the corporation around, the government gave it a big "thank you" for a job well done. Not only did the government grant the corporation operational autonomy through successive performance contracts (2000–2009), they also agreed to settle the arrears which had accumulated over the years, and, henceforth, to settle new water bills on time. As a result, the spectre of privatization that had haunted the corporation began to recede. This provided time to consolidate performance improvements through even more ambitious change-management initiatives. By 2005 the government had become a supportive owner in implementing internal reforms, which goes to show that governments are flexible enough to back the winning horse when they see one. This has been a useful lesson for NWSC managers as an organization.

Figure 10.2 A NWSC branch office in Kampala city.

Similarly, at the beginning of change-management programmes, donors, especially the World Bank, had grave reservations about the chances for success of home-grown capacity-building reform initiatives at NWSC. For example, the 100-days Programme was dismissed as an ingenious gimmick to hoodwink the public and to delay divestiture and privatization. Fortunately, actions always speak louder than words. Therefore, as the change-management initiatives began to transform the corporation into a viable entity, donors realised that there was the utmost seriousness and warmed up to the reforms. In addition, instead of seeing conditionalities as a suffocating yoke around our neck, NWSC managers saw them as challenges to overcome. Surely, conditionalities such as improving revenue collection, reducing UFW and arrears, and increasing staff productivity could only be seen as meant for the corporation's own good, rather than that of the donors.

In making the case for public enterprises, it is important to cultivate the support of the print and electronic media, for example by providing information, keeping an open-door information policy, publicising activities and achievements, and participating in phone-in programmes. Needless to say, the media are the eyes and ears of the public. The media raise public issues, shape the nature and direction of public debate, mould public opinion and can make or unmake the reputations of public enterprises. Since 1998, NWSC managers have done their best to cultivate media support and to win their confidence. There was extensive use of the media as the channel of communication and feedback. Through the media, NWSC publicised capacity-building activities, programmes and achievements, and responded to public criticisms and complaints. As a result, the public has become more sensitised to what NWSC has been doing and, by so doing, the corporate image has been boosted in the popular imagination. The corporation has become a favourite household name and an oft-quoted example of what other public enterprises and authorities should be doing to improve their public approval ratings. Managers who ignore the media do so at their own peril. They miss opportunities to make the case for their organizations.

10.6   FINAL THOUGHTS: LOOKING AT THE FUTURE

Upon looking back, one sees that NWSC managers have made progress over the last ten years by transforming the corporation into one of the most successful public enterprises in Uganda. The NWSC has indeed become a model public utility in the water industry and has begun to share its experience with sister utilities in Africa and other parts of the world through its External Services unit. In this respect, consultancy services have been rendered not only to sister water utilities in Zambia, Tanzania and Zanzibar, Rwanda, Kenya and elsewhere in the world, but also to private enterprises in Uganda.

For example, in April 2005, NWSC External Services signed a contract with the Alam Group of Companies in Kampala to "provide a wide range of professional support in respect of development and implementation of organization behaviour change management programme". Similarly, in partnership with ARD Inc. of the United States of America, and under financial support from United States Agency for International Development (USAID), the corporation won the contract to provide water utility technical and managerial support services to two towns in Northern Uganda: Kitgum and Pader. In addition, NWSC External Services was contracted by Kampala City Council to help review and recommend strengthening measures in revenue generation activities for the city. The External Services unit has also been actively involved in providing various forms of capacity-building programmes to the local private operators in Uganda. All these activities clearly point to a "management revolution" which has been carried out through local initiatives and programmes, without prompting or guidance from the outside world. Managers have revolutionised the corporation's organizational behaviour and transformed its work environment.

According to Muhairwe (2008), the corporation has built a professional team of managers and staff with the capacity, skills, brains, knowledge and experience to match their peers around the world. The NWSC experience has proved that it is possible to transform public enterprises using home-grown solutions. It also demonstrates that it is possible to apply private-sector principles, techniques and motives to public enterprises through delegation,

Figure 10.3 NWSC external services staff making a presentation to a Nigerian delegation.

decentralisation and operational autonomy. The emphasis on home-grown change-management initiatives does not mean that development partners have not made useful contributions to the transformation of NWSC since 1998. Indeed, donors have provided significant technical assistance in terms of money, input and expertise.

Without donor support to some service delivery activities, the corporation would not have accomplished as much as it did. NWSC could not have done everything on its own without a helping hand from outside development partners. They have been useful in the past and the present, and no doubt will continue to be so in the future. However, the point is: charity begins at home. The NWSC experience shows that change management must be conceived, borne and nurtured from within. The lesson to draw from this is that whatever they do, outsiders cannot be a substitute for local management initiatives.

Of course, it should be noted that the NWSC has not yet won the war of the quest for excellence or corporate perfection. The change-management case marks the early stages of a long journey toward the ultimate realisation of the corporate vision of 'being one of the leading water utilities in the world' through a corporate mission of 'providing efficient and cost-effective water and sewerage services, applying innovative managerial solutions to the satisfaction of our customers in an environmentally friendly manner'. NWSC ought to continue to aim at perfection in its service delivery, and not be oblivious to the challenges, or even to the internal and external threats, to the sustainability of its performance in the years ahead. In any case, experience has shown that every success is invariably pregnant with new challenges that require new solutions. After all, enhancing performance to improve service delivery is bound to be a process without end, for perfection is always elusive and shifting even in the most efficient of business enterprises. However, a strong, solid foundation has been built through a series of change-management initiatives, and gained sufficient experience, confidence, capacity and momentum to forge ahead into a prosperous and sustainable future for NWSC. The current and future managers of the corporation must take care because of the possibility of sliding back if hard work is not sustained.

REFERENCES

Chandan, J.S. 1987, *Management Theory and Practice*, Vikas Publishing House PVT Ltd, 576, Masjid Road, Jangpura, New Delhi-110014.

Hoffer, J. 1995, *The Challenge of Effective Urban Water Supply*, Ph.D Thesis, University of Twente, The Netherlands.

Kelman, S. 2006, *Downsizing, Competiton, and Organizational Change in Government: Is Necessity the Mother of Invention?*, Journal of Policy Analysis and Management, Fall, Vol. 25, No. 4, 875–895.

Matta, N. and P. Murphy. 2001, *When Passionate Leadership Stimulates Enduring Change: A Transformational Capacity Development Anecdote from Uganda*, Capacity Development Brief, World Bank Institute, Number 13, October 2001.

Mugisha, S. and S.V. Berg. 2008, *State-Owned Enterprises: NWSC's Turnaround in Uganda*, African Development Review, Volume 20(2), page 304–334.

Mugisha, S. and S.V. Berg, G.N. Katashaya. 2004a, *Short-Term Initiatives to Improve Water Utility Performance in Uganda: The Case of the National Water and Sewerage Corporation*, Water 21, June: 50–52.

Mugisha, S., S.V. Berg, H. Skilling. 2004b, *Practical Lessons for Performance Monitoring in Low-Income Countries: The Case of National Water and Sewerage Corporation, Uganda*, Water 21, October: 54–56.

Muhairwe, T.W. 2008, *Making Public Enterprises Work: A Case of NWSC, Uganda*. A Book in the Making, NWSC, Uganda.

Richards, T. 2003, *Global Trends in Private Sector Participation in the Management of Water Supply and Sewerage Services and the Impact on Sector Reforms in East Africa*, National Water and Sewerage Corporation, Uganda Working Paper.

Rugumyamheto, J.A. 2004, Innovative Approaches to Reforming Public Services in Tanzania, Public Administration and Development, 24, 437–446.

Savedoff, W. and P. Spiller (eds.). 1999, *Spilled Water: Institutional Commitment in the Provision of Water Services*, Inter-American Development Bank, Washington D.C., 1–248.

Chapter 11

LESSONS LEARNED FROM CAPACITY-BUILDING APPROACHES IN THE UN-HABITAT WATER FOR AFRICAN CITIES PROGRAMME

Graham Alabaster
UN-HABITAT

ABSTRACT

This paper describes the Training and Capacity Building (TCB) Programme of the UN-HABITAT Managing Water for African Cities (WAC) Programme Phase 1. The TCB programme was conducted between May 2002 and August 2005. The programme was carried out by UNESCO-IHE, supported by Network for Water and Sanitation (NETWAS) International, based in Nairobi, and the Centre de Formation Continue (CEFOC), based in Burkina Faso. It brought together participants from 6 cities in Western, Eastern and Southern Africa. The cities selected in West Africa included Accra (Ghana), Abidjan (Cote d'Ivoire) and Dakar (Senegal). The cities selected in Eastern and Southern Africa were Addis Ababa (Ethiopia), Nairobi (Kenya) and Lusaka (Zambia). The city of Johannesburg (South Africa) contributed as a resource organization.

The programme sought to build capacity in the participating cities in the water sector through:

- Making information available on best practices in urban water management, and
- Linking sector professionals with each other and with other networks, institutions, governments, municipalities, NGOs and the private sector.

The Training and Capacity Building Component specifically focused on the three inter-linked thematic priorities of the WAC I Programme, which were:

- Improving efficiency of water use through water demand management
- Preventing the negative environmental impact of urbanization on fresh-water resources
- Enhancing public awareness in African cities.

The programme was implemented over a period of three years. It started with a survey in the different cities to identify on-going pilot projects that could be used as training materials and potential local training institutes/trainers. Following this, the TCB component began with a training of (local and regional) trainers (ToT) who would lead the training in each of the cities. During the ToT, the trainers gained knowledge and expertise in the field of water demand management, pollution prevention and control, public aware-ness and planning, knowledge needed to train the target groups. The target groups included middle level managers (MLM), senior level managers (SLM) and top level managers (TLM) in the water sector of each of the six cities. The training courses were carried out in two cycles with a 6-month period of break time. In the first cycle, action plans were defined and developed for implemen-tation during the period between the two cycles. In the second cycle of train-ing, evaluation of the progress made in the implementation of the action plans were made and recommendations were suggested on experience learnt from the actions. During the training process, discussions and recommendations from the lower level were used as input and learning points for the higher level training. The training strategy was important to create a conducive envi-ronment for the lower and higher management levels to interact and discuss openly the important issues of water management. The course contents were designed to provide the participants a holistic approach on integrated water management.

During the training of trainers, a team of 22 professionals was trained from the participating cities that carried forward the process at the local level. These professionals contributed to the training of about 130 managers (MLM, SLM, TLM). Action plans were developed after discussions in the various levels and at different periods of the training. The action plans resulted in measurable achievements, e.g. in the field of reduction in unaccounted-for-water, envi-ronmental protection and public awareness campaigns. Moreover, training materials and training reports were prepared for each training activity that would be used for further reference.

Overall, the participants evaluated the programme as successfull, how-ever, there were some limitations and challenges with respect to roles and responsibilities of the different partners and actors, as well as to the budget. The lessons learnt from WAC I, if addressed adequately, would provide a strong platform for future training and capacity building programmes in the region.

11.1   INTRODUCTION

11.1.1   PROJECT BACKGROUND

The goal of UN-HABITAT's Water and Sanitation Programme is to contribute to the achievement of the water and sanitation related MDGs/WSSD targets in urban areas, with particular focus on Africa, by supporting the creation of an enabling environment for pro-poor investment.

The strategic vision of the UN-HABITAT Water for African Cities Programme is to reduce the urban water crisis in cities through efficient and effective water demand management, build capacity to reduce the environmental impact of urbanization on freshwater resources, and boost awareness and information exchange on water management and conservation.

The Water for African Cities Programme (WAC) has been addressing the need for improved management of urban water and sanitation resources in urban areas of Africa since 1999. Phase 1 of the programme consists of 3 main components: pilot studies, water education, and training & capacity building (TCB). The programme was initially funded by the United Nations Foundation for International Partnerships (Turner Foundation) and other agencies[1]. Matching counterpart support is also obtained from the participating cities/countries.

Phase 1 of the programme involved 3 cities from West Africa, and 3 cities from Eastern and Southern Africa. The cities from West African included Accra (Ghana), Abidjan (Cote d'Ivorie) and Dakar (Senegal). The corresponding cities from Eastern and Southern Africa were Addis Ababa (Ethiopia), Nairobi (Kenya), Lusaka (Zambia). Johannesburg (South Africa) is contributing to the programme as a resource city.

In the first phase of the TCB programme, three specific groups are targeted – middle level managers, senior managers and policy and decision-makers. Separate and sequential training is carried out at the three levels, with the aim of enabling the individual trainee to understand the subject area and his/her own contribution in realising an appropriate strategy to address it.

UN-HABITAT contracted the UNESCO-IHE Institute for Water Education, based in the Netherlands to advise UN-HABITAT in the supervision of the training and capacity-building component of the WAC programme. UN-HABITAT contracted two regional resource centres to carry out the training and capacity-building activities: Centre de Formation Continue (CEFOC)

---

[1] World Bank, UNDP/UNV, the governments of The Netherlands and Finland and the European Union.

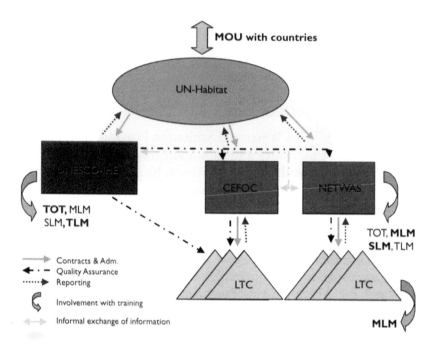

Figure 11.1 Relationship between the partners involved in the TCB program.

for the West Africa region, and Network for Water and Sanitation (NETWAS) in the Easter and Southern Africa region. The range of activities and the relationships between the different organizations involved in developing and delivering the TCB programs are shown in the flow chart in Figure 11.1.

The pilot studies and the TCB specifically focused on the following inter-linked priorities:

- Developing realistic and effective Water Demand Management (WDM) strategies in seven demonstration cities for efficient water use;
- Building capacity at the city level to monitor and assess sources of pollution loads to freshwater bodies, and to set up early warning mechanisms for timely detection of emerging hotspots of urban pollution;
- Enhancing region-wide information dissemination and awareness-raising on water conservation in African cities.

The TCB component was designed not only for the training of African professionals in water demand management and pollution control but also as a tool for capacity building of regional institutions to independently lead training of a similar scope and scale in the future.

Water sector capacity building is defined as the process of providing individuals, organisations and other relevant institutions with the capacities that allow them to run the water sector with optimal performance, now and in the future. In the context of a water sector that often performs poorly under severe resource constraints, capacity building provides a holistic approach that takes into account the environment, the managerial system and institutional as well as human resources development. Capacity building is a process that is characterised by:

- being supportive of local capacity builders and water sector institutions,
- integrating the relevant technical, economic, behavioural and management disciplines,
- being both issue-centred and output focused, and
- aiming at the creation of effective and learning-oriented organisations.

The subject of training and capacity building was discussed subsequently by the Programme, during several meetings of the City Managers. During this meeting the City Managers were consulted and reached agreement on the concept and approach of the intended training and capacity building programme. Preceding the consultation, a training needs analysis was conducted by means of a questionnaire circulated to the City Managers.

11.1.2   THEMES IN THE TCB

In order to effectively achieve the aims of the WAC Programme, the TCB component focused on the following important themes: Water Demand Management (WDM), Pollution Prevention Control (PPC) and Public Awareness (PA): crucial themes in the understanding and implementation of integrated water management in cities and regional scales.

11.1.2.1   Water Demand Management (WDM) component

Historically, the dominant approach to water resources development has focused on developing new supplies and structures to exploit available supplies in order to meet perceived water needs. The results of this strategy are often characterized by large dams, rivers, and diversions, as well as major water supply, waste treatment and other related infrastructure. Although these structures may be necessary in some cases, this conventional approach to water supply management tends to define water usage as a requirement that must be met, and not as a set of demands that are variable and changeable. This can and often does lead to overuse of water resources, overcapitalization, waste and other problems.

Governments are now beginning to understand that solving these types of problems requires fundamental change. Thus increasingly the water demands themselves, not structural supply solutions, are becoming the focus of policy and decision-making. This alternative policy strategy, focusing on water demand management, relies on proven, cost effective approaches for modifying water demand patterns, and in many cases lowering these demands substantially.

Water demand management emphasizes the socio-economic characteristics of water use. Water users are seen as consumers who can be influenced and governed by incentive structures, technology modifications, public education and regulations. Through a mix of these instruments, significant savings and better management of water resources can be achieved. Only after opportunities have been fully analysed for lowering or mitigating water demands should new capital-intensive supply systems be considered. (source: Strategic Plan (2001–2005) of the International Environmental Technology Centre (IETC) of the United Nations Environment Program (UNEP) – Nov. 2000.)

11.1.2.2   Pollution Prevention Control (PPC) component

Urbanisation is causing a continuous increase in the population size and density of all major African cities. Utilisation of resources within these cities generates wastes. Both the population size and density cause waste and pollution generation to exceed the natural carrying capacity of the environment in and around the cities. Therefore, waste management is required. Various stakeholders are involved in waste management, ranging from government institutions and NGOs to individual citizens. A proper mix of appropriate infrastructure and its operation and maintenance is required to deal with the wastes. Depending on the socio-economic, climatic and geographical situation, the emphasis in a pollution control strategy may be on centralised large-scale infrastructure development or on decentralised waste management by local stakeholders and their organisations. A proper mix of measures should be designed, which is affordable, acceptable and sustainable. For sanitation systems, the required water consumption to operate these systems and the treatability (costs, technical complexity of treatment systems required, reuse potential) of the generated wastes are important. Proper urban waste (water) management is an important step towards the development of sustainable urban centres.

11.1.2.3   Public Awareness (PA) component

It is believed that community participation is an important tool towards achieving the goals of the program. In this regards promoting Public Awareness among the community becomes essential and the TCB program considers this an important theme.

11.1.3   OBJECTIVES OF THE TCB PROGRAM

The overall objective of the TCB program was to improve the efficiency of water use and management through the training of professionals and to increase public awareness of the communities in the selected African cities.

Moreover, it aimed at building the capacity of local training institutions to promote the process. Each of the activities had specific aims set.

The TCB program consists of a training of trainers (ToT) followed by the training of various management levels. The objectives of the ToT course were to prepare future trainers on content and methodology for the MLM (and in some cases for the SLM) training course. Moreover, this aimed to equip and update participants with tools, skills and necessary knowledge for the development and implementation of the MLM (and SLM) training course and action plans.

The MLM training was planned to help participants understand the relationship between WDM, PPC and PA and the importance of an integrated approach to water management. The training was also designed to improve cooperation and coordination between the different role players within the utility, to develop an action plan and monitoring framework, and to present and share action plans with senior and top managers.

The SLM training was prepared to give the utility heads the opportunity to survey the action plans prepared by the MLM and evaluate their feasibility as well as to define their own tasks in implementation of the action plans and ensure timely monitoring. It also aimed to discuss and prepare an awareness-raising plan and to suggest changes in policies that should be discussed between the TLM.

The objectives of the TLM training were to enhance the capacity of the top WAC city managers and decision-makers in integrated urban water management with particular emphasis on key challenges, as well as to review achievements in action plans and make necessary adjustments.

11.2 METHODOLOGY

11.2.1 TARGET GROUPS

The programme developed separate training approaches in terms of content, training methodology, duration and implementation for three target groups of professionals from the six cities. The target groups were identified as follows (see also Figure 11.2):

- Top Level Managers (TLM) – This category included policy and decision-makers, managing directors and administrators, as well as political figures. The training was of the seminar/exposure/workshop type, and it addressed policy, strategy and programme development. This group consisted of about three people from each city.
- Senior Level Managers (SLM) – This category included heads of technical and financial departments. The training was of the seminar/exposure/workshop type, and addressed project planning, resource allocation and coordination. This group consisted of three people from each city.

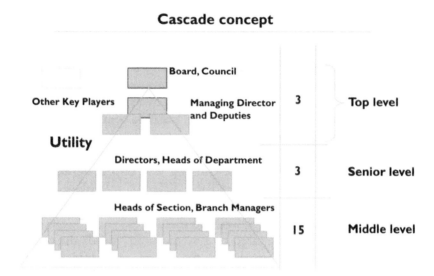

Figure 11.2 Relationship between the target groups – the cascade concept.

- Middle Level Managers (MLM) – Included heads of distribution districts, plant managers, etc. The training was of the workshop/exposure type and it addressed project preparation and implementation in the areas of water demand management and pollution control. This group consisted of about 15 persons from each city.

In total, 130 professionals (approximately 21 from each city) participated in the program.

11.2.2   TRAINING STRATEGY

This training and capacity-building programme was implemented along two parallel tracks. As its first objective, the project sought to enhance the capacity of managers and decision-makers in water utilities, other service providers and regulatory bodies in integrated urban water resources management with particular emphasis on water demand management and water quality management. Secondly, the programme sought to anchor the enhanced capacity within the region and ensure sustainability of this effort by institutionalising the training programme through capacity building in the form of a Training of Trainers programme and technical support to existing local training institutes and regional resource centres.

The TCB was designed to include the three upper levels of management in urban water delivery. It was planned in a sequential manner: to first train the MLM, then the SLM, and finally the TLM. This sequence was consciously selected with the following rationale: During the implementation of the training programmes for the MLM, participants would work on their action plan and identify limitations beyond their control in terms of resources at their disposal to undertake the work. During the training of the SLM, the resource constraints at the MLM would be introduced as an input to be discussed and addressed at that level. The SLM training would identify shortcomings in policy, regulations and instructions that were in turn an input to the training for TLM.

The TCB was carried out in two cycles, with a period of six months in between to implement the action plans developed in the first cycle. The first cycle of training focused on content such as WDM, PPC, PA, the exchange of experiences between the cities and planning tools and it resulted in the formulation of action plans. The second cycle of training provided opportunities to evaluate the progress made in realising the action plans, identifying bottlenecks, improving skills related to these bottlenecks, and it produced revised action plans. The aim of these training activities was to make the individual trainee understand the subject area and what his/her own contribution should be in realising appropriate water demand and water quality management objectives in her/his city.

Local trainers implemented the training for the MLM with support from the regional resource centres. The regional resource centres implemented the training for the SLM with support from UNESCO-IHE. The training for the TLM was implemented by UNESCO-IHE with support from the regional resource centres.

The proposed approach for the TCB program was developed with close cooperation between UN-HABITAT, the WAC city managers, UNESCO-IHE and the two regional resource centres. The two regional centres NETWAS and CEFOC covered the cities in the Eastern and Southern African region and the West African region, respectively. The local trainers covered the cities in which they were situated.

11.2.3 TRAINING DELIVERY

The training program started with the training of trainers. A team of trainers from the resource centres and each of the various participating countries was trained to conduct the training for the MLM at the various demonstration cities.

In the first cycle, duration of the training sessions were 3 weeks for the MLM and 1 week each for the SLM and TLM. The training was structured in three parts: a general part (20% of time), a WAC-specific part (30%) and an action planning part (50%).

The general objectives of the trainings were mainstreaming and operationalising of the WDM, PPC and PA through participants from the various organizations.

The general part aimed at:

- Increasing the insight and inter-relationships between the WDM, PPC and PA
- Educating the trainees with the approaches and methods for integrated water management.

The aims of the WAC-specific part were to:

- Make familiar strategies being used in the WAC programme;
- Understand related WDM, PPC and PA activities, demonstration projects and results (being implemented in the WAC cities);
- Examine own position and role, expected contribution to WDM, PPC and PA and required levels of cooperation.

The objectives of the action planning part were to:

- Increase the understanding and ability to design a project matrix using Log-frame Analysis and a schedule using the software MS-Project, or an equivalent,
- Enable the trainees to prepare a situational analysis of their work situation; to identify and evaluate alternative WDM/PPC/PA projects; to select and implement most beneficial project(s).

The content of the second cycle of trainings was fixed after evaluation of the first cycle and after analysing feedback from and needs expressed by the participants.

In general, the training content focused on key issues and challenges in urban water resources management, which were partly location specific. In order to realize the listed objectives, the training focused on topics such as the reforms at sector and utility levels, MDGs, water governance, water and poverty, institutional and organizational reforms, communication, and information sharing and personal management skills.

11.2.4 SELECTION OF REGIONAL TRAINING AND RESOURCE CENTRES

In order to select the best available training and resource centres in the selected cities, consultations were made with WAC program staff and WAC city managers in March 2002. UNESCO-IHE conducted surveys in each of the participating cities in the region to identify qualified resource centres and selected trainers from the centres that would conduct the trainings. This initial survey was conducted in the period June 2002 to August 2002. A meeting and telephone conference were conducted in June 2003 to help prepare and finalize the course curricula with the WAC city managers and to prepare the terms of references for the recruitment of the regional centres. The initial city surveys were used to identify potential regional and local training institutions (and trainers), and to get familiar with WAC strategies and activities in the different cities.

11.2.5 CAPACITY BUILDING OF RESOURCE AND TRAINING CENTRES

Strengthening the capacity of the local training centres and the regional resource centres was an important component of the programme. This capacity building activity focused on key technical components such as content (WDM, PPC and PA), as well as curriculum development, course development, and the development of didactical and managerial skills. The services of UNESCO-IHE were required for capacity building at the regional centres and local training centres. The regional centres recruited local trainers to carry out the job.

In terms of institutional capacity building, the efforts of UNESCO-IHE were directed towards a team of 4 regional trainers and 21 local trainers. A limited budget was deemed necessary to upgrade the training facilities of the regional and local centres. A workshop was conducted with the selected regional and local training centres to share the TCB project information, and to decide on the capacity-building agenda, input and coordination (June 2004). The workshop brought together relevant staff in managerial positions from the participating institutes, and one staff member from UNESCO-IHE.

11.3 ACHIEVEMENTS

11.3.1 PROJECT COMMISSIONING

A short brochure describing the project, its goals and the training methodology, was prepared. The printed brochure was used by the WAC city managers to look for and recruit trainees, and eventually to look for local and regional training centres.

11.3.2 SELECTION OF LOCAL AND REGIONAL RESOURCE CENTRES

The initial survey (March 2002) identified local training institutions that had full training facilities and convenient venues for the purpose of the WAC training. Moreover, the survey identified trainers (although experienced trainers were not immediately available at the training institutions), which included professionals and academics. This was assisted by the WAC city managers.

In June 2003, in response to UNESCO-IHE's assessment of local trainers and training centres, UN-HABITAT decided to contract only the two regional resource centres, letting both organisations the responsibility of sub-contracting the local training centres. UNESCO-IHE's assessment was made available to the regional resource centres to help them prepare their offer. Because of budgetary constraints, both regional resource centres proposed to make use of local trainers without any connection to local training centres.

The regional centres were selected based on the following criteria:

• Qualifications
• Documentation and training experience in subject areas of water and environment
• Expertise in the multi-disciplinary WDM and PPC subject areas.

The tasks of the regional training institutions were following:

• Carry out the regional seminars for senior (SLM) and local mid-level (MLM) training/seminars
• Assist in the development of training materials for all trainings
• Contribute to the ToT
• Assist in TLM training.

In the period September 2003 to March 2004, UNESCO-IHE supported UN-HABITAT in the selection process of the regional resource centres and made comments on the proposal presented by three organisations. The technical evaluation clearly showed that NETWAS, for Eastern & Southern Africa (Addis Ababa, Lusaka, Nairobi), and CEFOC, for West Africa (Abidjan, Accra, Dakar) were the best candidates to carry out the project in the regions.

When the resource centres and the name/profile of the trainers were known, the training of trainers programme was finalized. The training was tailor-made, focusing on specific topics that would augment the knowledge and experience of the selected trainers. The information collected during the initial city survey, such as strategies and pilot activities in the programme, have been made available to the ToT trainees.

UNESCO-IHE & UN-HABITAT held a workshop with the regional resources centres (NETWAS and CEFOC) to finalize the details of the training programme.

11.3.3 TRAINING OF TRAINERS (TOT)

A two-week ToT course was carried out by UNESCO-IHE in Nairobi between 21 June and 2 July 2004. There were four trainers from UNESCO-IHE and an additional four from the regional training centres. To increase the quality of the outputs of the training, UNESCO-IHE added one trainer to the team initially proposed.

The course brought together a total of 22 professionals, 11 of whom came from the three participating cities in Eastern and Southern Africa, and the remaining participants came from the corresponding cities in West Africa.

Topics addressed during the workshop included:

- Getting to know each other
- Understanding the TCB components of the WAC programme, such as organisation and relationships among ToT, MLM, SLM and TLM trainings
- Discussion about the ToT based on the concept proposed by UNESCO-IHE

  - topics on WDM, PPC and PA
  - guest lecturers, trainers, training materials
  - date, location, travels
  - selection of participants.

- Discussion on the content of the MLM, SLM, TLM trainings
- Discussion on the monitoring of the effects of the trainings
- Identifying roles & duties of the different partners
- Outlining the general planning of the programme
- Arranging administrative and contractual matters
- Agreement on the way forward to prepare the training materials.

At the end of the ToT the trainees were equipped with enough skills and materials to be able to develop training materials for the MLM. Approaches and methodologies, as well as program and some content of the MLM training, were prepared. A large majority of the participants agreed on the usefulness and relevance of the course in the preparation of the MLM training. Evaluation of some French-speaking trainees was difficult as expression in English during the ToT was limited.

11.3.4   TRAINING OF THE TARGET GROUPS

11.3.4.1   General

The training for the MLM was arranged at the local level due to the large number of participants (15 from each city), and it was decided that the training should be given in English in English-speaking countries and in French in French-speaking countries. Training centres were identified in each of the 7 cities where the two cycles of 3-week training seminars would be conducted. Local trainers, who participated in the ToT, conducted the trainings. Local trainers developed the training materials for the MLM after the ToT, based on basic materials made available during the ToT. The MLM training was given by the local trainers with full supervision of the resource centre's staff and with partial supervision by UNESCO-IHE.

The target groups of SLM and TLM in each city were too small (3 to 5) to justify local level training for each city, so training for the SLM was carried out at the regional level, and training of the TLM in Delft, the Netherlands, gathering delegates from all cities. This approach had the added advantage of allowing the exchange of views and experiences between the various cities. The regional centres and UNESCO-IHE jointly prepared the training materials for the SLM and TLM, and the courses were conducted by relevant staff members from the three organisations.

The training workshops were carried out as planned, in two cycles. The first cycle trainings were designed to develop action plans by each of the participants in the MLM, SLM and TLM. The implementation of the action plans (at the MLM level) were to be monitored by the local trainers and communicated to the regional trainers and the monitoring of the SLM and TLM action plans by the resource centres.

An assessment workshop was conducted in March 2005, in preparation for the second cycle of the MLM, SLM and TLM training. During the workshop, an evaluation of the first cycle training was carried out. Evaluating the experiences gained at all levels provided a valuable platform for discussion and preparation of the second cycle of training workshops. The second cycle of training was then designed and organized with all the feedback and output from the first track.

In the second cycle of training, participants came up with revised action plans. However, the implementation of these action plans was beyond the scope of the project.

11.3.4.2   Training for the MLM

Two series of MLM courses were held in each of the demonstration cities: the first in the period September/October 2004 (depending on the city) for three weeks, and the second series in April 2005 for one week, simultaneously. The key facilitators for both cycles of the training course were the local trainers who had attended the ToT course in June 2004 supported by trainers from the regional centres NETWAS and CEFOC, and UNESCO-IHE. Table 11.1 below displays the participants of the different trainings.

At the end of the first cycle of training, it was possible to understand and apply:

*   WDM, PPC and PA theories and practice in the WAC context,
*   The ins and outs of the WAC programme,
*   Overall logical framework of utility and logical frameworks of departments and sections

Table 11.1 Number of participants in the MLM training from the participating cities.

City	Cycle 1	Cycle 2	Distribution of Participants
Nairobi	14	13	Nairobi Water and Sewerage Company; Nairobi City Council Environment Department; Ministry of Water and Irrigation
Lusaka	12	11	Lusaka Water and Sewerage Company; Environmental Council of Zambia
Addis Ababa	15	15	Addis Ababa Water and Sewerage Authority
Abidjan	9	11	Ministère de la Construction et de l'Urbanisme, Ministère d'Etat, Ministère de l'Environnement, Ministère d'Etat Ministère des Infrastructure Economiques, Société de Distribution d'Eau de la Côte d'Ivoire
Accra	12	10	Ghana Water Company Limited
Dakar	11	8	Office National de l'Assainissement du Sénégal, Sénégalaise des Eaux, Société Nationale des Eaux du Sénégal, Direction de Gestion et de Planification des Ressources en Eaux du Sénégal
Total	73	68	

- Action plans and project schedules
- Monitoring framework
- Overview of resource constraints, including shortcomings in policies, regulations and instructions
- SWOT and Log-frame techniques

Action plans that addressed the key issues of concern in each of the cities were developed and implemented by the trainees in between cycle 1 and cycle 2 of the training programme. These action plans were reviewed and further refined in cycle 2 of the training programme. Implementation enabled participants to:

- sharpen focus on their day-to-day activities,
- clearly define individual contributions to action plan achievement,
- identify the driving factors and limitations for their achievements,
- identify the required support for attaining them.

The second cycle was conducted for one week in all of the participating cities (Abidjan was held together with Dakar in Senegal, due to the unstable political situation in the Ivory Coast). The course aimed to enhance the capacity of the MLM to implement and manage their activities better, to review progress made, to identify achievements and constraints, to share lessons learned and to determine the way forward. The second cycle training was completed with:

- Mainstreamed personal plans in specific, measurable, attainable, relevant and time-bound (SMART) terms,
- Clearly defined messages for senior level managers,
- Clearly defined messages for UN-HABITAT.

11.3.4.3 Training for the SLM

Two training courses were carried out for the SLMs from the various participating cities. The first cycle of training was organised and hosted by NETWAS International in Nairobi, in November 2004. The second one was organised and hosted by CEFOC in Ouagadougou, Burkina Faso in June 2005. Facilitation of both courses was done jointly between NETWAS, CEFOC and UNESCO-IHE.

About 20 participants from all the participating cities successfully completed both phases of the training activity. The key outputs of SLM training were increased knowledge of participants on specific content issues and a design of action plans (first workshop), a review and further refinement of the action plan and acquisition of knowledge in terms of personal management skills, etc. The action plans were made more specific by assessing the favourable factors for and limitations of their achievements. This training also reviewed and refined key messages from the MLM and defined their own messages to the TLM for support.

11.3.4.4 Training for the TLM

The first and second cycles of training workshops for the TLM were hosted by UNESCO-IHE and they were carried out in Delft, the Netherlands. The first one was conducted between 22 and 25 November 2004, and the second one was between 27 June and 1 July 2005. The course was facilitated jointly by UNESCO-IHE, NETWAS and CEFOC. About 18 participants attended the second cycle training. In the first cycle, there were only 13 trainees (Abidjan was not represented due to travel difficulties from Cote D'Ivoire).

The course aimed at enhancing the capacity of the top WAC city managers and decision-makers in integrated urban water resources management with particular emphasis on the key challenges in WDM, PPC and PA from the top level. The course particularly aimed at bringing the participants together to exchange experiences and ideas, as well as to make firm commitments in the implementation of action plans developed by the MLM and reviewed by the SLM.

During the first cycle of training, there were various activities including brainstorming sessions, which used various means such as metaplan cards, group discussions, presentations by various speakers, video-conferencing with an African-based water utility company (SONES, Senegal), as well as field visits to a water utility in the Netherlands (Amsterdam). In the second cycle of training, field trips were made to WMD (another water supply company) and the municipality of Utrecht. Participants also developed their own action plans that they would implement upon return to their workstations. These were the major outputs from the TLM training.

Other key outcomes of this workshop were:

- Participants shared their experiences, the key challenges they were facing, as well as key strategies to address the challenges
- Participants appreciated the key challenges in WDM, pollution prevention and control, as well as public awareness
- Participants also learnt some key strategies on how to deal with policy, regulations and reforms
- Participants understood the integration of and relationships between some key themes as related to the urban water resources management, for example water and poverty, and the Millennium Development Goals (MDGs), among others
- Participants also developed their own key action points that they committed to implement, as well as prioritised key actions from their city action plans for implementation within the next six months.

11.3.5 DEVELOPMENT AND IMPLEMENTATION OF ACTION PLANS

A number of action plans were developed and carried out in the different cities. A number of encouraging results have been noted, and some of the more important ones are discussed below. All the cities reported that the increased levels of awareness on water demand management are now yielding fruit in the reduction of UFW. In Addis Ababa and Nairobi, massive awareness campaigns were launched and these were coupled with improved revenue and billing systems. This resulted in an increased number of customers paying their bill. As part of the Value-Based Water Education Programme, the Addis Ababa Water and Sewerage Authority has embarked on construction of water education classrooms that are about 80% complete. Efforts in water management have included reforestation programmes (for example, in Addis Ababa over 20,000 trees have been planted) while also working with partners to improve catchment management. Sample studies on pollution from industries were carried out in Addis Ababa and in Accra, solid and wastewater disposal sites were identified.

In an effort to address the pro-poor issues, various efforts were put in place and, as a result, there is now an increased supply of water to the informal settlements in Nairobi and Lusaka. In Accra, the process has helped to establish the multi-stakeholder Densu River Board. The Public Private Partnerships

(PPP) efforts in Lusaka have been identified as a resource for strengthening the work of the WAC initiative.

There were a number of factors that contributed to the success of the action plans. These included:

- High commitment of staff who are also qualified and competent
- Regular evaluation of the action plan implementation, e.g. in Addis
- Interest at the various levels in the WAC programme in general
- Success of the WAC pilot studies related to the action plans
- Achievement of results from long-term planning (beyond the duration of the project phase)
- Consistency in implementation of the plan as part of a mainstream company work plan
- Improved interaction between the company & community, e.g. LWSC has opened an office in an informal compound
- Benchmarks set by regulators on reducing UFW as a prerequisite for the water company to raise tariffs – motivator
- Better communication channels between MLM & SLM on issues pertaining to the program.

11.4 LESSONS LEARNED AND SUGGESTIONS FOR THE FUTURE

There are clearly two kinds of lessons learned in the TCB component of the WAC programme: one related to the content of the programme itself and the other one related to the process.

11.4.1 LESSONS LEARNED ABOUT THE CONTENT

11.4.1.1   Content of the training workshops

As the training programme was implemented, some adaptation from the initial TCB methodology was required to fit the local needs and requirements of the WAC city managers and the participants, while still staying within the defined framework.

Cycle 1
The participants evaluated the training component of the WAC I programme in a very positive way. WDM, PPC and PA were considered very relevant topics for training. There was a good balance between these topics, the exchange of experiences between the participants from one city (at MLM level) and between the cities (at SLM & TLM levels) and the planning process. The planning process was found to be very interesting and useful. Suggestions were made to train more people from their own and other organisations, creating a

critical mass of people aware of and knowledgeable about WDM, PPC and PA topics. Creating contacts between organisations in one city is seen as very valuable, not only for the WAC city activities, but also for cooperation in general.

Inter-cycle period
As expected, participants faced difficulty implementing the action plans. In some organisations, a higher managerial level (or even boards, with respect to expenditures) did not approve the action plans. Organisations have their own annual plans in which MLM action plans should have been taken into account and within the proposed six months. The MLM action plans should be better adapted to provide technical and management guidance. For their success, MLM action plans should be incorporated in the long-term plans. As stated, lack of communication, delegation, and non-availability of time for carrying out the plans were expected to occur and are fundamental experiences in the learning process of the participants.

Cycle 2
Suggestions were made to have more time at the MLM level to increase the exchange of experiences (also from other cities) and to increase a better understanding of issues (WDM, PPC and PA) at the policy level. Comments were made that the work plans of MLM, SLM and TLM were far from optimal. As the programme is not mainstreamed and as the available budgets are often limited, the programme did not produce enough results, as it did not get enough priority from already over-loaded staff members.

Even then, the concept of 2 cycles of training was very much appreciated by the participants that joined both trainings.

11.4.1.2 Target groups

The training programme was considered by participants as very positive, in terms of human resources development. The programme increased the motivation of staff members at all levels, reinforced the linkage between MLM, SLM and TLM, and improved communication between peers and between hierarchical levels intra- and internationally. The second cycle of workshops, in particular at SLM and TLM levels, created a very trusting environment, allowing the participants to share experiences in a very participatory and open way. Informal networks were developed that created a critical mass to tackle issues.

In some cases the WAC city manager was expected to play a more prominent role and was missed in promoting the WAC programme. There were, in some cases, problems in selecting the participants, and in others the selection happened at the last moment due to lack of communication. The process of selecting the participants was not always optimal as a result of existing gaps in the MLM, SLM, and TLM chain. For example, in one city, staff from one organisation was selected for the TLM and MLM level but not for the SLM level. In the first cycle, some people were not able to come and were replaced by other staff members. This was due to a lack of understanding of the concept of the training. Furthermore, in some countries, reorganisation and staff transfer contributed significantly to the discontinuity between cycles 1 and 2, in particular when people were replaced and there was no transfer of information about the WAC programme.

Nevertheless, one of the results of the trainings is that motivation of the MLM has been increased and that motivation of the SLM and TLM has been reinforced.

11.4.1.3   Communication

The cascade model for training was identified as a good model for training development by the participants. Nevertheless, it became obvious during the training that strong communication was needed for such a model to give optimal results.

The main comment expressed by the participants was that when they had communicated with the higher hierarchical level, timely feedback should have been provided (on MLM training and plans by the SLM, and on SLM training and plans by the TLM). Moreover, the lower levels indicated that they were not well informed by the higher levels. In many cases, the roles of the WAC city managers were not clear for the participants and, even in some cases, for the WAC city managers themselves.

As an important lack of communication between hierarchical levels, intra- and international, was brought to light, the second cycle of training provided a clear focus on this issue, giving tools for improving communication and dialoguing between MLM, SLM, and TLM for organising and developing communication systems to remove any communication gap. Each participant was encouraged to identify the most suitable skills to be used within his/her organisation and culture.

Lateral communication between organisations at one given level (MLM) and between cities (SLM and/or TLM) has also been improved through the training, giving the participants the opportunity to reflect on other experiences and ways of acting. The groups of trainees at MLM, SLM and TLM levels can be considered informal knowledge networks.

In general, the TCB programme has shown that many participants had unfulfilled expectations, most of the time because the expectations were not shared with others. This tendency was discussed during the TLM 02, opening good possibilities to improve communication for WAC II.

11.4.1.4   Financial aspects

In most cases, the trainees did not develop plans that fit into the annual plans of their own organisations. Mainstreaming of the planned activities was the major bottleneck for successful implementation. The most successful city was

Dakar, where all activities planned were developed in the broader context of the ongoing World Bank Project. The financial support of the WAC programme was considered seed money to develop innovative approaches (pilot projects). The conceptual idea was that pilot projects could lead to full-scale actions with financial support from other sources. In the other cities, the mainstreaming of WAC activities was less successful. In Abidjan, the positive results are seen to be related to the relative isolation of the country in the last months, allowing staff of different organisations within the city to really carry out the activities with the resources locally available and to have more time for communication with peers (no external "disturbance" from many visiting donors, consultants, etc).

MLM and some SLM trainees indicated that they had lobbied for increased budgetary and resource support during the entire programme so that their plans could be developed. The request was to their superiors to allocate the required resources (time, money and materials) for achieving the action plans. But in most cases, they did not succeed: their activities were not mainstreamed into the planned activities of their departments or were not provided in time (for example, budgets for 2005 plans were already fixed in fall 2004).

11.4.1.5   Timing

The timing of the training was a very important factor. The initial approach was to have some pilot studies already running to serve as material for the training. During the process, and due to delays, in most cases the training was carried out after the pilot was completed. It is much better for the learning approach to have the action plans developed when the pilot is starting, to combine both WAC activities and create synergies.

11.4.2   LESSONS LEARNED ABOUT THE PROCESS

11.4.2.1   Planning/Training schedule

Some delays in the implementation were experienced, such as the initial delay in contracting the resource centres. To avoid more delays in the completion of the programme, UN-HABITAT requested that UNESCO-IHE review the overall planning of the TCB programme to see what was possible, in order to complete the programme as quickly as possible. UNESCO-IHE reviewed the planning in such a way that the TCB component could be carried out as quickly as possible without too much loss of quality. In particular, the following changes were recommended and adopted:

- The period for MLM training was shortened.
- The MLM trainings took place at several locations at the same period and not sequentially (instead of one city after the other) as initially planned.
- A shorter feedback period between the trainings of the different levels (between MLM and SLM and between SLM and TLM) was used.

- For financial reasons, UN-HABITAT decided not to make use of video-conferencing for de-briefing the trainers after the different trainings.
- The limited time period for classic feedback turned out not to be optimal due to technical communication problems between the different partners (mainly Internet problems).

From the moment that the ToT was given, the methodology was followed as initially planned, allowing the completion of the two cycles of training within 1 year (initially 1.5 years).

The duration of three weeks for the MLM 01 was experienced as too long. The content of the programme was interesting but the participants found that they were out of office too long. It was suggested to split the training MLM 01 into two parts. Another recommendation made at the MLM level was to have the training given outside the city to prevent participants from being disturbed by their daily work, since the training was given at a location too close to their office.

11.4.2.2   Staff involvement

UNESCO-IHE

UNESCO-IHE required more time inputs than originally budgeted for. The proposal for the selection of the regional resource centres and national training centres was reviewed and commented on several times, each time by a new project leader. This delayed the final selection of the resources.

Further, the quality control problems of contractual arrangements with the resource centres, UNESCO-IHE took the initiative to enhance its own control over the total quality of the TCB programme delivered by the regional resource centres and local trainers.

Due to the limited capacity available at the local level, in terms of trainers in the relevant fields (WDM, PPC, PA), UNESCO-IHE decided to reinforce their team involved in the training of trainers (ToT). More staff were made available for the ToT to increase the interaction between the UNESCO-IHE trainers and the trainees (future trainers of the MLM).

As the MLM trainings were carried out in parallel and not sequentially, and as the feedback period between the trainings was shortened, UNESCO-IHE took some measures to increase the quality of the training. Initially, one UNESCO-IHE staff member was supposed to travel to only one city for three weeks and to support the staff of the resource centres during the training, if needed. Then, the staff at the resource centres would be able to replicate to other cities. In fact, after reviewing the plan, each MLM had been visited for one week by a UNESCO-IHE staff member (total 6-week mission) as the resource centres was not able to have the same staff visiting all cities. This was compensated for by shorter MLM trainings in cycle 2.

Last, but not least, UNESCO-IHE contributed to two workshops with UN-HABITAT and the regional resource centres (before each cycle of training): one was initially planned, the other one was considered part of the general coordination of the programme.

Resource centres
Two staff from each resource centre were involved in the ToT. But the follow-up of the MLM and SLM involved (in Western Africa) many more people, some of them being consultants hired to fill the role of supervisor without appropriate preparation. In some cases, the fact that some staff members from CEFOC were not fluent in English created problems.

Local trainers
Not all local trainers had the same level of quality. In general, the teachers were experts rather than teachers. For this reason, more on didactics and moderation techniques should have been given at the ToT training. As local trainers were hired by the resource centres on a short-term basis to do the job, the level of commitment for continuity was non-existent. For this reason, it is impossible to talk about capacity building at the local level as long as the trainers are not aware of some continuity of work with other partners, based on a trusting relationship.

11.4.2.3   Contractual aspects

For the implementation of the TCB component, the chosen contractual model appears to be complex and not fully satisfying. UN-HABITAT has a contract with UNESCO-IHE as the international capacity-building institute and a separate contract for both regional resource centres. Those centres were given the responsibility to subcontract local training centres. This contractual model has major limitations, as stated below:

UNESCO-IHE had no contractual power to influence the way to work with the regional resource centres (NETWAS, CEFOC). The contractual arrangement which was used could have worked nicely if the contract had indicated clearly the roles and duties of all partners and had specified clearly the delegations. In particular, there was no contractual relationship between UNESCO-IHE and the regional resource centres (NETWAS and CEFOC). UNESCO-IHE was to be responsible for the overall quality control of the programme, yet without a contractual relationship with NETWAS and CEFOC requiring them to follow the requirements from and report to UNESCO-IHE, this was impracticable. UNESCO-IHE was not able to carry out quality control as expected by UN-HABITAT as the training materials were not made available following the procedure proposed by UNESCO-IHE.

Due to financial constraints imposed on the regional resource centres, the expenses for the local training centres were significantly limited. Freelance consultants and professionals were hired as trainers. These trainers, even when qualified, had no long-term relationship with the regional resource centres. For this reason, the TCB of local training centres was not possible. The conditions of the contracts for the local trainers and the contracting procedure applied by CEFOC and NETWAS implied that it was not possible to have adequate training materials designed as proposed by UNESCO-IHE. Trainers were contracted too late with insufficient time made available for the development of materials. A follow-up of the plan in the period between cycle 1 and 2 was very limited (for NETWAS trainers) to non-existent (for CEFOC trainers).

In general, the training materials used for the MLM 01 training was limited to the materials made available by UNESCO-IHE during the ToT. Adaptation and implementation to fit the local needs were not sufficient. For the second cycle of MLM training, some of the training materials were developed by NETWAS, the rest were made available by UNESCO-IHE. For both cycles of the SLM and TLM, the materials were mainly developed by UNESCO-IHE with some specific input mainly from NETWAS and a very limited input from CEFOC.

The effects of the language issue were underestimated in the design of the programme. The official language of the WAC programme was English. But two cities are French-speaking, three are English-speaking and one Portuguese- and English-speaking. This issue was not really a problem at the MLM level, since French- and English-speaking trainers were available for their respective trainees. But at the SLM and TLM levels, simultaneous translation was deemed necessary. The materials were supposed to have been made available in French and English at the MLM level (which was done partially) and in English at the SLM and TLM levels. The ToT was conducted in English and materials were prepared in that language. Several comments were made by participants about the non-availability of French training materials at ToT, SLM and TLM levels.

For the future, the main recommendation is to make the roles and duties of the different partners clear in the contract with specific, measurable results. This could be done in several ways, the easiest being for UN-HABITAT to contract a reliable consortium of international, regional and local training organisations to carry out the work. With this approach UN-HABITAT leaves the responsibility of administrative matters to the international TCB institute with a clear schedule of control points for itself. Another alternative is for UN-HABITAT to contract independently all partners (international, regional and local centres) with clear contractual arrangements between the partners, in terms of responsibilities. In the case of the WAC I programme, this approach would have led to nine contracts (UNESCO-IHE, NETWAS, CEFOC, and six local training centres). To avoid too many contracts, UN-HABITAT decided to reduce it to three contracts (UNESCO-IHE, NETWAS and CEFOC), requesting CEFOC and NETWAS to contract the local training centres.

11.4.2.4   Financial aspects

The contract between UN-HABITAT and UNESCO-IHE was signed in April 2002. All financial values were agreed on in US dollars. Most of the activities were carried out in the year 2004 and 2005. This means that the contract lost more than 20% of its value due to unfavourable currency exchange rates. These kinds of risks were not taken into account in the contract and no revision of prices was made.

The financial arrangements with the regional resource centres showed a discrepancy between what UNESCO-IHE considered was necessary for the development of *quality* training materials and delivery and what was available in the UN-HABITAT budget to pay the regional resource centres for preparing the materials. The consequence was that training materials were a collection of good PowerPoint presentations and relevant background information, but not real lecture notes as such.

11.4.3   SUGGESTIONS FOR TCB FOR FUTURE PHASES

11.4.3.1   General suggestions

The WAC programme as a whole needs much more political support than was the case at the time of implementation, and the TCB component will benefit from it. Political support was initially very high but decreased because of the delay in delivery (time span was too long to see the first results). In order to improve the outreach of the TCB component, it needs to be much better embedded in the actions planned by the WAC programme. Further, it is very important to implement the programme in the designed timeframe. Commitment from all partners to work along this line will make the follow-up of the TCB programme successful.

11.4.3.2   Suggestions for content

Training on relevant topics and skills
The educational backgrounds of participants were demonstrated in the varying skills of the participants, a few of them presenting very good managerial skills (from education) and others being very experienced water experts. Some participants showed weaknesses in formulating a report or presentation. The difference in managerial capacity was the most visible at the SLM level. This means that the target group is not homogeneous, and this will always be the case. For this reason it is very important to design training that is interesting for all participants.

Improved interaction between the different components of the programme
It was agreed that the WAC II programme would have a greater impact on the performance of the participants if there were greater interaction between the training and capacity-building workshops, the pilot studies within a city and between cities, etc.

Training materials
The development of training materials should be given greater emphasis in WAC II. The way this will be done needs to be reviewed as it appears unrealistic to expect that the local trainers can produce them without direct interaction with UNESCO-IHE (takes too much time, lack of experience in structuring the contents in English, insufficient supervision of resource centres, etc.). As the local trainers (who are in charge of a specific topic) in different cities are not working together, the quality of the results is inconsistent and efficiency is limited. In theory, it would save much time and improve quality when trainers develop a common basic product per topic (i.e. training manual), and afterwards each trainer could develop his/her own teaching material, based on this common training manual. But taking the general level of the local trainers into account, it should be considered that UNESCO-IHE, together with the regional resource centres, should make a major contribution towards the preparation of a standardised basic training package. The regional and local trainers can enrich the basic package with, for example, local case studies that could be exchanged between the cities.

11.4.3.3   Suggestions for process

Training form
It is proposed that the cascade system of training (MLM-SLM-TLM) be replicated. The training in WAC II should focus on selected topics (pro-poor governance, etc), on the development of action plans to continue the work done in WAC I, and on personal management skills. More consultations between MLM/SLM/TLM should take place in order to synchronize the action plans. Recommendations were made to organise a city workshop with MLM, SLM, TLM and WAC city managers at the start of the WAC II, to define the strategy of the city or, when it exists, to identify priorities to work on. This will allow a better coordination of activities within the city. It was also recommended that UN-HABITAT and UNESCO-IHE are represented. The role of the city manager in the TCB process should be defined in a better way. The city manager should be the coordinator and provide a central point to collect and share the information given by WAC.

Knowledge exchange
Another important point is the opportunity to increase the knowledge gained through the WAC programme and share it with others by exchanging experiences between the different cities, UNESCO-IHE, UN-HABITAT and the regional centres. It is very important that pilot studies and activities developed in the WAC II are well documented to be used as a base for training materials.

Exchanges between cities should be organised (e.g. study tours) to improve dissemination of knowledge. Furthermore, participants of the WAC I programme could be involved as resource people in the training of WAC II cities, where needed.

Make better use of existing materials of the WAC programme
Pilot studies (of WAC programme and others) should be used for taking better advantage of what has been done. In the same way, some experiences from others who have already implemented the programme could be added.

Translation
For the first cycle of SLM training, translation was insufficient. Translation at SLM and TLM levels was not planned and was added (at the last minute, in the case of the first SLM). Some participants with excellent English further assisted the simultaneous translation. For the second cycle of SLM training, simultaneous translation was very useful. For the TLM trainings (workshops in Delft), translation was available and useful but turned out to be very expensive.

Recommendations were also made to ensure the availability of training materials in French and in English for all levels of trainings (MLM, SLM, TLM) and not only for the MLM, as initially planned.

Cooperation in training chain
The cooperation between UNESCO-IHE, the resource centres, UN-HABITAT and the WAC city managers was a key issue in the programme. Participants had a lot of expectations that were not expressed at all or not communicated to the relevant organisation. A clearer definition of roles, tasks and duties will be needed in the future, maintaining as much flexibility as possible for a high level of efficiency.

Improved press coverage
As stated earlier, the WAC programme as a whole needs much more political support than is the case now. For this reason, press coverage has to be improved. Prior to this, it was very important to define a clear message, depending on the target group.

Improved gender balance
Unfortunately, gender aspects were not a priority in the first phase of WAC project. In the second cycle of the training, some time was dedicated for this issue but mainly limited to traditional approaches of gender awareness. This was very appropriate to help the participants understand that gender is a major issue in their own action plans. But at the level of participation, it was obvious that too few female participants were involved in the training. In the future, gender should also be taken into account, such as how to promote more women becoming active in the water sector.

Target group
In future rounds of training, it would be very valuable to include the representatives of finance and administration along with the technical staff. In principle, both should come from the same company. A consistent approach should be applied in selecting the participants, for example target one organisation/ company and pass through all the structures. It can also be acceptable to mix

technical people from organisation A and financial people from organisation B. Mixing backgrounds is very enriching for discussion, as well as gathering peers from different organisations.

It is obvious that, for one's own personal development, the staff involved in the first phase of WAC training was interested in being involved in the second phase of WAC. Continued training of these people will increase the sustainability of the actions of the WAC programme. At the same time it is very important that additional people are trained to create a critical mass for achieving results in different cities.

Local trainers
In general, the local trainers were experts rather than teachers. More on didactics and moderation techniques should be given in the next round of ToT. To really build capacity of regional centres, it is fundamental that local trainers be involved in the activities of the resource centres in a more structured way, to ensure sustainability of the efforts. Hiring staff is not the way to get continuity between the trainings and the dissemination of knowledge.

WAC project management
Both regional centres had difficulties carrying out the programme. Major reasons were limited human resource capacity at the resource centres, limited capacity to pre-finance activities, and limited capacity to invest in local trainers in a structured way. As the WAC programme is not giving a guarantee of full-time work for several months, the resource centres were obviously not willing to commit themselves to the long term, by recruiting the staff.

Fixed team of trainers
In the first phase of WAC, enough experience was gathered and some trainers appeared to be better than the others. It was recommended to compile a standard Anglo-French team to support the capacity-building activities. Creating continuity of the trainers between the MLM-SLM-TLM levels was fundamental to improving communication between and across the three levels. These core trainers will develop the training materials for all the groups.

Improved commitment of participants
A suggestion has been made to increase the commitment of the participants by asking them to sign a kind of training contract in which it is clearly stated which indicators will be used to evaluate the performance of the training.

Platform
Making use of an interactive online platform will be a useful tool to support the programme's activities. Such a platform will centralise the available training materials and all kinds of background documents, and will improve the exchange of information between the cities. The platform will be useful for the participants of the training and also as a source of information for external participants (not involved in the WAC programme). This will improve the dissemination of

knowledge and experiences. The platform is also very useful for the programme management, by creating another community of users (gathering UN-HABITAT and the resource centres, the trainers, the WAC city managers, etc.).

Tele- and video conferencing
Telephone and video conferences could be considered as ways to facilitate communication between the cities to avoid travelling. It was obvious, for example, that some TLM participants had to travel often out of their city/country for project work, donor conferences, upon ministerial request, etc., and it is not possible to assemble all TLMs at any given time for a short feedback (for meetings less than two days). Making use of telephone and video conferences at some crucial points in the planning would improve communication, with higher flexibility during the follow-up of the planning process in between the trainings.

11.5   CONCLUSIONS

Achieving the MDGs in water and sanitation will require improved capacity at both national and local level in many aspects of water and sanitation management. The WAC-TCB project clearly shows how multi-level capacity-building initiatives can contribute greatly to the improved delivery of services. The multi-level interlinked programme enables organisation-wide objectives to be reinforced, which builds synergy and improves the longer-term sustainability.

The choice of technical components on water demand management and catchment protection, linked with information and advocacy, was recognized by all stakeholders as efficient, effective and targeted. The topics could be expanded to include components on improved revenue collection and business development to further improve sustainability, particularly of the smaller utilities. The focus of the WAC-TCB was on utilities. Future components could also focus on other key stakeholders in the sector including government policy-makers and those who allocate funding from related sectors, and community-based and non-governmental organisations.

The capacity of locally-based institutions is still weak, but this is because there is little sharing of locally-developed ideas and approaches rather than a lack of technical capacity. The implementation of such capacity-building initiatives still needs more focused investment in follow-up actions plans if it is to achieve a major impact.

The vast majority of the training material and curricula benefited from local knowledge and solutions; without the participation of local institutions in the development of training materials, there would have been a reduced impact.

The identification of local institutions which are strong enough to deliver well-structured training is greatly lacking. Although highly-skilled experts and knowledge are available locally, there is little incentive to promote "centres of excellence". Many experts would rather focus their efforts on lucrative personal contacts. It is, indeed, extremely important that promoting centre of excellence would have a dramatic effect on water and sanitation-sector training.

In order to improve the impact in WAC, it would have been better to more closely tie the capacity building to investment packages. That way, the long-term aims and aspirations of personal action planning approaches could be could optimised.

In terms of overall coordination and next steps, more efficient contractual processes need to be developed. Quality assurance of local training institutions cannot be ignored, especially in terms of multi-source funding arrangements.

A future phase of the WAC-TCB is currently being considered by UN-HABITAT. In the meantime, there has been some evidence that the local capacity-building centres have sought to re-run the training programmes, however, no details were available at press time.

Overall, the final conclusion is that there is an urgent need to follow initiatives such as WAC-TCB to support more focused investments and to identify and promote locally developed solutions to local problems. There is also considerable scope for the twinning of utilities and sharing tools and approaches more readily.

At the outset of the programme in the early 90s, the use of web-based training was in its infancy. At the present time, the opportunities for web-based training and e-learning present new and exciting areas for future consideration.

Chapter 12

WATER EDUCATION:
Bridging divides for future generations

John E. Etgen
Project WET Foundation

Teddy Tindamanyire
Uganda Ministry of Water and Environment

Dennis L. Nelson
Project WET Foundation

Amy C. Fuller
University of Pennsylvania

ABSTRACT

The question: "How can youth and educators in schools and communities help address global water issues in local settings?" is addressed in depth from a regional and global perspective. These audiences play a key role in every community and are often overlooked in their ability to affect change and implement solutions to water issues. Educational systems are as diverse as watersheds – basic principles apply to all, while each has its own unique identity and characteristics. Survey results demonstrate that water education is not applied consistently across the globe and within countries, and that there are numerous barriers to water education. Successful methods of surmounting these barriers come from case studies including Project WET's global reach, Uganda's comprehensive approach, and discussions of other water education programs. The most successful programs combine the scientific and technical expertise of water professionals with formal and non-formal educators. Collaborative development and distribution channels can help scale-up the dissemination of materials, which can lead from awareness to action that creates positive changes and ultimately solutions to water issues. The water and education sectors have a unique opportunity to design, develop and deliver water education materials and training to countless people through new and innovative systems. Formal and non-formal educators are an underused resource and high-quality, hands-on educational materials and the networks to deliver them are the key to water education being a stronger tool in the water manager's portfolio.

12.1   WATER EDUCATION AS A KEY WATER MANAGEMENT TOOL

Over the past quarter-century, global water problems have continued to escalate. Every day, the health and well-being of the 6.72 billion people on our planet is affected by the availability of clean water resources. To face these challenges, governments, international and non-governmental organizations, private institutions and many communities are making efforts to build capacities in multiple water-emerging issues. Nevertheless, water education often remains a forgotten priority. There have been many disperse, isolated efforts to strengthen environmental and even water education, but few have had success or continuity.

To meet the water challenges of the 21st century, educators, students, business professionals, political leaders and others will require an understanding of water resources. Through water education, learners identify their watershed address, discover their role in the hydrological cycle, and recognize that water flows through and connects us all. Through water education, students, teachers, parents, business and community leaders are empowered to take action in their local communities; these ActionEducation™ projects contribute to a healthier local environment and economy. On a worldwide scale, water education helps learners recognize the relationship between the availability of clean water and global stability.

Figure 12.1 A student at the Children's World Water Forum (held in conjunction with the 5th World Water Forum) in Istanbul addresses the audience about his past week and what he learned about water.

Box 12.1 What is ActionEducation™?

ActionEducation™ is defined as education that empowers learners to take positive and appropriate actions to solve a local water resource issue. AE was initiated by the Project WET Foundation after being asked by a major philanthropic foundation what a water education program like Project WET can do to reduce the number of deaths caused by waterborne diseases. Through the ActionEducation™ initiative, Project WET has evolved from learner awareness to empowering students to take action leading to sustainable solutions for community water resource issues.

This initiative is strengthened by the emphasis of the global water community on local actions highlighted at the Fourth World Water Forum and a key emphasis of the Fifth World Water Forum, especially in the Global Water Education Village.

Box 12.2 Why do we need water education?

Escalating global water problems
Global water problems continue to intensify and affect the quality of life for billions of people.

An elastic global economy
The global economy of the 21st century will grow or stall based on banking, industry and commercial professionals and others understanding sustainable water management.

Disproportionate water distribution
Because of inequitable distribution of water and increased demand, the expansion of existing technologies, as well as innovations still to be imagined, it will be necessary to ensure clean water for the predicted population of eight billion in 2025.

Transboundary water conflicts
International stability depends upon nations protecting and sharing transboundary water resources, both above and below the ground's surface.

Figure 12.2 Ugandan students participate in Project WET's Blue Planet activity, learning what percentage of Earth is covered by water.

12.2 CHALLENGES AND BARRIERS TO EFFECTIVE WATER EDUCATION

Considering the importance of water education for students and community leaders alike, it is critical for water professionals to be aware of the current state of educational programs. Project WET created a survey to address the lack of information surrounding the prevalence of water education in communities across the globe. The following sections are challenges and barriers to water education as cited by local water officials, teachers and community leaders.

Funding

The current global recession has tightened most state budgets. In the United States, for instance, state deficits have affected teacher salaries and access to materials. Countries in the European Union have also suffered from the troubled economy, with Irish schools increasing class size to cut back on the number of salaried teachers. Nearly half the water education survey respondents listed funding as a major barrier to providing water education in schools. Monetary resources went to teaching required subjects, leaving little for additional topics such as water conservation. Materials could not be purchased unless distributed by the government, and there was no money for teacher training or field trips.

Time

Despite the importance of water education programs and materials, survey respondents revealed a lack of cohesiveness for water curricula around the world. Water education has been determined a voluntary subject behind such necessities as math and reading. The 2002 No Child Left Behind Act (NCLB) is a United States of America law based on "stronger accountability for results, more freedom for states and communities, proven education methods, and more choices for parents" (Department of Education, USA, 2001). Though NCLB has the potential to improve student achievement by setting standards for educational programs in schools nationwide, it is currently under review by Congress. Opponents have commented that NCLB's standards have created an incentive for teachers to "teach to the test," only covering the material that will appear on the standardized tests by which the school systems are evaluated. Topics such as water education are often not covered because they do not fall under the standard curriculum. In fact, when asked why water education is not taught in schools, many educators cite time constraints in the classroom as the primary factor. Though many curricula include some water education in science programs, there is not enough time to cover the material in depth when other standardized test topics are required by the state.

Lack of knowledge of water issues

In developed countries such as the United States, water education focuses on conservation of the natural resource. Water professionals have cited ignorance of water issues as one of the greatest barriers to water education: how can the public be educated about a problem if they do not believe one exists? However, as evidenced by expanding drought conditions in Georgia, South

Carolina, Texas and California, water conservation is imperative (Edwards, 2009). Globally, the United Nation's Commission on Sustainable Development (CSD) has stated that eight percent of the world's population lives in an area with high water stress and twenty-five percent with moderate stress. CSD warns that two-thirds of the population will live in a country with moderate water stress by 2050 if current usage trends continue. A lack of required curricula (as mentioned with NCLB) increases ignorance of the need for water conservation: omission of water education teaches students that the problem isn't important, while a quick overview of the topic does not properly educate students about what action needs to be taken. The introduction of required curricula into study plans could solve this problem.

School attendance
While water education faces curriculum and time-constraint barriers in developed nations, developing countries also suffer from the burden of not having children attend school at all. In some nations, children help support the family instead of attending school. Even if work conflicts don't occur, parents of large families may not be able to afford the school fees for all their kin. This situation becomes graver when considering the effect of civil unrest. In Zimbabwe, for instance, the United Nations believes that school attendance rates have fallen from ninety to twenty percent. This problem has been exacerbated by the August 2008 cholera outbreak, the increased cost of necessities from the global food crisis, and a backlash from economic recession.

Lack of sanitation also hinders school attendance. The World Health Organization estimates that 2.6 billion people worldwide lack access to improved sanitation: 75 percent of this population resides in Asia, 18 percent in Africa, and five percent in Latin America (UNICEF, 2006). Girls are particularly hindered by the lack of sanitation services, unable to attend school during their menstrual cycles or if there are no toilets separate from the boys' facilities. Furthermore, the absence of water and sanitation services forces girls to collect water from local sources, often walking miles per day, which severely impacts the time allotted for their daily education.

Literacy
The World Bank (2005) estimates that half of the children who do not attend school globally do not speak the language of their local school. For instance, in Togo (the Togolese Republic) the official language is French, though three other major languages are also spoken throughout the country: Ewe, Mina, and Kabye. Water professionals in the country have cited language as a barrier to education in schools. Materials given to schools may be in French while teachers lecture in a local language, or teachers may even be unable to communicate in the native tongue of their students. The Save the Children Fund has promoted mother tongue-based multilingual education (MT-based MLE)

to help address this problem. Under this model, education begins in the child's native language while a second language can be added as a subject (Alidou, 2006). Teachers may eventually switch entirely to the new language, but only after six years of instruction.

12.3 METHODOLOGY AND APPROACH: EMPOWERING YOUTH

Project WET's global program
To help students, their teachers, parents and community leaders meet 21st century water challenges, Project WET, an award-winning nonprofit global program, believes that water education has never been more important.

Initiated in 1984 as a local water education project in the USA, Project WET was developed by teachers, for teachers, with the guidance of water specialists and scientists. It soon became a national program in the United States and then grew internationally. Today, there are Project WET Coordinators in all fifty US states, and in more than forty countries. Project WET materials have been localized and translated into English, Spanish, Vietnamese, Arabic, Hungarian, Italian, French, Korean, and Japanese, with Urdu, Thai, and Chinese in progress. Over 400,000 educators have been trained in hands-on workshops with 40 million students reached.

Methods
Its curriculum and main tools include over 300 educational activities or lesson plans to be used with children, young adults, educators, and business and community leaders. Key program features include:

1 **Lesson Plans:** The cornerstone of Project WET is its methodology of teaching about water resources through hands-on, investigative, easy-to-use lesson plans. Project WET lesson plans are carefully developed and are designed to be:

   • *Interactive*: Learners participating in Project WET lesson plans are not passive observers. Engaging students through questioning and other inquiry-based strategies, educators become facilitators involving students in hands-on lessons and encouraging them to take responsibility for their own learning. For example, students design investigations to seek answers to real-world problems; play games to explore scientific concepts; reflect; debate; seek common ground to resolve conflicts, and creatively share their findings through songs, stories and dramas.
   • *Accurate*: All lesson plans are content reviewed by experts and are field-tested by educators with students.

- *Multi-sensory*: Activities engage as many of the learner's senses as possible. Research has shown that stimulation of multiple senses enhances learning.
- *Adaptable*: While adaptable for any environment, many Project WET lesson plans are ideal for outdoor settings and encourage children to be physically active.
- *Contemporary (21st Century Skills)*: Project WET lesson plans help students develop skills necessary for success in the 21st century. In most lesson plans students work in small, collaborative groups; many activities engage students in higher level thinking skills requiring them to analyze, interpret, apply learned information (including problem-solving, decision-making and planning), evaluate and present.
- *Relevant*: Information is not delivered in isolation; educators are encouraged to localize lesson plans to give them relevance.
- *Solution-oriented* (ActionEducation™): Through education, Project WET empowers students, teachers, parents and leaders to take action and find solutions to water issues in local communities.
- *Measurable*: Project WET lesson plans provide simple assessment tools to measure student learning.

2   **Interdisciplinary materials appropriate for both formal and non-formal educators:** Lesson plans can be used by teachers of different disciplines such as art, physics, language, literature, biology, history, geography, government, mathematics, chemistry, environmental science, ecology, health and so forth. The lesson plans can also be used by non-formal educators representing community organizations, government agencies, NGOs (non-government organizations), corporations, museums and parks.

Figure 12.3 Project WET Jamaica facilitators participate in Project WET's 8–4–1 one for all activity.

3   **Materials complementary to existing educational programs:** Project WET lesson plans are designed to satisfy the goals of educational programs by complementing existing curricula rather than displacing or adding more concepts.
4   **User-friendly lesson plan format:** Designed to respect the time and resources of educators, the lesson plan format was extensively tested in the development of the original Project WET Curriculum and Activity Guide.

Development of the Project WET international network
Project WET has proven to be a sound, creative, useful and accurate program to educate people on water issues regardless of country, culture or language. The rapid growth in organizations adopting Project WET as their key water education program is evidence of its success. Project WET works through strong local partners who request the program. After a series of initial communications and meetings to determine that the host institution and Project WET are aligned, key country leaders are trained in a Project WET Leadership Workshop. Project WET materials are localized and translated for use by the host institution. These local partners then set up train-the-trainer networks to reach key audiences and regions with customized water education lessons.

Adaptation process
Confirmed by educators' evaluations at many international workshops, Project WET methodology has been successfully adapted and applied. Many of the curriculum activities do not require adaptation at all; these include activities highlighting physical and chemical characteristics, water and Earth systems, water and life, and water as a natural resource. Nevertheless, some adaptation is required when activity topics are related to water management or water and social and cultural constructs. Some countries, like Canada, adopted the original **Project WET Curriculum and Activity Guide** and produced a supplement with country-specific information. Mexico translated and

Figure 12.4 Project WET's water, sanitation and hygiene materials.

adapted the **Guide** by including Mexico's history, water issues and teaching methods; in addition, three new activities and an appendix about water in Mexico were generated.

Japan translated the **Guide** and is currently adapting it. The Philippines used the original **Guide** for some years while they developed their adapted guide. For countries in Latin America and the Caribbean, the publication, **Water and Education**, adapted from the **Project WET Curriculum and Activity Guide** is available in Spanish and English and has been modified for Argentina and Costa Rica.

Summary
The mission of the Project WET Foundation is to reach children, parents, educators and communities of the world with water education. The Foundation accomplishes its mission by producing water resource publications; training educators, business and community leaders; establishing an international network of educators, water resource specialists, scientists and citizens interested in their local water resources; conducting events such as community Make a Splash with Project WET festivals or the Global Water Education Village™. Its experience is already an example of international cooperation – promoting an understanding of water resources through education, irrespective of political boundaries, to empower communities to take action to solve local water resource challenges.

12.4 UGANDA CASE STUDY: WHY YOUNG PEOPLE'S CAPACITY TO MANAGE WATER RESOURCES IS PIVOTAL

Young people have an important role in source water protection and water research based on traditional experiences and oral folklore. Individually, as members of a family and community, and as future citizens, the youth of today have the power to act as catalysts to initiate change and guide the development of future activities to ensure the health of the nation's drinking water sources. Yet, little has been done to motivate them, and their potential remains untapped.

Youth make up 70 percent of the global population and are the natural bridges between generations. They aid in relaying water education and information about water sources, thereby helping to mold future users, leaders and consumers of water throughout their communities. These bridges include a growing interest in the future of their water resources, familiarity with technology that makes information available on demand, connections to their community and a broader world through the Internet and multi-lingual experiences in addition to a thirst for making a difference. The participation of young people in water education is not only critical but also an inevitable venture that the global partners need to consider.

Education and capacity development targeting children and youth is the best strategy for protecting the waters of the world. Inspired action can be inculcated at an early age.

Education as a tool to improve water management has become a focus of water managers. Youth, as a category within the population, still have a low profile in water education and water management, yet they are the future users and managers of water. Young people need to be recognized as major stakeholders who can extend education and provide ActionEducation™ in their environments.

In Uganda, for instance, an analysis of the water and sanitation sector over the last year shows that youth are responsible for fetching water for domestic use at the household level. They are often left out of the capacity-building process because the focus is on professional expertise aimed at solving supply and water quality issues. Young water users are not given the attention they deserve and too little has been done to empower Ugandan youth.

Youth as water education facilitators
Young people need to understand water quality in a language that matches their environment and situation. Project WET's approach to water education provides educational materials that are interactive, practical, enjoyable, relevant and adaptable to situations in and out of the school environment.

The curriculum provides an integrated approach that has inspired the stewardship of water by children in schools in addition to outreach that encourages revitalization of civic life as a methodology to help youth focus on water education. This is an education that shapes the youth culture, and

Figure 12.5 Youth at a protected water source in a peri-urban setting in Masaka.

values local 'distinctiveness' and 'richness.' The methodology emphasizes education that is based on understanding the neighborhood as an important, useful and unique place.

Children in school need to be a focus of water education with the hope that they will duplicate the information and the experience not only with the school community but also with parents and peers, thereby influencing the larger community.

Background

At a glance, the hydrological overview of Uganda portrays the country as a well-endowed basin in surface water resources. The country's surface area of about 241,500 square kilometers is made up of 15 percent open water, 3 percent permanent wetlands and 9.4 percent seasonal wetlands. The open surface water bodies include lakes such as Victoria, Kyoga and Albert and rivers like the Nile, Katonga, Kagera and many others. Groundwater exists in both the fractures and weathered aquifers. The major input into the national water resources, apart from the Nile flow from the upstream countries, is the annual rainfall, which ranges between 600 mm to 1600 mm.

However, spatial and temporal distribution of rainfall is not uniform on the national level, and despite the seemingly abundant water resources of the country, their uneven distribution both in space and time poses big challenges to the water resources managers and decision-makers. There is increasing pressure on the already vulnerable water resources from factors such as rapid

Figure 12.6 Wetland ecosystem-based water source for the urban community in Masaka district.

population growth averaging three percent per annum, increased industrial activities, environmental degradation, soil erosion, drainage of wetlands and pollution of groundwater, rivers and lakes. Consequently, there is a need for strategic resource planning and management that will ensure the protection of the available water resources, and satisfy conflicting demands while ensuring sustainable water resources development. A number of stakeholders are critical in this endeavor.

Water is vital for sustaining life, promoting development and maintaining the environmental safeguard for water resources. The provision of safe water supplies and sanitation facilities, and their proper management and utilization, are necessary conditions for health and economic development and are vital for the welfare of society. The Water Sector is, therefore, one of the government's priority areas under the Poverty Action Fund (PAF).

Rural water supply and sanitation
The majority of the population in Uganda lives in the rural areas. They live in rural *communities* of up to 500 people and *rural growth centers* (RGCs) with a population of up to 5,000. The average water used per capita in Uganda is only half the recommended amount, and some 30 percent of constructed facilities do not function properly.

On the sanitation side, national household latrine coverage is estimated at 48 percent. There is, however, wide variation of coverage from district to district (with a low of four percent in Karamoja District and over 80 percent in districts in the southwest part of the country). Coverage of public latrines is also very low (19 percent), with most of these latrines located in schools, markets and health units. In terms of water supply, the number of constructed facilities does not necessarily reflect actual usage by all intended users (e.g., due to facility cleanliness and/or cultural taboos and beliefs). Presently, about

Figure 12.7 Some community members still use such muddy water for domestic consumption.

Figure 12.8 The quality of water and the surrounding environment are at stake. Children have no water alternative.

40 percent of the population lives without access to clean water while about 57 percent is without a safe place to go to the toilet. Women in Uganda often have to walk over 20 km a day to collect potentially unsafe water from rivers and muddy holes, thereby carrying disease. This is the only option open to them as, otherwise, their family would die from dehydration.

Getting Ugandan youth involved in water education
Uganda's youth are categorized as those in school and those out of school. In a country where the literacy levels is 84 percent among the girls and 88 percent among the boys, the approach of using interactive methods and easily adaptive lessons that are enjoyable and easy to grasp is ideal.

A case study of Uganda water educators reveals new experiences that are moving knowledge out of traditional classroom teaching and rhetoric-rich community presentations. The traditional education from those who know to those who don't know is moving towards interactive, practical, hands-on experiences that can be seen as a new step towards achieving wider knowledge about water issues and the discovery of multiple methods of handling water access issues, water diseases and water use in a variety of audiences. Everyone involved has a role to play in relating knowledge about water to the community.

There is a need to motivate youth to educate other youth and the community with a focus on local water problems and issues. This will encourage young people to think about potential outcomes for youth water education as well as to design relevant programs and develop skills important to water education. By giving youth opportunities to express themselves, they will be motivated to think of youth programs that are characterized by voluntary actions aimed at maintaining a healthy environment; they will

Figure 12.9 A guided youth of such status can be different if educated and mentored in water quality testing.

be motivated when they understand how their choices can improve their lives and communities, and they will become valued partners in local water problem-solving initiatives. The level of influence in the community that youth can provide has not been fully examined in the water education experience in Uganda.

Progress in capacity building for children and youth in Uganda
Uganda is among the African countries that have participated in different water management activities in line with the Millennium Development Goals and the UN Year of Water and Sanitation. An African, tailor-made curriculum is one such endeavor, in line with Sobel's principle of place-based education. This is a timely strategic move to address the unique African water issues as part of the global initiative and part of the Poverty Reduction Agendas that focus on children in school and synergies with the community.

For many years, Ugandan youth and students have been learning through the teacher-based education system. Interactive learning and hands-on activities have gained popularity and, no doubt, these will have more impact than the traditional methods for sanitation and water education in schools. The exposure of teachers in Uganda to Project WET activities has shown that teachers and students are positive about field-based, interactive learning mixed with classroom-based programs. A few monitoring exercises indicate that changes in students' attitudes and knowledge have generated behaviors to manage water and related ecosystems within their reach. It is, therefore, hoped that this approach will have an even greater impact on students, youth and other learners in advancing better sanitation and water stewardship.

Figure 12.10 Ugandan students read Project WET's kids in discovery series activity booklet: healthy water, healthy habits, healthy people.

Use of Project WET's materials, tested so far in about 12 schools, has indicated that young people can impact the future management of water. The motivation of learners resulting from their participation in enjoyable, interactive and relevant lessons is unequaled. This is the kind of learning that would help Uganda both now and in the future.

It has been fascinating to observe Envita School teachers build synergies between the schools and the community in peri-urban and urban school community environments – where community sanitation and school linkages have been strengthened through the education materials that have been pre-tested since February 2008.

The strategic issues that Uganda needs to address
Youth and children are the focus and core of water education at all levels. There is also a potential to involve the whole community in the education of children through targeted educational materials that both groups can take pride and ownership in.

Currently, Project WET is music to the ears of Ugandans, but the vision for the future is to see evidence of water education beyond classroom walls. The desire is to see youth and children do something important for themselves, for the community and for the country as a whole, as they contribute to better sanitary conditions and better health. The challenge is to see:

- students engaged in real water improvement projects at the community level.
- a water educator in every school providing direct teaching and support.
- teachers 'connected' to water and sanitation education materials, organizing interactive, relevant learning and helping to use the schools grounds and community to teach water issues in the framework of the African Curriculum (developed in February 2008).

- teams that would provide vision, guidance and support so that these new efforts result in a change that will help the local community to enjoy clean water and protect the catchments at local levels.
- seed teams to connect the community and school projects around water use, protection and sanitation improvement at the local level. Each seed team would have a teacher, youth facilitators and students whose main task would be to build connections between schools and communities so that outreach efforts emerge as a scaled-up process.

12.5 MEETING WATER CHALLENGES AROUND THE WORLD

The combined efforts of governments and NGOs have formed collaborations meant to increase water literacy both in school and throughout the greater community. In developed nations, officials have created materials and distributed them through government agencies, such as the Environmental Protection Agency in the United States. Despite a lack of a required curriculum in the country, the United States has resources for teachers who would like to introduce water education into their classrooms. NGOs assist the government by providing materials to more rural areas or serving as presenters in special classroom sessions.

Because of the hardships placed on governments in developing nations, water education often begins as a project conceived by NGOs. Collaboration usually occurs with Ministries of the Environment and Ministries of Education, but resources are often provided by aid organizations. For instance, the Southeast Asian Ministers of Education Organization convened a meeting in 2005 with UN-HABITAT, the Asian Development Bank, and the Society for Preservation of Water to create a proposal requesting that the Ministries of Education in Southeast Asia include water education in their national education plans.

The following sections describe similar efforts of NGOs, aid organizations and government agencies to educate youth on important water issues:

2020 vision for water
The Ministry of Water Affairs and Forestry of South Africa (1996) answered the need for water education in schools by creating the 2020 Vision for Water and Sanitation Education Programme (VFWSEP), a collaboration between the Ministry and the Stockholm Water Foundation. 2020 VFWSEP oversees three projects: curriculum support, the Baswa Le Meetse Awards and the South African Youth Water Prize. Curriculum support helps educators throughout South Africa by creating water materials, including Project WET, for students and providing training opportunities in collaboration with the Ministry of Education. The Baswa Le Meetse Awards and South African Youth Water

Prize encourage grade six and high school students, respectively, to engage in water research. It is hoped that these incentives will invite further research on water topics of interest, as well as encourage students to enter careers in the water sector.

Juenes volontaires pour l'environnement
Thanks to the efforts of Juenes Volontaires pour l'Environnement (2008) (Young Volunteers for the Environment, or JVE), in collaboration with the Regional Center for Water and Sanitation (CREPA), Togo launched its Water and Sanitation Network in November 2008. JVE was created in 2000 by a group of Togolese youth who were upset by environmental degradation in their communities. In 2005, JVE adopted Project WET as their primary water education curriculum. With steady growth, the NGO has grown to an organization with UN accreditation and the largest membership in the country. JVE is currently working on water relief projects as well as environmental education in classrooms. It is through their guidance that many Togolese teachers have received training in and materials for water education.

12.6   CONCLUSIONS AND FUTURE ACTIONS

The research conducted by the Project WET Foundation over the past twenty-five years has clearly documented the value and importance of educating children and young adults about water and its use, management and protection. The most successful water education programs combine the scientific and technical expertise of water professionals with formal and non-formal educators. When water and education sectors work together, children win.

The water and education sectors have a unique opportunity to design, develop and deliver water education materials and training to countless people through new and innovative systems. The first step in the design process is to ask: "What does every person need to know about water and its use, management, and protection?" A global survey of a broad cross-section of people is suggested. The results will be used to identify priority content for educational materials targeted at user groups (children, young adults, lay people, school educators and community educators). The universal nature of water will allow education and program developers to publish materials that can be used around the world – a water education content template. Project WET has documented that its water science methods can be translated, adapted (localized), and successfully used by educators worldwide.

The question now becomes one of delivery. There has never been a unified effort to scale-up water education with a goal of annually reaching large percentages of children and young people. The concept of World Water Forum events focused on children, such as the Global Water Education Village and

Water School, is recommended as one unifying idea to scale-up. A series of WaterCourses™ ranging from general water science and water management to more focused and in-depth courses on specific priority topics such as groundwater, wetlands or sanitation will be designed and made available to education providers and learners of all ages. The materials and programs will be available for use by any government, agency, organization or business interested in educating people about water and thus support scaling-up.

Finally, water and education sector leaders should establish a certification program to give credibility and value to the educational programs and to the people who take their courses. The key question is, "Does learning occur and can learners use the information in their daily lives to better manage and improve water conditions?" If the answer is yes, then the job is done.

REFERENCES

Alidou, H., A. Boly, B. Brock-Utne, Y.S. Diallo, K. Heugh, H.E and Wolff, 2006, *Optimizing learning and education in Africa - the language factor: a stock-taking research on mother tongue and bilingual education in Sub-Saharan Africa.*

Department of Education, USA, 2001, *No Child Left Behind*, http://www.ed.gov/nclb (Accessed 25 January 2009).

Department of Water Affairs and Forestry, South Africa, 1996, 2020 *Vision for Water and Sanitation Education Programme*, http://www.dwaf.gov.za/2020Vision/default.asp (Accessed 25 January 2009).

Edwards, L. 2009, *U.S. Drought Monitor*, Western Regional Climate Center, http://drought.unl.edu/dm/monitor.html (Accessed 20 January 2009).

Juenes Volontaires pour l'Environnement. 2008, *Water and Sanitation Network*, http://www.cooperationtogo.net/jve/info/112014.html (Accessed 25 January 2009).

Ministry of Water and Environment, Uganda, 2006, *Water Strategic Plan.* UN-HABITAT, 1999, Water for African Cities, http://www.unhabitat.org/content.asp?id=2154&catid=460&typeid=24&subMeuId=0 (Accessed 25 January 2009).

UNICEF, 2006, *UNICEF and WSSCC Launch WASH Partnership*, http://www.unicef. org/media/media_31476.html (Accessed 25 January 2009).

World Bank, 2005, *In their Own Language: Education for all*, Education Notes series. http://siteresources.worldbank.org/EDUCATION/Resources/Education-Notes/EdNotes_Lang_of_Instruct.pdf (Accessed 25 January 2009).

Chapter 13

AN OVERVIEW OF CAPACITY BUILDING ON GENDER EQUITY IN THE WATER SECTOR

Prabha Khosla
Women for Water Partnership (WfWP)

ABSTRACT

Capacity building on gender equity in the water sector covers almost all water sub-sectors. However, capacity building to enable equality and equity between women and men requires initiatives to focus on both – women and men, as well as women-only constituencies. This paper provides a brief review of capacity-building materials on gender and water. It identifies the production of guides, manuals and checklists as the dominant materials which encompass multiple functions, such as for information sharing, education, advocacy, and as lobbying instruments. The paper identifies resource materials on specific subjects and those geared to particular population groups. It also briefly profiles some water-sector organizations specifically working on building capacity on gender equity as well as organizations that also focus on women's groups and women and men in communities. Capacity-building materials on gender equity in the water sub-sectors are identified for a range of sub-sectors as well as for institutional change processes. Some closing comments offer a reflection on the experience of gender training and provide recommendations for strengthening capacity building on gender equity in the water sector.

13.1 INTRODUCTION

As with capability-building interventions in other disciplines and sectors, capacity building on gender[1] equality and equity in the water sector takes many forms and methods, and interventions take place at all levels of the sector – from the global stage to small villages and towns. However, what makes capacity building on gender equality[2] and equity[3] different from capacity building in other sectors is the importance of focusing initiatives on both women and men, and on women-only constituencies. Additionally, capacity building on equality between women and men requires trainers who are committed and engaged in advancing women's rights and gender equality. This is particularly so, as it is important that the trainers have experienced and engaged in the complex dialogues about (in)equality of power in gender relations and their implications for women, men and children with regards to access to resources and decision-making.

Interventions in capacity building include influencing and shaping the development of gender and poverty-focused participatory national water policy-making to gender-appropriate and affordable technologies, such as water pumps for women farmers working on small land holdings. Indeed, the water sector is vast and includes many sub-sectors, such as irrigation and agriculture, coastal zone management, wetlands, river-basin management, fishing, transboundary water, aquifers, potable water, sanitation, hygiene, grey water management, water and disasters, and water and climate change. Notwithstanding the newer and holistic approaches as articulated in Integrated Water Resources Management (IWRM), which is theoretically also committed to integrating gender analysis[4] as a cross-cutting issue, many of the institutions of the water sector continue to operate in silos of old and are slow to make the political and institutional paradigm shift required for gender equity and IWRM.

[1] Gender is the culturally specific set of characteristics that identifies the social behaviour of women and men and the relationship between them. Gender, therefore, refers not simply to women or men, but to the relationship between them, and the way it is socially constructed.
[2] Gender equality means that women and men have equal conditions for realising their full human rights and potential to contribute to national, political, economic, social and cultural development, and to benefit from the results. Gender equality is about the equal valuing by society of both the similarities and differences between women and men and their work in social production, and production of goods and services.
[3] Gender equity refers to the process of being fair to women and men. It recognises that different measures might be needed for women and men where: (a) they express different needs and priorities; or (b) where their existing situation means that some groups of women or men need to be supported by additional measures to ensure that they are on a 'level playing field'. This may require specific actions to enable equality of opportunity between women and women, or men and men, or women and men.
[4] Gender analysis is a process that assesses the multiple and differential impact of proposed and/or existing institutions, legislation, policies, and programmes on women and men. It is a systematic way of looking at their different situations and relations and is a tool for understanding social processes, and for developing equitable options.

Not surprisingly, the capacity-building methods and interventions for gender equity in the water sector reflect the diversity of the societies, cultures and their water organizations and institutions. Capacity building on gender equity in water focuses on all aspects of the sector, such as international bodies; national, regional, and local governments, institutions and organizations; non-governmental organizations (NGOs); environmental non-governmental organizations (ENGOs); women's groups; schools and health services; as well as legislation, policies, budgets, programmes, project implementation, operations and maintenance, data collection, monitoring and evaluation.

Before going further, a caveat is in order. While this paper presents a critical lens in its review of capacity building on gender issues in the water sector, let us recall the international climate on women's rights and gender equality and equity. For decades now, the international community, national governments, global water organizations and institutions have made countless declarations, conventions, resolutions and other pronouncements about the importance of equality and equity for women and especially poor women, men and children in water and sanitation, governance and decision-making, and for sustainability. However, the political will, commitment, and funds necessary to translate these noble words into deeds have not been forthcoming to the same extent; and, this is especially so in capacity building on gender issues and for the transformative action implicated in this development of human resources and institutional capacity. Capacity building in gender analysis and gender mainstreaming are critical to realizing the above-mentioned pronouncements.

This paper is based on a brief and non-comprehensive review and assessment of existing capacity-building materials, as well as the author's experience with and engagement in women's rights, gender, water, sanitation, environmental and capacity-building issues over the years.

A limited review of capacity-building materials on gender and water indicates there has been extensive development of certain kinds of materials and less so of others. This also implies that some sub-sectors of the water sector have had a greater encounter with questions and challenges raised by feminists and gender activists, than have others. The paper begins by assessing guides and manuals on gender and water. Some materials are overviews, and others are specific to certain constituencies or to particular areas for enabling gender analysis and gender mainstreaming, such as for gender-sensitive policy, gender audits, and for gender mainstreaming water-sector organizations and institutions. In the following sections, this paper highlights some capacity-building materials created for engineers and water managers, and discusses the work of some key organizations in the 'water world' working explicitly on gender equality and women's empowerment. A section is devoted to a discussion about sex-disaggregated data collection, and monitoring and evaluation, as these are two areas that are essential for gender analysis and gender mainstreaming and need a much stronger focus and investment of resources and expertise than has been the case so far. Some reflections and remarks about issues and debates on capacity building on gender equity and water and thoughts about the way ahead conclude the paper.

13.2   GENDER AND WATER GUIDES, TOOLS AND CHECKLISTS

The greatest production of materials is what has been called and classified as guides, i.e. guides to gender and water in the broadest sense of these terms. Numerous guides have been produced over the years. These guides were largely funded by bilateral institutions, the UN, or international financial institutions, and have some common objectives. They were developed for a wide range of audiences and serve multiple functions, such as for information sharing, education, advocacy, and as lobbying instruments. They provide cogent arguments as to why a gender analysis is needed in the water sector and the implications of gender-blind policies and projects. Examples primarily focus on the drinking water and sanitation sub-sectors. Most of the guides also include practical tips, checklists or a step-by-step process demonstrating how to integrate, for example, a gender perspective into the project implementation cycle. For a sample of guides see Baden (1999), Swiss Agency for Development and Cooperation (2005), Thomas et al. (1996), and UNDP (2003). For guides on agriculture and irrigation water, see Chancellor et al. (2000) and FAO (1999). Additionally, see the 'Minimum agenda for effective gender mainstreaming[5] in water management in agriculture'.[6]

Guides and manuals also exist for other water sub-sectors such as hygiene, the environment, climate change and water-related disasters, coastal zone management, etc. For an example of watershed management from a gender equity perspective see IUCN and HIVOS (2004). See Gender in Integrated Water Resources Management (GWA and UNDP, 2006) for extensive listings of guides, studies, books, on-line resources, and training materials in 13 water sub-sectors.

13.3   MANUALS AND GUIDES FOR 'ENGENDERING' POLICIES, ORGANIZATIONS AND INSTITUTIONS

Another set of tools include manuals and guides for the development of gender-sensitive policies, and for the engendering of water-sector organizations and institutions.

---

[5] Gender Mainstreaming is the process of accessing the implications *for women and men* of any planned action, including *legislation, policies and programmes* in all areas and at all levels. It is a strategy for making women's as well as men's concerns and experiences an *integral dimension* of the design, implementation, monitoring and evaluation of *policies* and *programmes* in all political, economic and societal spheres, so that women and men *benefit equally* and inequality is not perpetuated. The ultimate goal is to achieve *gender equality* (ECOSOC, 1997, emphasis added). For the full document see: http://www.un.org/documents/ecosoc/docs/1997/e1997-66.htm

[6] http://www.iwmi.cgiar.org/assessment/Synthesis/minimumagendagender.htm

On the policy front, the Gender and Water Alliance has produced both an excellent Policy Development Manual (see Derbyshire, H. et al. 2003), as well as an overview to inform those interested in the development of gender-sensitive water policies.[7] Additionally, see the Policy Brief produced by the Global Water Partnership (GWP, no date).

Today, the two gender-inclusive national water policies – of South Africa and Uganda – have themselves become capacity-building tools. They offer much food for thought, both from the policy analysis point of view as well as from the point of view of policy implementation. A comprehensive critique of the policies and their implementation would have much to offer *all* actors in the water sector.

The process for the development of gender-sensitive and poverty-focused water policies needs to be accompanied by a parallel process for the engendering of relevant water-sector organizations and institutions. Transforming the institutional culture so that it is informed by the principles of equality, equity, and sustainability – especially inclusiveness, transparency and accountability – is critical to changing the culture of water management. However, the water sector is far from showcasing its success in this area. There are some institutions that are attempting this transformation, such as the South African Department of Water Affairs and Forestry; however, few others can demonstrate their commitment to this indispensable transformational change.

Nevertheless, the Gender and Water Alliance (GWA) has attempted to engage the Asian Development Bank (ADB) in an initiative to assess the gender-sensitivity of its policies and operations. In partnership, they identified a methodology for the assessment – a Gender Scan. The Gender Scan was structured into four areas of activity: a review of key documentation, particularly the Policy on Gender and Development, "Water for All" (the water policy of the ADB), The Poverty Reduction Strategy and progress reports on policy implementation; a review of a 25% sample of loan documents approved since 1995; field visits to two countries [Sri Lanka and Viet Nam] to review four projects under implementation; and, key informant interviews and questionnaires (Derbyshire, 2005, p. 1).

Briefly, the recommendations include: a review of the water policy to strengthen gender mainstreaming as a cross-cutting theme in the context of poverty reduction and IWRM; increased attention to mainstreaming gender effectively in the water sections of the countries' strategies and programmes and the water sector road maps; the formation of a committee with staff from the Regional and Sustainable Development Department (RSDD) and the gender and water networks to elaborate priorities and actions arising from the Scan; the production of a 'gender typology chart' of the different water projects to monitor and support gender mainstreaming; production of an

[7] See the following section of the Gender and IWRM Resource Guide http://www.genderandwater. org/page/2859

advocacy package for policy dialogue in the executing agencies to deepen their understanding about implications of gender relations in water policies and activities; the collection of gender-analytical information by engaging gender consultants in loan fact-finding, appraisal, inception, reviews, etc; and the identification of ways of supporting the executing and implementing agencies for the effective implementation of their gender action plans.

Another example from and experience of engendering institutions worth noting is from the International Labour Organization (ILO). The ILO began its first gender audit in 2001 and has been consolidating and refining its gender mainstreaming process since. Guidance for engendering policy and institutions is readily available and the experience of mainstreaming gender in development organizations will be instructive for water-sector organizations and institutions.

13.4   CAPACITY BUILDING ON GENDER EQUITY FOR ENGINEERS, WATER-SECTOR MANAGERS AND POLICY-MAKERS (AND OTHERS)

That gender equity capacity-building initiatives should focus on engineers and water-sector professionals and managers – professional categories dominated by men – was recognized by women's rights and gender equality activists early on. Recent years have seen the development of materials specific to different water professionals, such as engineers. During the last decade, the Water, Engineering and Development Centre (WEDC) of Loughborough University in the UK produced some very useful and accessible materials geared to engineers in the developing world. For further reading, see Reed and Coates (2007), Reed, Coates et al. (2007), and Reed and Smout (2005).

The formation of the Gender and Water Alliance (GWA), the Women for Water Partnership (WfWP), and various Women for Water networks all testify to the urgency and mobilization to put women's rights and equality between women and men on the front page of the water sector. All these networks of women and women and men devote considerable resources and time to capacity building, evidence-based policy development and programmes, and knowledge networks. They work with water-sector professionals as well as communities of women and men.

The Gender and Water Alliance (GWA) was created to build the capacity of its members and partners in gender mainstreaming. Over the years, it has produced a tremendous amount of material advocating for gender equity in the water sectors, Training-of-Trainers (TOTs) courses, and collaborative research initiatives with water sub-sectors. Today, it is creating regional hubs for building capacity on gender and water. Perhaps, an indicator of their success is that the demand on them and their expertise for capacity building on gender and water far exceeds their own capacity and resources.

The Women for Water Partnership (WfWP) is an alliance of women's groups and networks working in almost all water sub-sectors. They are committed to linking macro- and micro-level interventions and creating conditions that enable women and their organizations to become partners in the development of their own environments. Every six months WfWP organizes Regional Working Conferences where initiatives of member organizations are developed into concrete projects. Relevant stakeholders are invited to participate in the joint development and subsequent implementation of these projects. The knowledge and experiences about partnerships, appropriate technologies, micro-finance, and up-scaling of successful initiatives are exchanged and put to practice.

Through WfWP's methodology of *Dynamic Networking*, all stakeholders arrive at a shared-needs perception and jointly develop the most appropriate solution. In the resulting project(s), all parties involved work together on an equal footing, contributing their specific expertise within the overarching principles of sustainability.

The members of GWA and WfWP encompass a broad range of stakeholders and build capacities of both women and men. WfWP member-organizations are primarily women's or women-headed organizations. They are committed

---

Box 13.1 NetWwater, Sri Lanka.[8]

NetWwater (Network of Women Water Professionals) is a voluntary group of Sri Lankan women who are committed to promoting the principles of holistic water management to meet the current freshwater crisis and to create awareness about the relevance of gender in water resources management. NetWwater was created in 1999 and its work engages both women and men in all aspects and sectors of water.

NetWwater incorporates capacity building on gender by building the knowledge and capacities of women and men in technical issues and incorporating a gender analysis in their technical capacity development programmes. They begin with Integrated Water Resources Management (IWRM) and link this to the Dublin Principles.[9] Depending on the constituency and the region of the country, they address issues such as floods, irrigation and farming, water pollution, climate change, river sand mining, sanitation in schools and rain water harvesting. Broadly speaking, their approach follows the following elements:

1 Create platforms where water issues can be discussed, including hot topics such as river sand mining, river ecology, sanitation, and the filling in of critical urban wetlands. Platforms are enablers for discussions between women and men, professionals and non-professionals, state and non-state actors, and provide opportunities for questions about gender differentiation in water resources use and management.

---

[8]This information is based on the author's interview of Kusum Athukorala of NetWwater. For additional information on NetWwater check http://www.womenforwater.org

[9]For information on the Dublin Principles see: http://www.gwpforum.org/servlet/PSP?iNodeID=1345

2   Foster a holistic approach and ecosystem analysis of water sectors and related activities.
3   Demonstrate gender analysis by linking it to what people are facing in day-to-day living with the relevant water issue(s) they are engaged in.
4   Link capacity building to activities and actions in specific water sectors and geographical regions.

This approach to building capacity in gender equity in water has worked for NetWwater as they do not always have the resources to put 30 people in a room for a given number of days of capacity building only on gender and water. Furthermore, it has also assisted them in engaging many women and men who would otherwise not be open to discussing gender and the unequal power relations between women and men that are inherent in the water sector and the society.

How does NetWater gauge their impact on capacity building in gender in the water sectors? The example below illustrates the difference ten years and a commitment to gender equity in the water sector can make.

"Ten years ago, we were considered a joke by almost everyone and every institution. Today, we are able to access the highest levels of government and water-sector institutions for the poorest women in the driest part of Sri Lanka. For example, last year when we had a school programme, we examined toilets in a school in the Central Province. The school had 1000 students and six toilets for girls and six toilets for boys. None of the girls' toilets were functioning. Only one of the boys' toilets was still working. We took photographs of the toilets and showed them to the people responsible for school sanitation in the Province.

Photo of girls' toilet.

The Sri Lankan database on sanitation speaks of extensive sanitary coverage for the country, with percentages from 70%–90% being quoted. However, the reality on the ground is very different. Toilets may be there, but they are not useable. Within five days of showing the photos, we got appointments with senior decision-makers and they immediately gave us access to all the schools in the district to carry out a sanitation sensitization programme."

to working on the ground directly with partners and communities of women and men to create solutions by working together. Their capacity-building efforts are given shape as they jointly create solutions to water problems. The example below of WfWP member, NetWwater from Sri Lanka illustrates one such approach.

There are innumerable international, regional, and national capacity-building organizations that focus on the water sector. It would be a very positive development if all of them were to include gender training as an integral part of *all* their trainings on the water sub-sectors. However, that is far from the case as the experience of Cap-Net, a partnership of autonomous international, regional, and national institutions and networks committed to capacity building in IWRM demonstrates.

Cap-Net[10] has attempted to raise the awareness of the need for capacity building on gender in its network and with professionals and practitioners. It has stressed the importance of selecting women participants in training and capacity-building courses. Table 13.1 provides a breakdown of female and male participants in their recent courses.

Table 13.1 indicates that there are fewer women participants despite there being many courses. The numbers above were a surprise for Cap-Net. This example also demonstrates the importance of the collection of sex-disaggregated data in the water sector. The Cap-Net secretariat is now attempting to improve the participation of women in the courses it directly facilitates and especially in the context of the development of materials. Additionally, they hope to highlight to the member networks the drop-off in the participation of women and support them in their efforts to recruit more women for their courses.

Yet another method under consideration by Cap-Net is to increase women's participation in the capacity-building networks by selecting women as resource people and facilitators in the training-of-trainers courses and other capacity-building interventions, such as in the development of case study research materials. It would be revealing to examine the sex-disaggregated

Table 13.1 Course participants by sex.

Year	No. of Courses*	Women	Men	Total
2007	21	273 (43%)	356 (576%)	629
2008	27	285 (32%)	593 (68%)	878
Total	48	558 (37%)	949 (63%)	1507

*This refers to courses where the participants' list specified participants by sex.

[10]Thanks to Simone Noemdoe from Cap-Net/UNDP for this information about Cap-Net.

enrolment statistics of other water sector educational and training institutes and see what they illuminate in terms of women's (in)equal access to educational and training opportunities.

SaciWATERs,[11] the South Asian consortium for interdisciplinary water resources studies, in their pioneering initiative 'Crossing Boundaries, regional capacity building for IWRM and gender and water in South Asia', works with academics and postgraduate students in the water sector. Crossing Boundaries is changing the technical and civil engineering curriculum by introducing subjects which address IWRM, gender, participatory management, equity, sustainability and environmental concerns. The initiative has successfully developed a new curriculum for targeted technical and engineering universities and colleges in South Asian countries and raised the number of women Masters and PhD students in engineering.

13.5   SEX-DISAGGREGATED DATA COLLECTION AND MONITORING AND EVALUATION

Two additional and related areas of importance for building capacity in gender equity in the water sector are the collection, dissemination and use of sex-disaggregated data and the development, dissemination and use of gender-sensitive and poverty-focused targets and indicators for monitoring and evaluation. Recognizing that the water sector is large and encompasses a vast array of sub-sectors, clearly it is not realistic or feasible to see only one system of sex-disaggregated data collection or one set of gender and water indicators. Sex-disaggregated data collection systems need to be developed in conjunction with national statistics offices and national and local systems of data collection and dissemination. Gender-sensitive and poverty-focused indicators are ideally developed to suit the change process they are intended to monitor and measure.

Capacity building in both these areas is of key importance to advance equality between women and men, to measure the impact of policy and project initiatives on women and men and especially on poor women, men and children, as well as for monitoring changes in organizations and institutions engaged in gender mainstreaming. There are some initiatives attempting to define and develop indicators, such as Van Koppen's work on the development of gender performance indicators for irrigation (2002). The Mainstreaming Gender Dimension into Water Resources Development and Management in the Mediterranean Region (GEWAMED) is a new initiative that also proposes to prepare a set of gender indicators for monitoring gender mainstreaming in water resources management.[12]

---

[11] To learn more about them see: http://www.saciwaters.org/
[12] For additional information about GEWAMED see: http://www.gewamed.net/index.php

In December 2008, the UN Department of Social and Economic Affairs (UNDESA) and the UN-Water Decade Programme on Capacity Development (UNW-DPC) jointly organized an Expert Group Meeting (EGM) on 'Gender-disaggregated Data in Water and Sanitation'.[13] The meeting had several goals; namely, to take stock of sex-disaggregated data on water and sanitation at global and regional levels, to identify obstacles to sex-disaggregated data collection and capacity, and to identify data needs and priorities. Recommendations were made on policies, practices and priorities to improve the state of sex-disaggregated data in water and sanitation. As this process evolves, it will have much to offer the drinking water and sanitation sub-sectors.

Frequently, it is difficult to obtain data at the local level, and specifically data that are disaggregated by sex and other social relations such as ethnicity, poverty, age, race, caste, etc. The Community-Based Monitoring System (CBMS)[14] is a unique experiment in the Philippines and was developed due to the lack of data at the micro-level which was essential to address poverty locally. The CBMS is an organized and participatory method of collecting information at the local level for use by local government units (municipalities), national government agencies, non-government organizations, and civil society for planning, programme implementation and monitoring. It is a tool intended for improving governance, and for greater transparency and accountability in resource allocation. It is currently in use in 161 municipalities and 13 cities covering about 4,438 *barangays* (neighbourhoods) in the Philippines.[15] The CBMS offers an enormous potential as a methodology for collecting data for many water-related programmes and projects.

13.6   ISSUES AND DEBATES IN CAPACITY BUILDING AND TRAINING IN GENDER EQUITY

The methodologies for gender training have perhaps not been a point of critical reflection and analysis compared with the feminist critiques of water and development. There is some debate about how much gender training reflects the pedagogical traditions of the dominant paradigm of patriarchal societies (i.e. those informed by positivist and linear approaches to knowledge acquisition and learning), compared with those that reflect a feminist epistemology of knowledge, or what is popularly referred to as 'women's ways of knowing', and that women's ways of acquiring and sharing knowledge is not necessarily linear and often based on learning that is experiential and relational.

[13] For information about the meeting see: http://www.unwater.unu.edu/file/EGM+ summary+ report.pdf
[14] For information on CBMS see: http://www.pep-net.org/NEW-PEP/index.html
[15] http://econdb.pids.gov.ph/index.php?option=com_content&task=view&id=25&Itemid=40

While not wanting to 'essentialize' women and water, this difference is particularly important for recognizing women's knowledge of their environments, their knowledge about water issues, and the solutions they present and can present. Additionally, women's decision-making approaches can be more inclusive than men's and women are and can be a powerful force for the resolution of water-related conflicts. Capacity-building efforts would benefit from valuing the differences women bring to decision-making, and their engagement in a holistic approach for natural resources management. For a larger and more nuanced discussion of feminist epistemologies and their relationship to methods of learning and knowing, readers are encouraged to explore the work of, for example, Harding (1987) and Smith (1987).

An inspirational example of women-specific and poverty-reduction training from Kerala, India, is provided below.

Box 13.2 New skills, New lives: Kerala's women masons – Thresiamma Mathew.

Socio-Economic Unit (SEU), Kerala, was established in 1988 to implement water and sanitation projects in Kerala through community participation from planning to monitoring. SEUs have been involved in the construction of 53,763 household latrines, 253 institutional latrines, and two pay-and-use latrines with full community-group, local government and user participation. To meet the shortage of skilled masons for such construction work, the Jeevapoorna Women Masons Society (JEEWOMS), an offshoot of SEU, was formed in 1989 and trained the first group of 12 women from Thrissur, one of the *panchayats* (local rural governments) in Kerala. As these women began to marry outside their villages, they dropped out of the programme and therefore, in 1990, the second group of 14 women had to meet certain criteria; namely, that they had to be married, below 45 years of age, from below-poverty-line households; that their children should be more than 3 years old; that women in greater need of extra income were preferred; and, that all candidates should have previous experience as a mason's helper. The women learned such skills as cement-block making, bricklaying, reinforcing steel work, and how to construct low-cost twin-pit latrines.

In the beginning, what proved difficult was persuading women that they could learn masonry, and overcoming their fears of violating the cultural norm. A participatory training programme was developed to build up women's confidence that they could become skilful masons, to strengthen team-building so that women could face opposition, and to awaken the women's obvious but latent potential. To reinforce this, trainees and the facilitator formulated the "Ten Commandments" which included participating attentively, being determined, persevering, being confident, cooperating, working hard, and being honest and loyal. Training also involved savings and money management and hygiene so that the women could be agents of change within their communities. The women were trained by male master masons and gained experience in different *panchayats*. Once trained, they earned equal pay to men. Over the next few years, the women of Thrissur became a crucial component of SEU's sanitation programme. Despite some local problems and opposition, they also received publicity and appreciation. In 1996, JEEWOMS ventured into machine-operated hollow-block production, which developed their skills and boosted their earnings. Twelve masons completed one-month training courses in house construction, and some trainees became trainers. Seeing this success, more of Kerala's district *panchayats* are training women in masonry as a main plank of their women's-empowerment initiatives. The women participants have gained confidence to speak in public, leadership abilities, managerial skills, dignity, and an improved status in the community.

Source: *Waterlines*, 1998. Vol. 17, No. 1, p. 22–24

13.7   SOME REFLECTIONS

A significant portion of gender training is focused on understanding language and concepts such as sex, gender, gender equality, equity, etc., but their actual cultural meanings and implications on the ground where the training takes place are not discussed and explored sufficiently. For example, see the PowerPoint on gender-sensitive training, monitoring and evaluation for MPAP.[16] Identifying the example of MPAP is not to isolate them as a special case, but to demonstrate that this approach is common to capacity building on gender equity.

Tools such as checklists have come in for particular criticism as they have been used in an instrumentalist manner, devoid of engagement with social relations on the ground or in the organizations promoting them. While they are useful educational tools and reminders, a deeper gender and diversity analysis is required in practice.

There has not been sufficient reflection on and assessment of our work as activists and professionals engaged in building women's capacities to engage on their own terms and with their knowledge as legitimate and of value in the water sector; and as gender trainers, training women and men in organizations and institutions in the water sub-sectors. There is a need to reflect on both the production of knowledge and training materials, as well as on the application of methods and tools.

Additionally, there is a need for trainers, whether in the development of curriculum or in the conduct of trainings, to reflect on their own location and position as the 'trainer' and the implications of this in water-sector institutions and with the actors with whom they engage.

The United Nations Educational, Scientific and Cultural Organization (UNESCO) has fairly extensive engagements in the production of knowledge materials for capacity building on gender equity, as well as on gender and water. UNESCO has also conducted numerous in-house capacity-building

---

[16] See: http://www.iwmi.cgiar.org/southasia/ruaf/CD/s26.html

trainings on gender. Their reflection on the practice of gender training is an informative assessment and is produced below.

Box 13.3 Tips and good practices for conducting gender-training
for UNESCO staff.

In addition to operational gender-mainstreaming tools, gender training is one of the key methods to support behaviour and organizational change. Providing training is, however, on its own not sufficient. Training is helpful if it is part and parcel of a comprehensive corporate culture of learning. This means that both the organizers and the trainees draw lessons from the training, and apply this new knowledge to improve their daily work.

Below you will find the lessons learned by those who have organized gender-training sessions for UNESCO staff both in headquarters and in field offices.

If you are planning to organize a gender-training workshop for your office, division, bureau or programme, we suggest you carefully read through these.

Planning
- External gender expertise was proven very useful to both facilitate and plan the training sessions. The voice of someone who is not UNESCO staff and who has worked with many different development institutions adds credibility to gender work. It also offers the Gender Focal Point who is co-organizing the workshop the opportunity to acquire new skills.
- It appears that UNESCO staff value learning from other staff more highly when the group is composed of staff from several offices or different sectors and services.
- The role of the Office/Division/Institute director in encouraging participants to attend and in giving a thought-provoking opening speech adds legitimacy to the training. This certainly helps participants justify investing their time.
- Having the Office/Institute/Sector Gender Focal Point co-facilitate the workshop also reinforces the 'value-added' expertise she/he can bring to programme specialists. It also creates a natural transition for their collaboration in following up the training. ˙
- HQ + BSP presence (i.e. representative of the Section for Women and Gender Equality in BSP) is helpful to establish the importance the Organization gives to gender work.
- Men should be encouraged to take part in the training. Their participation is not only necessary to promote the integration of gender issues in UNESCO's programmes; it brings new perspectives to the issues being addressed.

Design and implementation
- Key gender terms and language should be clarified in a non-threatening way at the start of the training. This creates a common foundation of 'gender understanding' among participants that is conducive to interaction.
- If the training is to be conducted in French, or will be run with non-English native speakers, time should be foreseen to discuss issues of translation. Language directly affects the way that people think about and see the world.

- Participatory + experience-based training is the most effective way to get participants to internalise new information. This approach builds comfort and competence with gender issues.
- Participants are actively engaged in the training when they are given the opportunity to develop, during the session, a 'product' that they will be able to use when they go back to their workplace. (examples of products: Gender Lens/Checklists, or post-literacy material, recommendations for future actions, outline for common project).
- The experience of working in small groups to review a UNESCO project from a gender perspective proved especially useful in workshops. This exercise appeared to be one of self-discovery for many participants. Several seemed surprised that there was so much to consider and equally surprised that *'they got it and got it so quickly'*. Many found they simply needed to clearly focus on gender. The same *'ah hah'* could not have been achieved through passive listening/learning.
- Participants were very appreciative of the handouts/resource materials provided. A number of these can be used in discussions or activities with partner groups.

Monitoring and evaluation
- Gender training increases demands and expectations of the Office/Institute/ Sector Gender Focal Point. Resources need to be foreseen to respond to the demands for gender technical assistance that gender training and advocacy generate.
- Maximizing the benefit of these training workshops can be done in many ways. Among them: refresher trainings or discussions; recognition of employees who perform well in advancing gender equality, institutionalising the use of the gender tools developed in the workshops.

Source: http://portal.unesco.org/en/ev.php-URL_ID=14405&URL_DO=DO_TOPIC&URL_SECTION=201.html

Another area in need of critical review is an assessment of the use and utility of the capacity-building materials that exist and, hopefully, are in circulation. While numerous guides, checklists, manuals on 'how-to integrate gender' in various aspects of the water sector and its activities exist, there has been very little evaluation of whether these reach the practitioner or engineer, or the irrigation, agricultural, fisheries and environmental extension workers; the managers of water utilities; the professors and directors of water training and academic institutions; the policy- and decision-makers; or, the women and men in rural and urban areas whom they are supposed to assist. A report by Udas (no date) examines the use of manuals and guidelines for gender mainstreaming in agriculture, water and the environment. The report indicates that only a limited number of people in institutions have access to the manuals and guides, and field staff in particular do not have access to such resources. Accessibility to documents is restricted to the few who have ready access to

Internet communication technologies. Other limits on the use and distribution of guides and manuals include language restrictions, use of jargon, and the lack of suitable examples and illustrations.

In a new book examining methods and methodologies for gender training, the editors, Mukhopadhyay and Wong (2007), explore the evolution of gender training and its 'development' and ideological underpinnings. Specifically, they raise the question of the need to revisit gender training by tracing its historical roots and its links to early development debates and the making and remaking of gender knowledge. Those who are directly engaged in developing capacity-building materials and organizing and delivering trainings on how to create equity in the water sector and society will find the reflections in this book useful.

13.8 THE WAY AHEAD: SOME THOUGHTS ON STRENGTHENING CAPACITY BUILDING IN GENDER ISSUES IN THE WATER SECTOR

The nature of power and power relations in our societies and in the water sector (i.e. the personal, public, and political) needs to be grounded in capacity building in the sectors. Attempts to keep away from conflict and not discuss implications of (in)equity in families, communities, societies, water-sector institutions, administration, management and implementation are compromising equality between women and men. While this is contentious ground, it is not new, and it is vital that the capacity-building materials strengthen the link between the personal, political and public.

There is an understanding by most trainers, and in the capacity-building materials, that the materials and the pedagogical approach need to be locally informed and relevant (i.e. in terms of culture, religion, language, politics, geography, etc.). Insufficient attention, however, has been paid to the integration of capacity building in gender with other social relations such as class, caste, race, ethnicity, age and sexuality that also inform gender relations and identity.

Capacity-building needs to be integrated into all activities in the water sector and should be tied to practice. It needs to become a core component of all water sector initiatives and be an on-going professional development process and not a one-time activity. Training on gender and diversity should be linked to the responsibilities of managers and staff so that the conceptual understanding is tied to practice. On-going reviews and discussions of implementation of a gender analysis in the mandates of managers and staff will assist in changing workplace practices and in deepening the understanding of gender and diversity equity and its implications.

Informal discussions between activists and trainers have raised the issue of an assessment tool such as a 'gender equity performance measurement tool' or perhaps a 'Gender Report Card' for water-sector organizations and institutions. The development of this tool is worth pursuing as it can serve multiple functions, such as monitoring progress on the implementation of gender action plans of institutions and mobilizing for accountability to gender equity in the water sub-sectors.

Recognizing the success of gender-responsive budget initiatives or gender budgets at the national and local government levels, time is long overdue for capacity building on gender-responsive budget initiatives (GRBIs) in water-sector budgets. This should be a key capacity-building focus of the coming years.[17]

Water-sector organizations and institutions need to hire gender specialists as part of their core staff and train existing staff members on gender and water as part of their on-going professional development and core job skills and requirements. This refers to capacity building on gender as an on-going programme and not a one-time activity.

It is vital that finances are allocated and programmes developed to expand the number of gender and water specialists and gender and water facilitators and trainers globally. Water-sector NGOs, ENGOs, utilities, ministries, organizations and institutions number in the hundreds of thousands and the current capacity of the water sector to train on gender equality and equity is severely limited and cannot meet this challenge.

13.9   CONCLUSION

Globally, the water sector is a massive sector – both in terms of the number of people engaged in it, as well as its total intellectual, financial, and infrastructural resources. Water-sector NGOs, ENGOs, ministries, utilities, public- and private-sector organizations and institutions number in the thousands, as witnessed by the number of participants at the World Water Forums and water-related international conferences. The planet's population is roughly half female and half male; still, the recognition that women and girls are half the global population has not dawned on those managing water resources (mostly men) as well as those (also mostly men) engaged in building capacity in the water sector. The profile of gender-focused guides, manuals, toolkits and organizations in this paper barely scratches the surface to provide the kind of capacity-building on gender equity that is needed in the water sector.

The women and men working for gender equality and equity in the water sector are but a drop in this vast ocean.

[17]The following website will answer your questions about GRBIs. http://www.gender-budgets.org/

Building an understanding of the implications of (in)equality between women and men and boys and girls and the ability to facilitate gender and social equity is essential for the institutionalization of IWRM, for building a culture of sustainability, and to bring about equality and equity in the distribution of a scarce essential and fast-depleting resource. A gender and poverty-sensitive approach in the water sector is the only answer for sustainable solutions.

REFERENCES

Baden, S. 1999, *Practical strategies for involving women as well as men in water and sanitation activities.*, Report prepared for the Gender Office, Swedish International Development Cooperation Agency (SIDA). BRIDGE, Institute for Development Studies, University of Sussex, Brighton, UK.

Chancellor, F., Hasnip, N. and O'Neil D. et al. 2000, Gender-Sensitive Irrigation Design. Developed by HR Wallingford under contract to the Department for International Development (DFID), United Kingdom.

Derbyshire, H. 2005, Gender Responsiveness in ADB Water Policies and Projects. Report on the Gender Review of ADB Water Operations jointly conducted by the Gender and Water Alliance and the Asian Development Bank. Technical Note # 3. Asian Development Bank, http://www.genderandwater.org/content/search/?SearchText=ADB+Report&SearchButton=Search

Derbyshire, H. et al. 2003, *Policy Development Manual for Gender and Water Alliance Members and Partners*, Gender and Water Alliance (GWA), http://www.genderandwater.org/content/download/191/1589/file/policy_manual.pdf

FAO, 1999, *Participation and Information. The key to gender-responsive agricultural policy*. High Level Consultation on Rural Women and Information. Rome, October, 1998, http://www.fao.org/docrep/x2950e/X2950e00.HTM

Gender and Water Alliance (GWA) and UNDP, 2006, Gender and Integrated Water Resources Management (IWRM), http://www.genderandwater.org/page/2414

GWP-TEC, undated, *Gender Mainstreaming: An essential component of sustainable water management*, Global Water Partnership – Technical Advisory Committee (TEC), Policy Brief No 3, http://www.gwpforum.org/gwp/library/Policybrief3Gender.pdf

Harding, S. 1987, *Feminism and Methodology: Social Science Issues*, Bloomington IN: Indiana University Press, USA.

Mukhopadhyay, M. and Wong, F. (eds), 2007, *Introduction: Revisiting gender training. The making and remaking of gender knowledge*, In Revisiting Gender Training – The Making and Remaking of Gender Knowledge, A global sourcebook, KIT Publishers, Amsterdam, Netherlands. http://www.kit.nl/smartsite.shtml?id=SINGLEPUBLICATION&ItemID=2074

Reed, B., Coates, S. and Parry-Jones, S. et al. 2007, *Infrastructure for All. Meeting the needs of both women and men in development projects – A practical guide for engineers, technicians and project managers*, WEDC, Loughborough University, UK, http://wedc.lboro.ac.uk/publications/details.php?book=978-1-84380-109-2&keyword=%infrastructure%&subject=0&sort=TITLE

Reed, B. and Coates, S. 2007, *Developing Engineers and Technicians. Notes on giving guidance to engineers and technicians on how infrastructure can meet the needs of men and women*, WEDC, Loughborough University, UK, http://wedc.lboro.ac.uk/publications/details.php?book=978-1-84380-110-8&key word=%infrastructure%&subject=0&sort=TITLE

Reed, B. and Smout, I. 2005, *Building with the Community: Engineering projects to meet the needs of both men and women*, WEDC, Loughborough University, UK, http://wedc.lboro.ac.uk/publications/details.php?book=978-1-84380-081-1&key word=%building%&subject=0&sort=TITLE

Siles Calvo, J. and Soares, D. with the collaboration of E. Alemán, 2004, *The Force of the Current: watershed management from a gender equity perspective*, IUCN and HIVOS, San José, Costa Rica.

Smith, D. 1987, *The everyday world as problematic*, Boston: Northeastern University Press, USA.

Swiss Agency for Development and Cooperation (SDC), 2005, *Gender and Water. Mainstreaming gender equality in water, hygiene and sanitation interventions*, SDC. Federal Department of Foreign Affairs, Bern, Switzerland.

Thomas, H., Schalkwyk, J. and Woroniuk, B. 1996, *A Gender Perspective in the Water Resources Management Sector: Handbook for Mainstreaming*, (Prepared in close cooperation with the Department for Natural Resources and the Environment), Swedish International Development Cooperation Agency, Publications on Water Resources, No. 6.

Udas, P.B., undated, *Report on Use of Manuals, Guidelines for Gender Mainstreaming in Agriculture, Water and Environment*, Both Ends, Gender and Water Alliance, and Comprehensive Assessment. http://www.genderandwater.org/content/download/7262/50055/file/070212_report_on_use_of_manuals.pdf

UNDP, 2003, *Mainstreaming Gender in Water Management: A Practical Journey to Sustainability*.

Van Koppen, B. 2002, *A gender performance indicator for irrigation: Concepts, tools and applications*, Colombo, Sri Lanka: IWMI. 42p., IWMI research report No 59.

RESOURCES – ORGANIZATIONS, INSTITUTIONS AND CAPACITY-BUILDING TOOLS AND METHODS

The following documents and websites offer rich perspectives and materials relevant to the topic of capacity building on gender equity and the water sector.

CAP-Net and GWA, 2006. Why Gender Matters.

Explaining why gender is important to water managers, this CD is in the format of a tutorial. It addresses the importance of a gender approach in agriculture, water supply, sanitation and water resource management.

http://www.cap-net.org/node/847

Food and Agriculture Organization of the United Nations. (FAO)

Portal on capacity building.

http://www.fao.org/capacitybuilding/

GWA and UNDP, 2006. Gender and Integrated Water Resources Management (IWRM)

This is a revised and updated version of the 2003 UNDP Resource Guide for Gender Mainstreaming in Water. This Resource Guide divides the resources among thirteen water sub-sectors to facilitate access for specific purposes and water uses. Introductions to the sectors describe current debates and gender issues. References, resources (including manuals and guidelines), case studies and relevant websites are all grouped by sub-sector.

http://www.genderandwater.org/page/2414

Gender Training Websites

http://www.un.org/womenwatch/directory/gender_training_90.htm

http://portal.unesco.org/en/ev.php-URL_ID=12001&URL_DO=DO_TOPIC&URL_SECTION=201.html

International Water Management Institute (IWMI)

Documents related to gender, land, water and livelihoods in Africa and Asia.

http://www.iwmi.cgiar.org/Research_Impacts/Research_Themes/LandWaterandLivelihoods/IrrSystem/Outputs.aspx

IRC International Water and Sanitation Centre. Gender and Equity

http://www.irc.nl/page/118

IRC Publications on Gender and Equity

http://www.irc.nl/page/7319

IUCN Gender and Environment

Extensive listing of resources of various environmental issues, including capacity-building materials.

http://generoyambiente.org/

Mukhopadhyay M., F. Wong (eds), 2007, *Revisiting Gender Training – The Making and Remaking of Gender Knowledge. A global sourcebook*, KIT Publishers, Amsterdam, the Netherlands

The book includes an extensive annotated bibliography of documents, books, articles, etc. related to gender training and available in the KIT Library. (p. 84–127) The book can be bought or downloaded from the following URL.

http://www.kit.nl/smartsite.shtml?id=SINGLEPUBLICATION&ItemID=2074

Socio-economic and Gender Analysis Programme (SEAGA)

http://www.fao.org/sd/seaga/1_en.htm

UNDESA

Gender and Water Reference Materials. 2006. June.

http://www.un.org/esa/sustdev/inter_agency/gender_water/referencematerials.pdf

UNESCO

Capacity Building and Training Programme on Gender Mainstreaming

http://portal.unesco.org/en/ev.php-URL_ID=11005&URL_DO=DO_TOPIC&URL_SECTION=201.html

UN-INSTRAW's Gender Training Wiki

http://www.un-instraw.org/wiki/training/index.php/Main_Page

Women for Water Partnership (WfWP)
Documents: http://www.womenforwater.org/openbaar/index.php
Partners and Networks: http://www.womenforwater.org/openbaar/index.php

Women's Environment and Development Organization (WEDO)
The site has documents related to gender and water, climate change, and sustainability.
http://www.wedo.org

World Bank, Gender and Development, Water and Sanitation
http://web.worldbank.org/WBSITE/EXTERNAL/TOPICS/EXTGENDER/0,,contentMDK:20205024~isCURL:Y~pagePK:210058~piPK:210062~theSitePK:336868,00.html

Chapter 14

KNOWLEDGE AND CAPACITY DEVELOPMENT AT RIVER BASIN LEVEL

Daniel Valensuela
International Office for Water, Paris, France

ABSTRACT

In the global context of putting in place an IWRM approach to face the current and future challenges in water resources, river basin management can be seen as a way to concretize IWRM principles, with basin organisations at the core of the institutional arrangement.

To be successful with river basin management, the river basin organisation has to have the necessary knowledge, data and information about water and water-related issues. This issue has to be addressed with the vision of developing an integrated information system. Based on an identification of capacity development needs, the basin organisation also has to put in place a capacity development plan, in line with the situation of the basin and the strategy adopted by the basin organisation. This plan should cover technical and non-technical needs, and include individual and institutional capacity building.

This paper analyses the kind of knowledge that is needed to ensure the basics of river basin management and how it has to be organised and managed in the basin context. Also, an analysis of the capacity-building approach is presented, showing the wide range of needs for managing the water resource and the basin organisation in an integrated way. The case studies of the Mekong River Commission and Niger Basin Authority illustrate the process for developing capacities in the complex transboundary context. Finally, key recommendations are delivered about knowledge management and capacity development at the basin level.

14.1   INTRODUCTION

To face the challenges related to water, in particular those concerning the MDGs, the improvement of water resource management and governance is essential. For several years, most countries have been developing Integrated Water Resource Management (IWRM), following the recommendations of the international community such as recommendations made by the World Summit on Sustainable Development in Johannesburg (2002). At the same time, to put IWRM into practice, many countries have set up basin organisations, within the country or at the transboundary level in the case that water resources are shared between several riparian countries. Indeed, the river basin is the relevant natural territory to ensure and organise efficient water management, provided that the existing basin organisations or those in creation adopt the IWRM principles. In that sense, countries have to think deeply about the basin organisation mandate, their roles, their responsibilities and their operation. Regulations, legislation and financial arrangements have to be decided and implemented to ensure the good functioning of the basin organisation. To properly execute their tasks, these new or revamped organisations need to have the right capacity, including an appropriate organisational structure, management and professional capacity in terms of numbers and quality, and the necessary equipment business processes and knowledge systems. This capacity is often not there or incomplete and needs to be developed.

This chapter focuses on both knowledge to be managed at the basin level and organisational capacity to be developed in basin organisations in order to get a real, effective and efficient IWRM. Knowledge and capacities are key aspects for water management in any situation. It is particularly true for water management at the river basin scale. Water resource managers, decision-makers, professionals and practitioners need to be able to get hold of reliable, up-to-date and relevant data and information on issues related to water in

---

Box 14.1 Capacity and knowledge.

Capacities can be seen as the knowledge, skills and other faculties in individuals or embedded in procedures and rules, inside and around sector organizations and institutions. These main capacity-building components are:

- The creation of an enabling environment with appropriate policy and legal frameworks;
- Institutional development, including community participation (of women in particular); and,
- Human resources development and the strengthening of managerial systems.

Source: Delft Declaration 1991 (UNDP symposium)

Table 14.1 Issues to be addressed at the basin level.

Issues about basin organisation management	Issues about basin management
Structure (organisation, mandate)	Water management
Statute of the organisation	River basin, aquifer or lake
Role in the institutional context	Land, forest or urban management
Governance of the organisation	Rules, Regulations
Finance management, Accounts	Institutional structure or organisation
Staff	Allocation, Quality
Human resource management	Planning

the basin. Knowledge and capacity are strongly linked, as using knowledge efficiently requires relevant capacities, competencies and skills that are based on accessible knowledge (Box 14.1).

At the basin level, knowledge and capacity have to address issues concerning both basin organisation management and basin management (Table 14.1).

14.2   KNOWLEDGE DEVELOPMENT

One of the key issues around knowledge is that necessary data and information are often dispersed, heterogeneous and incomplete, and are rarely comparable and adapted to meet the prerequisites for objective decision-making and management. Moreover, it is a fact that a lot of public, semi-public and even private organisations produce and manage some of the necessary data but lack sufficient means and guidelines for exchanging, gathering, standardising, summarising and capitalising on it among themselves.

At the same time, a key aspect in this issue is the notion of the need to manage river basins as a whole and in an integrated way, departing from sectoral and isolated water management in order to reach a higher level of integration. Institutional arrangements also have to be in place, such as a basin organisation, basin commission or an authority. To achieve a higher level of sustainable water development and management, capacities have to be developed at basin and sub-basin levels. Unfortunately, capacities at these levels are very often weak or even absent. Capacity building closely supports and helps guide the required institutional strengthening, the reform programmes and the development of management tools, thus making effective integrated water resources management operational at the river basin scale.

14.2.1   WHICH KIND OF KNOWLEDGE IS NEEDED?

Two types of knowledge are necessary for water management: actual data about water resources and uses, and knowledge in the sense of information, know-how, practices, institutional arrangements and tools.

Managing a river basin is not only managing the water itself. It is managing all things impacting the water resources within the limits of the hydrological basin. Therefore, there is a wide range of types of data and information that the basin organisation has to tackle (Table 14.2).

The first type of knowledge is obviously about the status of the surface water resources in the river, the tributaries, associated lakes and wetlands, and associated groundwater, from both a quantitative and qualitative viewpoint, and their seasonal and yearly fluctuations. As water in the basin has to be managed as a unit, we also need to know about the availability and use of the groundwater resources, since it impacts surface water usage.

It is also crucial to have reliable data about all uses: surface and ground-water uses, withdrawals for industry, drinking water and irrigation, point or non-point sources of pollution and discharges, and nutrients coming from various human activities on all surfaces of the basin. Biophysical characteristics of the basin (e.g. soils, topography) are also key elements to know, as they influence infiltration, the recharge of groundwater or erosion, and the amount of sediment in river water. The status of aquatic ecosystems and wetlands and their degree of sensitivity are other areas which need accurate information. In particular, we need to get the real needs for the ecosystems, in river beds or wetlands for instance, and the trend of their evolution or degradation. The people in charge of water management have to know the risks of recurring extreme phenomena, such as floods or droughts and accidental pollution (helped by a spatial map view and forecasting that takes into account climate change scenarios applicable for the basin). All economic indicators like cost, price, taxes and charges are another set of data needed for management.

Beyond that, managing water at the basin level also means getting reliable information on land use, such as the status of forestry, agriculture, industry, economic activities, infrastructure and urbanisation related to population changes; demography impacts water resources. Social and health indicators are also key data to be collected. For instance, such is the case for OMVS (Office for the Valuation of the Sénégal river), the basin organisation for the Senegal River, which has developed a specific structure for managing all kinds of information within the basin (see Box 14.2).

Table 14.2 Various types of data and information.

Information on water	Other information
Water resources quantity	Land use: status and trends
Water quality	Population: status and trends
Uses of water resources	Social indicators
Status of ecosystems	Health indicators
Risks: floods and droughts	Economical indicators
Regulation on water resources	

Box 14.2 Environment observatory for Senegal basin.

The role of this institution is to monitor the environment in the whole basin and to provide information needed to evaluate the impacts of the dams and other hydrological installations. Based on the information, decision-makers and population on the ground can decide activities to be developed to mitigate the impacts. The system has been designed with the rationale of the Local Information System and Decision-Support (MERISE Method). The tool called BASE SOE-OMVS allows the institution to manage information, data and actions coming from the ground. This format allows easy updating. A first outcome is a series of maps showing the zero-status of the basin for many topics related to water (maps for 2003).

www.omvs-soe.org

Examples of maps: Rainfall Agriculture and Paludism Development

 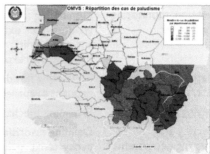

Source: website OMVS / SOE

The responsibility of managing relevant knowledge and data at the basin level must be taken care of in accordance with the prevailing institutional arrangements and the functions of the basin organisation (Table 14.3).

14.2.2   KNOWLEDGE FOR WHOM?

Access to and the sharing of knowledge by individuals and groups is critical for addressing water-related problems. Knowledge is a key input for water management, for instance regarding resource allocation, environmental impact assessment of a project and decision-making. Knowledge is also a base for informing all stakeholders and, more widely, the whole population in the basin. In that way, it is a critical element for developing a consultation process and the real, constructive participation of people.

The use of knowledge concerns a large number of target groups: decision-makers (directors of river basin organisations and administrations, basin committee members, representatives of local authorities), associations of users, NGO representatives, and the population. On the other hand, in the case of an intra-national basin, knowledge is also used at the national level, both for

Table 14.3 Range of possible functions for a basin organisation.

Collecting data	Collecting, managing and communicating data regarding water availability and water demand (including environment) to support basin functions
Planning	Formulating medium- to long-term plans for developing and managing water resources in the basin
Allocating water	Defining mechanisms and criteria by which water is apportioned among use sectors, including the environment
Constructing facilities	Designing and constructing the hydraulic infrastructure
Maintaining facilities	Maintaining the hydraulic infrastructure
Operation and management	Ensuring proper operation of dams and distribution infrastructures; ensuring conjunctive surface and groundwater management
Prevention, monitoring and enforcing	Monitoring and controlling water pollution, salinity levels, and groundwater extraction; ensuring that they remain within accepted limits; and, enforcing relevant laws and regulations to prevent overexploitation, restore ecosystems
Preparing against water disasters	Preventing floods and developing emergency works, flood/drought preparedness plans, and coping mechanisms
Resolving conflicts	Providing mechanisms for negotiation and litigation
Protecting and conserving ecosystems	Defining priorities and implementing actions to protect ecosystems, including awareness campaigns
Coordinating	Harmonizing policies and actions relevant to land and water management undertaken in the basin by state and non-state actors
Mobilising resources	Ensuring financing for other functions, for example, by collecting water user fees or water taxes

Source: Brief IWMI/IRD- INBO – GWP "Developing, managing river basins" 2008

monitoring and control, and for guiding the national policy and its regulations. The database and knowledge base at the basin level also have to be seen as a simultaneous combination of various levels of action: local, basin, regional, national, international, etc.

14.2.3 HOW TO MANAGE KNOWLEDGE AT THE BASIN LEVEL

Considering the features of knowledge, it is important to organise and manage it, particularly at the basin level, in order to use it efficiently and at the right time.

In many countries, however, the accessibility of a water resources and services knowledge base has often been limited due to budget constraints (the emphasis often being on developing a new infrastructure), a lack of professional education, language barriers, and the view that such knowledge represents strategic information which is better not shared with other stakeholders. This results in a decreased capacity to translate available data into usable knowledge for water management in an integrated manner. Basin managers need reliable, accurate and up-to-date information about the state of natural resources and uses in a basin and how they change over time. This means that basin organisations need to create a Basin Information System or a Water Information System (WIS) that will both manage the data collected and deliver analyses of data to different groups of users in formats they can understand and make use of. This knowledge base can be utilised to better analyse and understand current changes in water management. A better understanding of these changes and the challenges that result from them allows for better-targeted and more appropriate intervention strategies. Generating this knowledge base requires accurate data describing the state of water resources and their management. The Water Information System is also the key tool for developing an information/communication flow towards the population, the base for a participatory approach and stakeholder awareness development. These information systems, then, often constitute one of the priority tools to be implemented in order to support an efficient policy for water resources management and risk prevention. The implementation of WIS is supposed to work first at an institutional/organisational level and then at a technical level.

In parallel, monitoring programmes should be organised to produce the necessary data on water resources and water uses, and to allow water management actions to be assessed.

One difficulty met by the basin organisation when starting the process of developing the basin's information system is that the information is usually spread out across many organisations and administrations; these have no willingness, at the beginning, to share as information is regarded as a source of power. The first task therefore is to develop an information system that regroups all pieces of data and makes it accessible to everybody. The first step, then, is to have a strong political will at both the national and basin levels to oblige institutions, NGOs and other bodies that are producers of data to share their knowledge and work together to build an efficient system that also exchanges data. Several experiences exist around the world about developing such a WIS at the basin level. One case is the Körös – Crisuri pilot basin shared by Romania and Hungary, where work led to the development of a database on water quality, among others (Box 14.3).

To be efficient, the system has to be alive: knowledge and data have to be updated with relevant frequency. It has to be seen also in terms of its potential evolution (forecasting). In particular, we need to know all projects well before the end of the decision-making process, in order to both assess the (possible) impact on the water resources and related ecosystem, and to eventually integrate that knowledge into the database used for water management. It is also important to anticipate the changes in land uses. The basin organisation has to establish and maintain a comprehensive data network, available to all stakeholders in ways which suit their needs and encourage them to develop and implement

---

Box 14.3 Management of the Körös – Crisuri pilot-basin
(Romania/Hungary).

The accidental pollution of the Tisza, the main tributary of the Danube, led to the strengthening of cooperation between Romania and Hungary. The Körös/Crisuri sub-basin, a main sub-basin of the Tisza, was chosen for testing the Water Framework Directive implementation with a sub-basin approach and coordination by the ICPDR. In the first phase, among various activities, the data management issue was addressed and allowed to develop; after 2 years of work, a catalogue of shared metadata was available, a website hosted by ICPDR (www.icpdr.org) was created, and methods for a Biological Quality Index were developed.

Source: Documentation of INBO

'knowledge-based' natural resources/water management policies and strategies. The basin organisation itself has to be established and develop procedures that ensure that information is relevant to users' needs, is available in good time for the purposes of planning and management, and that information exchange is accessible, appropriate, equitable and affordable. Data and information systems must meet the particular needs of all stakeholders. Monitoring can be very costly and, before setting up the systems, an estimate of the costs and the availability of funding should be assessed. Often basin managers are under pressure from government, from donors, and from academics to provide a wide range of information. However, the funds available often do not match these demands. The basin manager must, therefore, prioritise what to monitor; the first priority will be that required by law, followed by that needed for the basin organisation to fulfil its mandate, and finally that which is needed for future planning. There will have to be trade-offs between data collection, monitoring and reporting to find the optimal use of the human and financial resources available.

In the case of a national basin, the Water Information System at basin level also has to be part of the National Information System. In Mexico, for example, national and basin information systems are connected to each other, allowing synergy (Box 14.4).

14.3 THE KEY TASKS FOR SETTING UP AN INTEGRATED WATER INFORMATION SYSTEM (WIS)

To set up a basin data and information system, the following tasks have to be undertaken:

- Make an inventory of the existing water-related information institutions, at the basin and sub-basin levels and at the national level, and evaluate the

Box 14.4 Links between national and basin
information systems in Mexico.

The legal and institutional context in Mexico (i.e. water law, laws on statistical, geographical and environmental statistics, as well as the role of various institutions) influences the links between the Mexican National Water Information System and the regional basin water information systems. The system was set up by creating a federal, topical water group to lead the project. Various training sessions were held on data exchange tools, language and information systems. *More information is available at www.conagua.gob.mx*

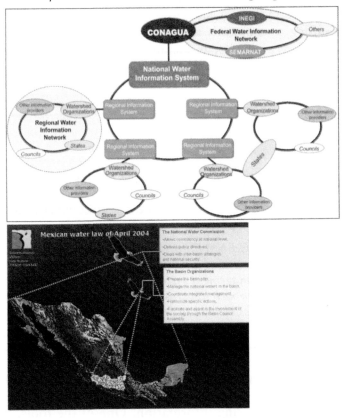

Source: Documentation of INBO

relevance of the information. Those institutions should work together to develop the WIS.

• Identify the relevant needs of the various groups existing in the basin, such as stakeholders and decision-makers, users, and institutions, and define the target groups.

• Define common standards, rules and tools to favour the homogeneous exchange of data between all stakeholders and allow the inter-operability of the information.

- Ensure stakeholders can access and use the data and information in ways that suit their needs.
- Use geographic information systems and other user-friendly means or tools to present the state of basin resources, and monitor changes.
- Make sure there is an interactive, accessible, affordable, appropriate and equitable basin information exchange programme.
- Set up a basin monitoring programme that coordinates information from state, federal, commercial and non-governmental organisations.

Based on those tasks, a basin water information system has to regularly collect and control data. Information must be put in databanks in sustainable ways and treated according to the needs and targets for dissemination and uses by various means, such as a website, as documents, etc. The WISE (Water Information System in Europe) is a good example of what can be done at the regional level (Box 14.5).

Box 14.5 Water Information System for Europe.

The Water Information System for Europe – or WISE – is "Your Gateway to Water". It contains data and information collected at the EU level by various institutions or bodies; data which have either not been available or only been fragmented across many places. The WISE project started in 2002 with the following core objective:

"...the European Commission (DG ENV, Eurostat and JRC) and the EEA are committed to continue the development of a new, comprehensive and shared European data and information management system for water, including river basins, following a participatory approach towards the Member States, in order to have it operational as soon as possible and to implement it, including all the various elements set out in this document, by 2010.

Decision support and modelling tools can integrate and rank information of different types to help decision-making. Modelling tools are particularly useful where there are gaps in existing information. One of the tools, the multi-objective decision support system (MODSS), builds scenarios of the outcomes of alternative land and water management strategies. These systems are most useful when stakeholders specify the size and importance of the variables and processes, and then rank the outcomes predicted by the models. The Geographic Information System (GIS) can be a useful tool to get stakeholder ownership and understanding, especially at a sub-basin level, as there is a real complexity between all elements impacting water resources.

Maps help stakeholders see the likely impacts of proposed changes, not only on the sub-basin, but on the basin as a whole. Presenting the bigger picture in an easy-to-understand format, such as with a GIS, gives basin managers opportunities to negotiate targets and priorities, for example for reducing pollution, for the entire basin ecosystem at the same time as for sub-basins.

Such a system allows information on basin management to be shared clearly and transparently, on progress made (or not), for example. Basin information exchange programmes do not have to be sophisticated. In developing countries, information exchange may first take the form of local meetings or newspaper articles. To be effective, a basin information exchange programme needs a stable organisational 'home', which could be the basin organisation, even though the costs of basin information exchange may be shared by stakeholders. Information needs to be appropriate to the task at hand, proven through research and development, tested in the field and presented to the institutions, practitioners and stakeholders so that they can understand and use it. Information needs to be affordable, preferably free, so that there is no discrimination between information providers and users because of a lack of funds. Information needs to be accessible to all practitioners through the channels they normally use, not dependent on major upgrades. Information processes should be equitable. This means they should respect cultural needs, gender issues and embrace stakeholders distanced from decision-making because of their location, or economic or social status.

14.4 CAPACITY BUILDING FOR ORGANISATION DEVELOPMENT AT THE RIVER BASIN LEVEL

Developing basin organisations is a necessity for applying IWRM efficiently at the basin level. However, the term 'basin organisation' covers a wide variety of organisations, including an administrative commission with a permanent or temporary secretariat, the owner of large infrastructure (such as dams, canals and lock gates for navigation), agencies for planning, financing, implementation, a basin council gathering of all ministerial departments, associations or consortia of users and NGOs to solve a targeted issue, or even a structure temporarily established to drive a project. The important thing is to put in place arrangements fitting with the needs of the basin users, which obviously depends on the scale (transboundary, national or local), the step of development, the major challenges that water management has to face, the economic, social, ecological and political contexts.

There is no universal model of basin organisation structure; however, it is possible to learn from the experience of the other basin organisations, particularly thanks to twinning programmes (Box 14.6) or basin organisation networks (Box 14.7).

Box 14.6 Twinning between basin organisations, a way for capacity building by exchanges of experience and know-how.

Direct exchanges seem to be the more reliable way of disseminating good practices and strengthening the human resources in basin organizations. Several programmes using the twinning method between basin organisations have been developed for several years. It is the case of STRIVER project, driven by NIVA, TWINBAS or TWINBASIN funded by the EU. In order to build the capacity of basin organizations, INBO promoted from 2004 to 2007 bilateral twinning between basin organisations to facilitate exchanges of knowledge, methods, tools, know-how and practices among qualified and identified staff. In five years, 41 twinnings, concerning 70 basin organisations in 42 countries, have been implemented. The main topics addressed were about stakeholder participation, information and databases, and institutional frameworks.

Source: INBO and STRIVER websites

Box 14.7 INBO.

The International Network of Basin Organisations (INBO) was established in 1994 to facilitate exchanges among basin organisations about institutional, operational and technical issues and to mobilise the experience of the professionals of administrations and organisations directly responsible for the implementation of IWRM at the river basin level. INBO and its regional networks are networking 185 members and are currently present in 68 countries.

A basin organisation is not a monolithic structure which puts all of its functions under one roof. It is a body which brings an added value to the existing institutional framework and which can carry out activities in an efficient manner, thanks to its proximity to the stakeholders and users. To develop a basin organisation, a wide range of capacities is needed. Those capacities depend a lot on the functions assigned to the organisation. Moreover, we need to consider the capacities, competencies and skills of the staff working within the basin organisation – namely, in the secretariat – and the capacities of the stakeholder representatives that exercise decision authority, like the basin committee. Globally, a river basin organisation requires staff with backgrounds in a variety of disciplines, not only in water engineering, but also in environment, economics, law and social sciences. A strong capacity in planning, in data monitoring, in impact assessment and in conflict resolution is also required. Specific skills in communication and pedagogy will be a basis for developing the involvement of each inhabitant in water management of the basin. A good understanding of the interactivity among the various sectors and domains in the basin is a key topic which has to lead to specific activities of capacity development for targeted people, in particular to improve the participation of the population and stakeholder representatives in the planning and decision-making processes. Whatever the role and mandate of the body in place for river basin management, capacities are required to develop an effective cooperation among the numerous institutions operating at various levels in the basin. To be efficient, river basin management has to be based on autonomy. So, specific competencies are needed in finance, budgeting, accounting, prioritising, human resource management, etc. Capacity development has to be an outcome of strategy, coming from an analysis of the baseline and the status of the basin, an analysis of the tendencies and analysis of the priorities to improve water resources. Therefore, the basin organisation has to be structured to improve its capacity for forecasting. Based on the basin strategy and then the priority objectives to be met, and taking into account the current capacities and skills in the basin organisation, a capacity-development plan can be developed.

14.5 HOW TO DEVELOP A CAPACITY-BUILDING PROGRAMME

14.5.1 CASE STUDY MEKONG RIVER COMMISSION

14.5.1.1 Context

The Mekong River Commission (MRC) was formed on 5 April 1995 by an agreement between the governments of Cambodia, Lao PDR, Thailand and Vietnam. The four countries signed The Agreement on the Cooperation for the Sustainable Development of the Mekong River Basin and agreed on joint management of their shared water resources and development of the economic potential of the river. In 1996 China and Myanmar became Dialogue Partners of the MRC and the countries now work together within a cooperative framework. The MRC is an international, country-driven river basin organisation that provides the institutional framework to promote regional cooperation in order to implement the 1995 Agreement.

Since the 1995 Mekong Agreement, the MRC role has expanded and the focus of basin development competencies has also shifted. New basin-wide activities require different skills, beyond an exclusive focus on technical issues. In order to create riparian ownership of the MRC and its programmes, the organisation has needed to focus on how to generate relevant capacities at all levels.

The MRC has recently developed a capacity-building programme to support the Mekong Programme, (the Regional Cooperation Programme for the Sustainable Development of Water and Related Resources in the Mekong Basin, established for 2006–2010).

Mekong Basin
- 6 countries
- 4,800 km
- 795,000 km^2
- Population: 55 million

Figure 14.1 Map of the Mekong river basin.

This programme is in fact, the outcome of a process started in 2002/2003 with the development of an "Integrated Training Strategy", to define various capacity-building needs through an extensive participatory approach. The process was assisted by the UNESCO-IHE Institute for Water Education. As defined in the Integrated Training Strategy, "Capacity Building is widely regarded as the key strategy in ensuring sustainable water sector development", and political will for capacity development was clearly stated (Box 14.8). The Mekong Programme has four goals: three are related to regional cooperation, sustainable development and strengthening of basin-wide environmental monitoring and impact assessment. The fourth goal is directly focused on the IWRM capacity and knowledge base of national and basin structures and stakeholders. Indeed, all aspects of organisational development envisaged in the Plan 2006–2010 rely heavily on adequate human resources, such as IWRM professionals, to develop policy and legal frameworks, river basin organisation professionals to manage and develop the necessary institutions and mechanisms at different levels (Mekong Commission, national, basin and sub-basins). Creating a critical mass in IWRM and basin management competencies was considered a critical step for the success of the Plan 2006–2010. The general objective of the Integrated Capacity-Building Plan (ICBP) is to ensure that various relevant entities (MRC, national entities, agencies and other stakeholders) will be able to manage water resources by applying IWRM principles and using basin-wide information and knowledge management systems.

Box 14.8 **Strong political will for developing capacities.**

The Strategy and Action Plan for "Riparianization of the Mekong River Commission Secretariat" endorsed by the Joint Committee in June 2007 put great importance on the Integrated Capacity Building Programme as quoted below:

> "Across the MRCS, the competency shortfalls have been observed, in the main, not to be related to technical qualifications, but to the cross-cutting or 'integrative' competencies necessary for effectively carrying out water resources management, river basin planning and environmental management, particularly in the areas of resource management policy and IWRM approaches".

Achievement of the MRC strategic goals for the next five years will depend greatly on the ability of the MRCS to provide neutral, balanced and objective advice to the Joint Committee and member countries for the benefit of the basin as a whole. Building the capacity of the MRC Secretariat is crucial in this respect. As stated in the Strategic Plan 2006–2010 "Riparianization must occur in tandem with capacity building and targeted training, and also in a phased way that allows riparians to become managers with the appropriate management skills, plus with the right degree and nature of technical support." Capacity building is therefore a priority activity of the MRCS to accompany the riparianization process.

Source: MRC Work Programme

14.5.1.2 *Elaboration of the Integrated Capacity-Building Plan (ICBP)*

For the elaboration of the Capacity-Building Plan, teams at regional and national levels were established to enable activities to be carried out efficiently and in a timely manner. A clear framework has been developed defining the responsibilities and necessary capacities of team members. To assist the programme management team in formulating the Capacity Building Programme document, consultants were commissioned. At the national level, coordinators were appointed to ensure activities at this level. As the training activities are supposed to involve various levels, partnerships with regional and national training institutions have been developed, particularly for the implementation of priority training courses. To address the function of the MRC, major cross-cutting (integrative) competencies have been identified, covering knowledge areas, skills and attitudes. These are:

- Integrated Water Resources Management;
- Integrated River Basin Planning and Management;
- Environmental Management;

- Flood Management and Mitigation; and,
- Management Support Tools.

A rapid needs assessment (RNA) was conducted in January 2008, with the active participation of staff from all entities (basin, national, sub-basin).

The RNA recommended that the MRC urgently address these cross-cutting capacity-building needs simultaneously on two levels: human resource development and institutional development. The RNA pointed out that, in the past, training activities were scattered, formulated independently into different programmes and not sufficiently coordinated. This led to overlaps between the various training activities and an overall waste in training expenditure. Moreover, it demonstrated the need for riparian officials to be more thorough in his/her area of expertise and to be equipped with basic competencies required by the organisation, meaning competencies in IWRM; river basin planning; environment management; flood management; and with skills in organizational development, networking, communication, and in political and cultural matters.

According to the RNA, the ICBP was designed to focus on the main priority training areas which are: IRB planning, programme/project planning, cross-cultural communication, networking, fundraising and leadership. The ICBP integrated three ongoing training projects: gender mainstreaming in water resource development, junior riparian professional development, and MRC staff training. In order to be efficient, the programme management team made a thorough search of lessons learned from previous and ongoing capacity-building programmes and projects within the MRC and other related river basin development programmes/projects, and also analysed various training modalities implemented in the past. In the whole process, many consultative meetings were undertaken at the MRC and national levels with staff and stakeholders.

To improve the trainings, training institutions are asked to provide an evaluation report on the trainings and provide recommendations, which are used as input to the development of the Monitoring and Evaluation System. Also, target groups representing the main actors in the Lower Mekong River Basin were identified and training courses will be adapted to the needs and features of each group. It is interesting to note that the priority is not only with the professional staff of the MRC Secretariat, but also with the professionals in the national committees, and the professional staff of national line agencies and training organisations involved in MRC activities.

14.5.1.3   Implementing priority capacity-building activities and evaluating their effectiveness

The priority activities proposed to cover the following areas, and are focused on professional knowledge, skills development, attitude change and institutional capacity building.

- *Integrated River Basin Planning and Management*

Integrated River Basin Planning (IRBP) is developed first by the training of trainers, with technical assistance from the Murray Darling Basin Commission. The training of trainers programme will address IRBP and IWRM concepts and tools, the participatory learning approach, training course development and facilitation, development of national training packages for the IRBP training course at national level.

- *Programme/projects planning and management*

Skills and attitude in programme/project planning have to be developed in depth, and a practical training course needs to be designed to enhance the skills of the planners of the programme/project planning, including fundraising skills. Results-based mechanisms for monitoring and evaluation of skills have to be developed also:

   - Networking and Coordination
   - English Proficiency
   - Report Writing
   - Cross-Cultural Communication.

- *Cross-cultural Communication*

This capacity is needed to help participants learn to accept cultural differences and to have the ability to adjust one's style of behaviour based on neutrality and openness. In addition, the capacity to recognise the differing interests and power spheres in the region, both within and outside the MRC organisation, is essential in the MRC context.

- *Leadership and team-building*

Leadership, coaching skills, consensus-building capacity, capacity to develop networking, to coordinate and to build teamwork have been considered critical areas to tackle immediately. These topics will be addressed by seminars delivered for executive level (directors-general, directors) to enhance their leadership skills.

Various training modalities have to be used during the implementation of the capacity-building programme, such as the training-of-trainers, an in-depth and practical training, a workshop and seminar according to the target and the topics.

All training courses or modules have to be tailored to the Mekong region and should focus on practical knowledge, respect adult learning principles, use real case studies, adopt participative approaches, use role-plays, and be linked to the daily realities of the participants. From the beginning, they should incorporate concrete actions for follow-up activities after training. In addition, special consideration and priority will be given to qualified women professionals who are involved in water resources development in

Mekong countries. In particular, the programme team, in coordination with the national coordinators, has to encourage female professionals in all entities to participate in capacity-building activities.

14.5.1.4   Monitoring and evaluating the training programme

The MRC supports monitoring and evaluation as the joint responsibility of the MRCS ICB programme management team and the selected training institutes. Their responsibilities may include obtaining agreements from all project operators and implementers on the milestones and targets to be achieved and the schedule of activities; reporting requirements; setting clear targets and evaluation criteria for each team of external consultants; holding working sessions to review and monitor progress; and, conducting an evaluation session. Key M&E activities to be carried out by the programme management team during this process are realised through meetings and reports.

14.5.2   CASE STUDY NIGER BASIN AUTHORITY

Created in 1980, the Niger Basin Authority (NBA) with its nine country members has developed a vast process of "Shared Vision" started in 2002 for making "the Niger Basin, a common area of sustainable development by the integrated management of the water resources and associated ecosystems for the improvement of living conditions and the prosperity of the populations". Within this approach, a Sustainable Development Action Plan for the period until 2025 and an investment programme have been elaborated.

In this evolutionary context, a strong political will has been clearly expressed for capacity-building activities. Indeed, capacity building has been considered one of the three priority domains in the strategic document (Action Plan for Sustainable Development) which defines and guides the process of integrated development shared by all country members (Box 14.9). Moreover, the basin charter states explicitly the need for promoting information exchanges: reinforcement of capacities into the basin, in particular as regards

Figure 14.2 Map of the Niger basin.

> *Box 14.9* Action Plan for Sustainable Development.
>
> The Action Plan for Sustainable Development (APSD) aims at elaborating a basin management master plan which accompanies the sustainable development of the Niger Basin, allows a better prioritising of actions for poverty reduction, basin environment protection, and reinforcement of cooperation between country members. The objective is also to ensure sustainable and efficient participation of the civil society and private stakeholders to implement the shared vision.
>
> Several actions developed in other programmes are integrated to the APSD, such as a master plan for reducing riverbed sedimentation and the formation of sand banks, a strategic action plan to reverse the degradation of soils and water quality, a development programme for water resources and ecosystem sustainable management.
>
> Source: website ABN/NBA

IWRM and the use of adequate technologies for the management of the Niger Basin hydrological catchment area.

To accompany the process, it appeared necessary to reinforce capacities to the level of the Executive Secretary. A capacity-building plan was prepared, based the outcomes of an organisational audit and using a participatory approach involving all senior staff of the Executive Secretariat. Workshops have been organised to think about baseline competencies, and the capacity- strengthening and training needs for all staff of the Executive Secretariat. Thanks to this process, major challenges to be faced and priority fields for intervention have been identified and seven priority themes have been targeted:

- Reinforcing administrative arrangements and developing efficient tools for financial accountability;
- Professionalizing human resource management;
- Developing a judicial framework to prevent conflicts and to secure the cooperation:
- Setting in place information and communication systems which guarantee better visibility of actions of NBA and contribute to improving collaboration within the organisation;
- Developing technical expertise within NBA for advising Member States on water issues;
- Improving monitoring and assessment tools for a better decision-making process; and,
- Evolving the management style to enhance leadership and responsibility.

Developing capacities in these areas will help the harmonisation and coordination of national policies, the elaboration and implementation of an integrated development plan, the design and operation of works, the control and regulation of navigation, the formulation of technical assistance and the mobilisation of financing.

According to the seven priority themes identified, a wide set of activities has been set up.

In administration and accounting, training sessions will be realized, for instance on using accounting tools, on increasing the capacity in procedural matters, and on financial control.

To address the capacity of leadership and management, several trainings for the top management are foreseen, on subjects such as strategy, gender approaches, teamwork, participative approaches, results-based management, and conflict resolution. Through the trainings, a new style of management will be promoted based on better listening, anticipation, a bottom-up approach and increased delegation of responsibilities, it being recognised this will improve the efficiency of the Executive Secretariat.

For human resource management (HRM) which is considered a key point of attention, a system for competencies and skills management will be developed. HRM staff will benefit from training on the use of HRM tools including the definition of job profiles. A handbook for HRM will be developed, and include recommendations on managing equity and gender issues. Internal workshops will be organised on procedures, on staff performance and evaluation (the process of dialogue, feedback, coaching, counselling and mentoring). In addition, a human resource development plan will be elaborated.

As the prevention of conflicts within the basin represents a major challenge for the "Shared Vision", a judicial framework will be developed. To support its implementation, an audit has been necessary. The four components of the judicial framework are: the Niger Basin Charter, the Environment Charter, the instruments for the prevention and resolution of conflicts, and the instruments for navigation management. All this will be accompanied by workshops for information, ownership and validation.

About communication, a new three-year communication plan will be implemented, including the development of a more efficient information system. All staff, including higher staff will participate in training sessions for gaining the capacity to use such an information system efficiently.

Of course, technical expertise is also a target for training activities. The priorities are about project cycle management (identification, formulation, assessment, monitoring). Capacity to design, plan, program and manage projects and programmes has to be reinforced by trainings and workshops. A set of trainings is also planned for managers on water quality, hydraulic models, awareness, participative approaches, formulation, coordination and the monitoring of projects/programmes, and planning and management by objectives and results. Assessment of socio-economic and environmental impacts of projects, the follow-up of environmental impact studies, elaboration of management plans for wetlands, or development of mechanisms for payment of environmental services are targeted for specific training sessions. In addition, methodological guides, synthesis documents, and action guidelines will be elaborated on and disseminated according to the analysis of relevant situations in pilot basins. National and regional workshops are organised for defining an action plan for IWRM at the basin scale, and for supporting the ownership of and strict application by country members, decision-makers and stakeholders. As some countries already have an IWRM plan, national and regional workshops to share the lessons learned about the preparation and implementation of those plans and good IWRM practices applied in the basin will be organised.

To improve the monitoring and assessment capacity for the NBA executive secretary, a set of trainings will be developed, focusing on Results-Based Management and understanding the role of M&E for effectiveness of the organisation. The top management will be also benefit from those trainings while an M&E handbook will be elaborated including training to use the M&E system.

As the implication of the civil society moving into the management and development of water resources within the basin is a priority for implementation of the shared vision, an NGO has been mandated to develop a study to identify and better know the stakeholders, their organisations, their features and their roles (current and potential) within the basin area. Since 2007, representative organisations (coordination) have been in place at the national level; the NBA executive secretary is developing awareness raising and training activities with them.

14.6 RECOMMENDATIONS

The two case studies presented above and the experience of other basin organisations helped determine some key recommendations regarding capacity development and knowledge management at the basin level.

- **Political will** is required for success in capacity building and for developing an information and knowledge system and communication flow. Political will has to be stated officially by top management and the political body of the organisation, for instance in the charter (Niger Basin) or in a Strategy of the Basin (MRC) document. Financial resources have to be allocated to fund these activities, showing that the first priority is with knowledge and capacity development. The political will be translated into decisions and choices which will facilitate the implementation of training activities (budget, time to develop the training plan, incentives to staff for training).
- **Commitment of countries** in the case of a transboundary basin: as shown in the NBA example, riparian countries have to be committed, particularly

because capacity development will be at all levels, including the national level. Such commitment should be about supporting implementing activities and transparency (the case for NBA).

- The basin organisation has to get enough **autonomy** for making key decisions about capacity building, without referring to a higher level (which would be rather inefficient and time-consuming).
- Capacity building needs to be in line with the **Basin Strategy Plan**: for success of capacity development, the training plan has to be designed as a support to the basin strategy and its objectives; strategy has to focus on information and training after identifying the priority needs.
- The organisation has to think about a **long-term training plan** based on a long-term (around 5 years) work plan or strategic plan. In the case of NBA, before the elaboration of the training plan, there was the phase of the elaboration of the action plan for sustainable development including a base line on capacities and definition of needs for reaching the priority objectives of the Basin Management Master Plan. Making a long-term plan avoids overlaps and action on an ad hoc basis and facilitates prioritisation of needs according to the objectives.
- **Implication for staff**: staff of the basin organisation secretariat, when it exists, have to be involved in the process of preparing the training plan so their needs and vision can be taken into account.
- While the capacity-building plan is being elaborated, it is important to get a **broad vision** of the potential topics. Indeed, as basin management refers to a wide range of people, stakeholders and subjects, including multiple links with issues such as land management, it (basin management) is a rather complex task. So many capacity needs are not directly related to water issues or technical matters. In particular, the notions of integration, sharing and ownership are very important to the success of basin management. The institutional capacity of the basin organisation is also critical. Therefore, HRM, financial management, leadership style, teamwork, information, and communication are themes for which the strengthening of capacities and skills are necessary. Those aspects will need to be included in the capacity-building plan.
- The capacity-building plan has to address **all levels**: country level in case of a transboundary basin, sub-basin level in case of national basin. Local bodies have to be involved according to their relevance. As CB is a real investment, it is also important to think about the retention in the organisation of those to be trained, particularly staff members at high level that tend to be subject to frequent inter-organisational transfers.
- Relevant lessons learned from experience with **other capacity-building programmes and projects** must be incorporated into the design of the capacity-building plan.
- In the case of a transboundary basin, training and workshop activities have to address the aspects of **regional and cultural differences, and the**

sensitivities between the riparian countries. This also pertains to large national basins where there may be large differences between up- and downstream areas, for example. In this connection, efficient communication and information tools are very important (WIS).

- The capacity-building plan has to concern the entire staff (both technical and administrative staff and higher top management), as well as decision-makers and elected officials, members of commissions or sub-commissions, representatives of users and stakeholders. Basic and continuing professional training often needs to be increased: this means assessing the training needs, creating/reinforcing capacities for training water professionals, and developing funding mechanisms for professional training.
- During the formulation of the training programme in a transboundary context, special attention needs to be devoted to several issues such as the use of local languages for the delivery of training, explicit use of local or national trainers, and the involvement of national commissions or sub-basin commissions, when they exist.
- Support by other basin organisations is particularly useful for developing knowledge, learning about the use of tools like water information systems and modelling tools, and developing know-how, for instance, about the participation of the civil society in basin management. Twinning between basin organisations can be a good process for improving the capacities of basin organisations.

Part 4

Evaluation and indicators

Chapter 15

THE ROLE OF EDUCATION IN CAPACITY DEVELOPMENT

Stéphan Vincent-Lancrin
OECD Centre for Educational Research and Innovation

ABSTRACT

Acquiring and producing new knowledge is a key component of capacity development. Whether this knowledge has been produced abroad or domestically does not really matter. This is true for all sectors of capacity development, including water. This chapter illustrates the need to strengthen education for any capacity development strategy, including water capacity development. The first section presents the concept of capacity development and shows its multi-level dimension and its links to knowledge and education. The second section stresses the different roles of primary, secondary and tertiary education in capacity building, as well as the different levels of efforts needed to strengthen them in different countries. The final section shows how cross-border tertiary education, that is the mobility of people, institutions and programmes, contributes to capacity development and argues that it should be an instrument for capacity development.

15.1 WHAT IS CAPACITY DEVELOPMENT?

The concept of *capacity building* or *capacity development*[1] appeared in the late 1980s and has become the buzzword of development in the 1990s.

---

[1] In this paper, the two terms are considered synonymous: although they are sometimes used with slightly differing meanings, this is how most people treat the two notions. "Capacity building" is

Rather than capturing a new idea, it embodies the criticism of development assistance by emphasising the need to build development on indigenous resources, ownership and leadership and by bringing human resources development to the fore. While the 1980s are typically described as the "stabilisation and structural adjustment" decade, in the 1990s a strong emphasis was placed on the building of human capital following advances in the "endogenous growth theory" in economics (Thorbecke, 2000). The shift from traditional development aid to capacity building is illustrated by the well-known proverb: "give someone a fish and he eats for a day; teach someone to fish, and he can feed himself for a lifetime". The concept of capacity building signals a shift from *assistance* to a less dependent "help yourself" attitude in the development community. It appeared in a context marked by a widespread (and possibly exaggerated) dissatisfaction with technical cooperation (Arndt, 2000) and, more generally, with aid effectiveness – the so-called "aid fatigue" of the 1990s. This new philosophy has started to be challenged with clear evidence that existing institutions or individual preferences do not necessarily make of poor people themselves the best drivers of development (Banerjee and Duflo, 2008).

The OECD has recently defined capacity and capacity development as follows:

> *Capacity is the ability of people, organisations and societies as a whole to manage their affairs successfully. Capacity development is the process whereby people, organisations, and society as whole unleash, strengthen, create, adapt and maintain capacity over time* (OECD, 2006).

And the United Nations Development Programme (UNDP) defines capacity and capacity development as follows:

> *Capacity is the ability of individuals, organisations and societies to perform functions, solve problems, and set and achieve goals. Capacity development entails the sustainable creation, utilisation and retention of that capacity, in order to reduce poverty, enhance self-reliance, and improve people's lives. [...] Capacity development builds on and harnesses rather than replaces indigenous capacity. It is about promoting learning, boosting empowerment, building social capital, creating enabling environments, integrating cultures, and orientating personal and societal behaviour. (www.capacity.undp.org)*

often used in the African context as well as in relation to trade and private sector development, whereas "capacity development" is more commonly used in aid development agencies. There is actually no clear cut distinction and both terms refer to the same idea: capacity development is more accurate, though, in that it recognises that there is always some initial capacity to be developed (whereas capacity building seems to imply one builds on a *tabula rasa*).

Capacity building is thus based on learning and the acquisition of skills and resources among individuals and organisations. The acquisition of skills should be seen in opposition to the transfer of technology or technical assistance, neither of which have necessarily led to individual and/or organisational learning in developing countries. While this process does rely on some imported resources, foreign capacity should be used as a knowledge-sharing device, which allows the strengthening and development of the local capacity (Figure 15.1). Capacity building is committed to sustainable development, to a long- rather than short-term perspective, and attempts to overcome the shortcomings of traditional donor-led projects – typically criticised for being too short-term rather than sustainable, and not always addressing the needs of the recipients. Development within a capacity-building context allows developing countries to identify their own needs, and design and implement the best suitable development strategy within the local context. Its ultimate aim is to make developing countries less dependent on aid. As a process, it builds on monitoring and evaluation in order to identify existing capacities, deficiencies, and the progress and achievements of development.

In Boxes 15.2 and 15.3, two sets of guiding principles for capacity development are outlined. They offer two different entries based on a shared philosophy. According to capacity development principles, ownership of development projects is transferred from the donor to the recipient community and mirror

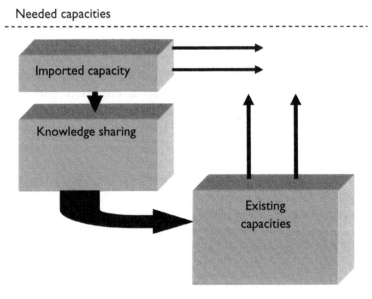

Figure 15.1 **Capacity development.**
Source: OECD (adapted from UNDP, 2003)

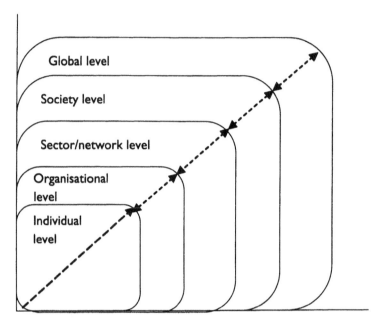

Figure 15.2 Capacity development: a multi-level conceptual
framework.

Source: OECD (adapted from Bolger, 2000)

recent aid-effectiveness principles.[2] For this reason, capacity development is
not necessarily linked to development aid but can also describe countries'
efforts to meet their development goals regardless of external assistance.

An important dimension of capacity building is found within its systemic or multi-level approach to development: capacity building acknowledges the need to consider several levels of interventions and understands
their interdependence in order to develop a coherent and sustainable
development policy. Adapting Bolger (2000), five levels of capacity are
considered in this paper (Figure 15.2): the individual, organisational,
sector/network, society and global level (OECD, 2001; UNDP/Global
Environment Facility, 2003).

At the individual level, capacity development refers to the acquisition of
skills, through formal education or other forms of learning. Although skills
and knowledge can be acquired in various settings, formal education systems
play a paramount role in this connection. In the case of water, formal education can be very important to learn some water management techniques

---

[2] Namely, that foreign aid should depend more on what the recipient needs than on what the
donor can provide; that foreign aid through foreign technical expertise is often unsustainable
once the foreign experts are gone; that aid should be managed by the recipient country and
untied, etc.

Box 15.1 Example of the CKNet-INA network: capacity development
at individual, organizational and sector levels.

CKNet-INA is a network of ten universities in Indonesia with the aim of improving water management and supporting the government policy to decentralize their water management responsibilities to the regions and to strengthen the local government institutes and their human resources (CKNet-INA, 2008). The objective of the project is to contribute to the improvement of the Water Resources and Irrigation Sector performance in Indonesia, through increasing the capacity of lead Indonesian universities to deliver training courses on water resources management and irrigation. The focus is on building the capacity in water resources and irrigation management (WRIM) of selected leading universities in Indonesia. Five WRIM universities have been selected on the basis of their regional coverage and their experience in delivering water resources and irrigation-related educational and training programs. The project provides an opportunity for the network partners to learn and to acquire experiences in collaborative networking and how to further develop and maintain their network in a sustainable way. Based on scheduled Capacity Building Needs Assessments and Training Demand Assessments in the sector, capacity building programmes are identified, developed and implemented to strengthen and/or develop capacity of the universities to offer and deliver demand responsive programs to strengthen the capacity of WRIM sector professionals and their institutions active in the sector.

In this way the national strategic component in professional learning is enhanced and the Government of Indonesia's policy in water sector reform is supported.

Source: CKNet-INA, 2008.

---

in agricultural settings, but also to master and absorb knowledge about water management and use that has been developed abroad or to invent new ones. The individual level is crucial as a more efficient use and management ultimately rely on individual behaviour.

At the organisational level, capacity development focuses on infrastructure and institution building, the availability of resources and the efficiency of processes and management to achieve effective and quality results within existing infrastructures. In education, this level signifies the improvement of domestic educational institutions (*e.g.* primary schools or universities), through additional resources and a better use of those already available. As for water management and use, this level refers to the development of organizational processes that lead to a more efficient use of water, be it within collaborative labour in villages, in production processes within companies, or in other kinds of organizations such as schools.

At the sector/network level, capacity development seeks to enhance the consistency of sector policies and promote a better coordination between organisations. In education, capacity development can, for example, aim at improving links between vocational and academic educational institutions, between research-intensive and teaching-only institutions or to improve the coordination of institutions across different academic fields. In the water area, this can relate to capacity development in a specific sector, for example a water industry or the use of water in agriculture, but also to the interaction between different uses of water in different sectors: ensuring that some industrial companies do not make water improper for effective use by farmers, or that people up a river do not compromise water down the river, can be an important capacity development area.

The society level refers to the human frameworks (conventions, habits, values, regulations, political regimes, policies, etc.) within which development takes place. The society level can enable or constrain development. Gender inequality, racial discrimination, corruption, lack of security and commitment

---

Box 15.2 OECD Development Aid Committee (DAC) guiding principles for sustainable development strategies.

Strategy formulation

- Country ownership and participation, leadership and initiative in developing their strategies.
- Broad consultation, including particularly with the poor and with civil society, to open up debate on new ideas and information, expose issues to be addressed, and build consensus and political support on action.
- Sustained beneficial impacts on disadvantaged and marginalised groups and on future generations are ensured.
- Convergence and coherence are enabled by building on existing strategies and processes, rather than by adding additional ones.
- A solid analytical basis, taking account also of relevant regional issues, including a comprehensive review of the present situation and forecasts of trends and risks.
- Integration of economic, social and environmental objectives through mutually supportive policies and practices and the management of tradeoffs.
- Realistic targets with clear budgetary priorities.

Capacity development

- Strengthening and building on existing country capacity – public, civil society and private – as part of the strategy process.
- Linking national and local levels, including supporting devolution, in all stages of strategy development and implementation.
- Establishing continuous monitoring and evaluation systems based on clear indicators to track and steer progress.

Source: OECD, 2001a.

Box 15.3 UNDP's 10 default principles for capacity development.

1 Don't rush – Capacity development is a long-term process. It eludes delivery pressures, quick fixes and the search for short-term results.
2 Respect the value system and foster self-esteem – The imposition of alien values can undermine confidence. Capacity development builds upon respect and self-esteem.
3 Scan locally and globally; reinvent locally – There are no blueprints. Capacity development draws upon voluntary learning, with genuine commitment and interest. Knowledge cannot be transferred; it needs to be acquired.
4 Challenge mindsets and power differentials – Capacity development is not power neutral, and challenging mindsets and vested interests is difficult. Frank dialogue and a collective culture of transparency are essential steps.
5 Think and act in terms of sustainable capacity outcomes – Capacity is at the core of development; any course of action needs to promote this end. Responsible leaders will inspire their institutions and societies to work accordingly.
6 Establish positive incentives – Motives and incentives need to be aligned with the objective of capacity development, including through governance systems that respect fundamental rights. Public sector employment is one particular area where distortions throw up major obstacles.
7 Integrate external inputs into national priorities, processes and systems – External inputs need to correspond to real demand and be flexible enough to respond to national needs and agendas. Where national systems are not strong enough, they should be reformed and strengthened, not bypassed.
8 Build on existing capacities rather than creating new ones – This implies the primary use of national expertise, resuscitation and strengthening of national institutions, as well as protection of social and cultural capital.
9 Stay engaged under difficult circumstances – The weaker the capacity, the greater the need. Low capacities are not an argument for withdrawal or for driving external agendas. People should not be held hostage to irresponsible governance.
10 Remain accountable to ultimate beneficiaries – Any responsible government is answerable to its people, and should foster transparency as the foremost instrument of public accountability. Where governance is unsatisfactory it is even more important to anchor development firmly in stakeholder participation and to maintain pressure points for an inclusive accountability system.

Source: UNDP, 2003

---

to development, inability to raise taxes, etc., are typically constraining factors. Stable political and economic environments, commitment, sound policies, etc., on the other hand, typically facilitate development. In some cases, capacity development has the ability to transform attitudes and values which are hindering development, for example through efforts to fight corruption, crime and insecurity, or other policies challenging socially unproductive behaviour, such as gender inequality. Transforming society, however, is a slow and uneasy process. The society level, whether this is a facilitating or constraining development (and regardless of whether this level is changing) forms the basis for capacity-building activities. The importance of appreciating the nature of this level corresponds with the principle of developing capacity-building activities within the local situation: one size does not fit all.

Finally, the global level needs to be taken into consideration, *i.e.* the international context in which the country operates. This includes multilateral agreements, international laws, but also geo-strategic considerations. At this level, capacity development seeks to improve a country's participation in, and utilisation of, international organisations, treaties and agreements. Water management, use, and scarcity are global challenges which can be the source of conflicts, wars, or the vehicles of diseases. That makes this level particularly important.

15.2　THE CENTRALITY OF EDUCATION AND HIGHER EDUCATION IN CAPACITY-BUILDING STRATEGIES

The education and higher education subsector play a significant role in any capacity-building strategy. The ultimate goal of a national capacity development strategy is to achieve progress and development, in specific sectors but also by becoming a more affluent nation with less pressure in the use and distribution of its public (and private) resources. According to their natural assets and constraints, to their already existing capacities, to their possible competitive advantages, and to their priorities, countries need to develop differing national development strategies. National development strategies build on a variety of complementary sectoral capacity development strategies. A country may need and want to develop capacity in education, in trade, in health, in engineering, in agriculture, in water management, etc.: each sector contributing to growth and to its development goals in a different manner.

Some sectors such as education, health or trade are cross-sectional or horizontal in the sense that they impact on all sectors in the economy. Even if agriculture is the main priority in a developing country's strategy, it should not neglect the horizontal sectors as its agricultural sector will be more competitive if it has a healthy labour force (health), if its peasants know and use the latest agricultural and water treatment and management techniques (education), and, possibly, if it can trade them effectively on the national and world markets (trade). In turn, better management and agricultural output will contribute to better health and education.

Education has a unique privilege as a built-in feature of any capacity development strategy. Whatever the sector, capacity building relies to some extent on the strengthening of individual capacity through training and learning, in order to raise the domestic stock of quality human resources.

This can be done by setting up or increasing access to specific educational programmes in the formal education system or by other forms of learning. Although some of the necessary skills would typically be acquired on-the-job or through learning-by-doing, developing countries characterised by less efficient organisations of work or by less advanced technologies might need to rely more on formal vocational education and training. What level of education (primary, secondary or tertiary) is required to achieve this goal depends on the kind of competence to be built.

15.2.1   THE CONTRIBUTION OF EDUCATION TO ECONOMIC DEVELOPMENT

Why invest and develop capacity in education and tertiary education as part of capacity development? One reason is that education and tertiary education can contribute to economic development. In the case of water, this can imply that some countries will have the resources to access cutting-edge water technologies and infrastructure.

As (basic) education for all, and more precisely "achieving universal primary education", is one of the internationally agreed upon Millennium Development Goals, the importance of education for development hardly needs discussion. Education is widely seen as a good in itself and one of the "primary goods" all people are entitled to in democratic societies. Understood as a road to freedom, development policies can certainly not neglect education and treat it as a luxury in the context of developing countries as it enhances people's "personal capabilities" which are seen as a fundamental objective of development (Sen, 1999; Sen and Williams, 1982). A host of basic ethical, humanistic and political reasons justify investment in education in all countries in the world.

But there is also a host of economic and social reasons for developing capacity in education. Education is widely considered a significant engine of economic growth. The estimated long-run effect of one additional year of education in the OECD area generally falls between 3% and 6% of GDP per capita (OECD, 2004a). The few economic studies which attempted to weigh the impact of different levels of education on economic development have shown that the impact of education differs according to countries' stages of development, although explanations of the differences differ (Pritchett, 2001; Hall and Jones, 1999; Hanushek and Kimko, 2000; Krueger and Lindahl, 2001). According to Gemmell (1996), tertiary education is more important in OECD countries, while secondary and primary education contribute the most to growth in the intermediate and poorest countries, respectively. This does not imply, however, that tertiary education does not play a role at all in developing countries.

At the macro level, recent advances in the economic growth theory have brought human capital to the fore. Two main mechanisms explain how the stock and/or growth of human capital can impact on growth and economic development (Aghion and Howitt, 1998; Sianesi and Van Reenen, 2003; de la Fuente and Ciccone, 2002).

First, a rise in education could have a once and for all impact on economic growth: it would lead to a rise in the level of output of the economy (Lucas, 1988; Mankiw, Romer and Weil, 1992). The output growth is then proportional to the growth of education. A developing economy could thus develop by increasing its quantity of quality human capital, defined, for example, as the educational attainment of its population. Quantity does not necessarily imply quality though, which is what matters most, as shown by studies linking growth to assessments of cognitive skills rather than attainment (Hanushek and Wössman, 2008). All other things remaining the same, a developing country would then catch up with developed countries once it has accumulated the same amount of human capital.

Second, a rise in human capital could have a permanent effect on economic growth. Human capital is seen as a determinant of the *growth rate* of the economy rather than just a determinant of its growth (or level of GDP). This implies that human capital allows developed countries to grow more rapidly than developing countries and that the gap between them could continue to widen if developing countries were not to catch up in terms of human capital. The underlying mechanism is the following: growth is driven by physical capital investment, which is in turn driven by innovation, by investment in research and development (R&D) generating ideas for new designs or goods (Romer, 1989, 1993; Aghion and Howitt, 1998). For this to happen, a country needs a population with different levels of education, but tertiary educational attainment is particularly important. Researchers and highly skilled workers drive innovation, and possibly technology transfer, but an educated workforce with lower educational attainment is also necessary to absorb the new technologies. Another close explanation views education (and more broadly human capital) as a facilitator of transfer technology from "innovating countries" to "imitating countries". The larger the stock of educated labour countries with lagging technological capacity would have, the easier for them to catch up on the more effective technologies and develop (Barro, 1991; Benhabib and Spiegel, 1994). Although basic and secondary education improves the returns of R&D activities, tertiary education and R&D activities are crucial in economic development. In line with this view, the World Bank has recently highlighted the role of tertiary education in developing countries to construct knowledge societies and create local innovation networks (World Bank, 2003).

Third, education (including tertiary education) can have positive social externalities modifying the country at society level: it contributes to *social* capital as well as *human* capital (OECD, 2000). Education is often associated with better health, higher life expectancy, lower crime, better parenting, better governance, enhanced trust, etc. (World Bank, 2003; OECD, 2007: SOL).

15.2.2   THE CONTRIBUTION OF PRIMARY AND SECONDARY EDUCATION

Primary and secondary education contribute to teach the knowledge and practices that can lead to healthier or more sustainable practices in particular areas, as well as to slowly change or challenge some societal practices that can hinder capacity development.

Developing capacity in primary and secondary education is important for water capacity development. Formal education about water use and management or the use of water as a pedagogical theme can help raise awareness of the value of water and improve people's basic knowledge about how to use and manage it in the best possible ways in its local context. Concretely, this can also take the form of water issues being used as a thematic thread in the teaching of different skills, especially in the frame of innovative pedagogies relying on collaborative project- or problem-based teaching and learning. These pedagogies are increasingly used in developed countries and are not necessarily more expensive than traditional ones, although human and financial resources can indeed be a limiting factor for their implementation in poorer countries.

Moreover, beyond the potential inclusion of water issues in school discussions, projects or curricula, better literacy allows people to be receptive to more information sources and more complex messages and might thus indirectly enhance the effectiveness of information campaigns trying to change people's behaviour in the use and management of water. Formal education alone is often not sufficient to change some habits or beliefs that shape people's identity and behaviour, and cannot replace an appropriate incentive structure. However, it helps to disseminate basic knowledge, including about water, its use and its value.

Most countries have made significant progress toward universal primary education over the past decade. There have been impressive increases in the net enrolment ratios in primary education in Sub-Saharan Africa (from 54% to 70% between 1999 and 2006) and in South and West Asia (from 75% to 86% between 1999 and 2006) (see Figures 15.3 to 15.6). However, about 75 million school-age children were still estimated as not attending primary school in 134 countries. Moreover, only 25% of children with the appropriate age to go to secondary school enrolled in Sub-Saharan Africa. There is therefore much room for progress before the largest share of the population of all countries reaches secondary education.

Further effort to improve the participation in and the quality of primary and secondary education will remain crucial in any capacity development

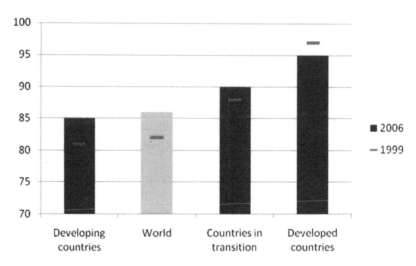

Figure 15.3 Net enrolment rates in primary education.

Source: UNESCO Institute of Statistics

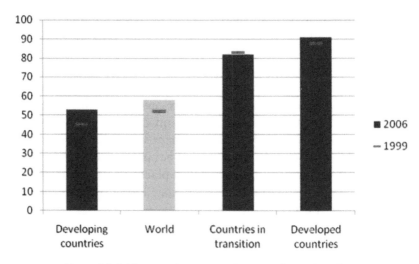

Figure 15.4 Net enrolment rate in secondary education.

Source: UNESCO Institute of Statistics

strategy. The quality of education needs to be strengthened as it is actually the skills rather than formal educational attainment that matters. Moreover, a lack of quality can be a factor of school dropout as some studies have shown (Hanushek et al., 2008). Enhancing gender equality and outreach to rural areas are also critical in many poor countries where insufficient participation of girls and children from rural areas explains low participation (UNESCO, 2008).

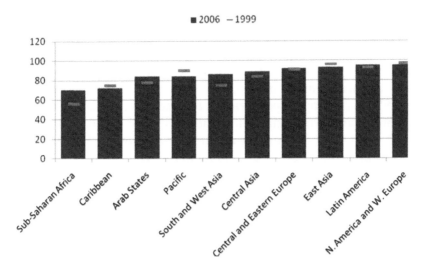

Figure 15.5 Net enrolment rates in primary education, by region (%).

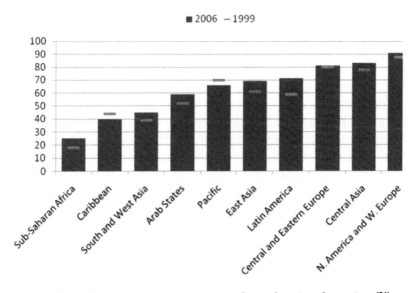

Figure 15.6 Net enrolment rates in secondary education, by region (%).

Source: UNESCO Institute of Statistics

Finally, developing technical and vocational education and training is important for facilitating the adoption of new technologies and having a strong absorption capacity. Organisational improvement constitutes the bulk of innovation in most countries: in many industries, including water, this requires the adoption, adaptation, and integration of existing technologies or making incremental changes in the production process. Technical and vocational education and training aims precisely to develop the skills required in this process. It is a significant instrument of technology transfer and diffusion. In 2006, vocational education and training represented about 16% of secondary enrolments in developed countries, 12% in countries in transition, and 9% in developing countries. It represented only 2% of school enrolments in South and West Asia, 3% in the Caribbean and 6% in Sub-Saharan Africa (UNESCO, 2008). In many countries, vocational education and training needs to be further developed and modernised.

15.2.3   THE CONTRIBUTION OF TERTIARY EDUCATION

Tertiary education contributes to the design of capacity-building strategies and to the construction of an information base for monitoring its progress. It can develop new knowledge in specific areas and sectors (e.g. water) through its research activities as well as contribute to the contextualization, adaptation and application of existing international knowledge.

The higher education sector, including research, plays a specific role in any capacity development strategy. First, domestic researchers and academics should help design the national development strategy by exploring the costs and benefits as well as the feasibility of alternative policies. Second, an essential feature of capacity development strategies lies in the establishment of continuous monitoring and evaluation systems based on clear indicators to track and steer progress. Here again, academics and researchers are well equipped to contribute to this task, as it is the case in many developed countries. When it is carried out in the higher education sector, this evaluation benefits from an open and contradictory scientific debate and allows for shedding light on many possible consequences of the policy. But even if it is carried out outside the academic sector, this policy assessment requires highly educated workforce people, typically domestic tertiary-level graduates. For example, according to Schultz (1999), an information base is lacking to set human resource priorities for Africa and allocate, on a firm foundation, public resources among human capital resource development programmes.

In the case of water, in spite of recent progress, developing countries often lack the capacity to monitor their policies and progress related to the use and management of water or to the impact of specific water policies (e.g. Suhardiman, 2008; Mollinga and Bolding, 2004). Poorer countries also often lack an explicit capacity development policy with solid knowledge and information management throughout the sector (or the country), which makes monitoring difficult. Foreign scholars and academics can help developing countries to develop their monitoring capacity, but this imported capacity is generally insufficient to gather data and design an evaluation framework. In some cases, even when they have the appropriate capacity, domestic scholars lack the academic freedom to undertake such tasks in a credible way because of the political implications of their findings for the government. To limit this problem, some development economists call for an enhanced role of international organizations in the monitoring of capacity development policies, using, for example, experimental methodologies (Duflo and Kremer, 2005).

Tertiary education is also important in an education capacity-building strategy because it supports the development of primary and secondary education. The training of teachers and school principals, the curriculum design and reform, the educational research and innovation are primarily the responsibility of tertiary education. A strong tertiary education is thus necessary for quality primary and secondary education sectors.

Many countries will have to continue to expand their tertiary education system if they are to catch up with richer economies. Gross enrolments rates in developed countries were, on average, at 67% in developed countries and 70% in Western Europe and North America, compared to 17% in developing countries, and 5% and 6% in Sub-Saharan Africa and the Caribbean, respectively (Figures 15.7 and 15.8). The expansion of tertiary education is not an end in itself but has to be proportional to the expansion of the other education sectors and relate to the needs of tertiary graduates of the local economy.

15.3   CROSS-BORDER HIGHER EDUCATION: AN ENGINE OF CAPACITY DEVELOPMENT

Cross-border flows of people, services and products also contribute to capacity development. Education and science policies that encourage the international

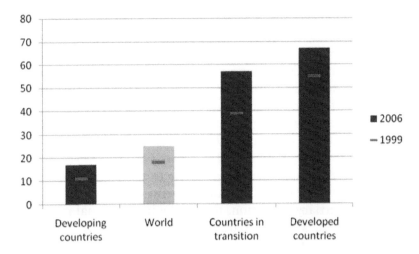

Figure 15.7 Gross enrolment rate in tertiary education by country category (%).

Source: UNESCO Institute for Statistics

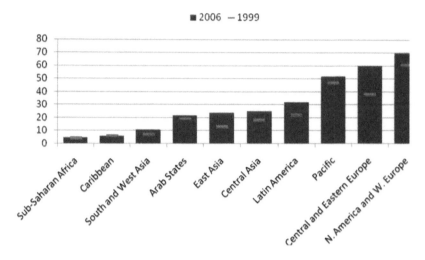

Figure 15.8 Gross enrolment rate in tertiary education by region.

Source: UNESCO Institute for Statistics

mobility of students, researchers and, more generally, highly trained human resources are increasingly seen as crucial to capacity development strategies (OECD-World Bank, 2007, OECD, 2008).

Learning-by-trading has become central to many contemporary treatments of trade and growth: the learning externalities of trade have become an important theoretical argument for considering trade liberalisation as an engine of growth, especially in developing countries. Some argue that trade is a means of knowledge circulation giving access to knowledge to all trading partners. Exports expose domestic firms to foreign knowledge and allow developing countries to reap benefits from foreign research and development (R&D): they may learn about new technologies and materials, production processes, or organisational methods. Imports of goods and services can also be seen as diffusing foreign R&D developed by trade partners: importing intermediate goods embodying foreign research and development corresponds to a use of this technology by the importing country, which could positively affect its productivity (Grossman and Helpman, 1991; Coe and Helpman, 1995; Bayoumi, Coe and Helpman, 1999; Romer, 1993). Cross-border flows of products and services under other arrangements than trade (e.g. technical assistance, partnerships or collaboration) yield the same benefits.

Cross-border tertiary education, that is the international mobility of students, academics, educational programmes and institutions, is a key component of knowledge diffusion and is increasingly seen as an integral part of capacity development strategies.

The number of foreign students within the OECD area has tripled since 1980, and increased by 54% between 2000 and 2006 to 2.9 million students

(Figure 15.9). While OECD countries host about 85% of all foreign students in the world, two thirds of these foreign students come from non-OECD countries, primarily China and India.

Cross-border education can typically help a country to expand its tertiary education system more quickly than it could with its domestic resources alone. Many developing countries lack the tertiary education places and staff to meet their domestic demand. There is some evidence that student mobility, has served capacity development. In 22 countries, domestic students studying abroad represented over 30% of domestic tertiary education enrolments in 2004, showing a clear use of foreign tertiary education systems to increase their tertiary education capacity (Figure 15.10). Cross-border education also plays a role in enhancing the quality, variety and relevance of domestic tertiary education systems, not the least by giving institutions a quality benchmark.

The cross-border mobility of tertiary educational programmes and institutions is an innovation that has surged over the past decade: from a relatively low starting point, it has grown very quickly, under a variety of contractual arrangements and "business models" (OECD, 2004b). Tertiary education institutions generally partner with local institutions when they deliver their programmes abroad, but they are sometimes just open branch campuses, or turn their institutions into a multi-campus institution, or network with other foreign institutions to operate abroad. This presence abroad in turn facilitates the mobility of students and academics abroad, as well as collaborative research projects. Cross-border higher education represents a potential capacity development tool for the host countries, by providing quality

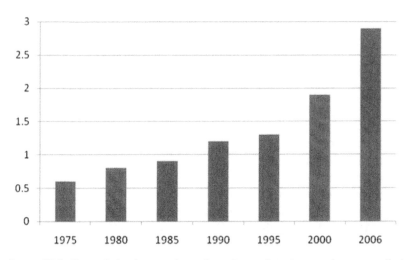

Figure 15.9 Growth in the number of tertiary education students enrolled outside their country of citizenship worldwide (millions).

Source: OECD and UNESCO Institute of Statistics

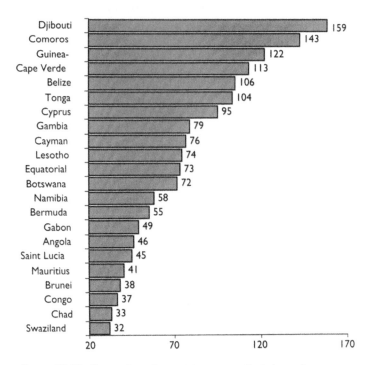

Figure 15.10 Countries where citizens enrolled abroad represent over 30% of domestic tertiary education students (%), 2004.

Source: UNESCO Institute of Statistics

benchmarks to their institutions, thanks to incoming and outgoing mobility, by supplementing local capacity in the training of qualified manpower, by helping students and academics benefit from cutting edge knowledge, or by inducing positive organisational or cultural change within their higher education sector (OECD/World Bank, 2007). Increasingly, it explicitly becomes part of innovation and capacity development strategies. These foreign institutions are, for example, grouped in clusters encompassing businesses, as part of regional innovation clusters (and strategies) with the ambition of creating a knowledge economy; the Knowledge Village in Dubai, the Education city in Qatar, or the Kuala Lumpur Education City in Malaysia are examples of these new innovation clusters.

Programme and institution mobility can provide another way of improving the quality of domestic educational provision and of knowledge in specific fields. At their best, such programmes are indeed able to link developing countries with cutting-edge knowledge and in this way assist in training an effective workforce, as well as a high quality faculty for the domestic system.

The growing international mobility of academics and of doctoral students is another part of the picture that contributes to the internationalisation of academic research. Flows of international scholars into the United States have, for example, doubled between 1998 and 2007 (IES, 2007), and there is scattered evidence of growth in many other countries.

Having a critical mass of high quality academics is important to develop capacity in tertiary education or in a specific discipline, for example water engineering. When this capacity is not available domestically, quality cross-border educational provisions can help reach it. Faculty and post-graduate students can, through international mobility, obtain a high quality education or develop their competencies before returning to the university sector in their home country. When their country is too small to have a "critical mass", it allows them to be part of international networks of knowledge and communities of practice.

It is widely recognised that student and scholar mobility allows developing countries to access recent knowledge and research methodologies. Indeed, mobility contributes to the creation and diffusion of knowledge. This is true for codified, but probably even more important for tacit, knowledge, that is knowledge that cannot easily be completely "written down" and is more easily transmitted by sharing a common social context and physical proximity.

Partly related to this mobility[3], international collaboration has grown significantly in academic research. This is reflected in the growth of internationally co-authored (or collaborative) scientific articles, meaning articles with at least one international co-author (in terms of institutional affiliation). Between 1988 and 2001, the total number of international articles more than doubled, increasing from 8 to 18% of all scientific articles. In the United States, the share of internationally co-authored articles in the total article output more than doubled between 1988 and 2001, and amounted, on average, to 23.2%. In Western Europe, international collaborative articles accounted for 33% of all articles in 2001, up from 17 percent in 1988 – the collaboration having a strong intra-regional component. In Asia, the percentage of international articles also increased from 11% of all articles in 1988 to 21% in 2001. Moreover, the breadth of countries with which each country collaborates for scientific research has increased. Between 1994 and 2001, all countries (for which information is available) have raised the number of countries with which they have jointly authored articles: for an OECD country, the average number of collaborating countries in scientific activities rose from 89 to 102 countries between 1994 and 2001. But this trend goes beyond the OECD area: emerging and developing countries have actually expanded the number of countries they collaborate with more than developed countries (NSB, 2004). Finally, foreign scientific articles are increasingly cited in the

---

[3] The US National Science Foundations notes a moderately high correlation between the number of US PhDs awarded by country to foreign-born students in 1992–96 and the volume of papers co-authored by the United States and those countries in 1997–2001 (NSB, 2004).

scientific literature worldwide: in 1992, foreign articles accounted for 55% of all citations, against 62% in 2001 (NSB, 2004).

A concern with cross-border education is that it could increase a brain drain rather than the circulation of skills between countries that is necessary to capacity development, especially in developing countries. There is no systematic data on the relationship between the mobility of students and researchers or the acquisition of a foreign degree at home and subsequent migration. What little exists does show that there is a link, though not necessarily a strong one. For some years, almost one-half of the candidates admitted under Australia's skilled migration programme hold an Australian degree. In Canada, it is estimated that between 15 and 20 percent of foreign students have stayed and are working in the country; in New Zealand, 13 percent of the foreign students entering the country to study between 1998 and 2005 had obtained a residence permit in 2006; in Norway, 18 percent of the foreign students studying there between 1991 and 2005, originating from outside the European Economic Area (EEA), stayed in the country (as against 9 percent of foreign students from the EEA) (Suter and Jandl, 2006). In the United States, where 'stay rates' for foreign students after receiving their degree are periodically compiled, the average stay rate for foreign recipients of science and engineering doctorates in the United States, four to five years after earning their degree, rose from 41 to 56 percent between 1992 and 2001. The figures skyrocketed from 65 to 96 percent for Chinese doctorate recipients, and from 72 to 86 percent for Indians. Stay rates in countries following the completion of studies vary considerably depending on country of origin and discipline, but there seems to be no systematic pattern in that regard.

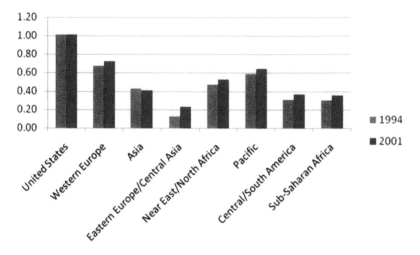

Figure 15.11 Relative prominence of citations of S&E literature, by region.

Source: NSB, 2004

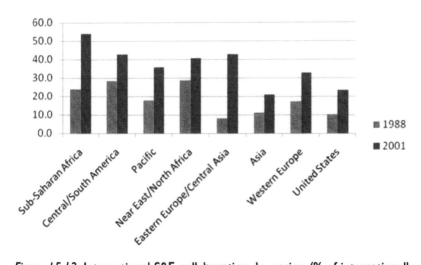

Figure 15.12 International S&E collaboration, by region (% of internationally co-authored papers).

Source: NSB, 2004

The OECD migration database shows that high expatriation rates for the highly skilled (in terms of stock rather than flows) mainly concern African and Caribbean countries: over 70% of Jamaican and Guyanese nationals holding higher diplomas are expatriates in an OECD country. On the other hand, despite their high stay rates in the United States after they finish their studies, Indian and Chinese nationals each account for less than 3% of the expatriates holding a higher degree in OECD countries, as is also the case for Brazil, Indonesia and Thailand. Out of the 113 countries for which information is available, 27 have expatriation rates of their tertiary educated people over 20%, including nine over 50%. Although holding a degree from a country certainly makes it easier to get residence, no consistent pattern can be found between high levels of student mobility and expatriation rates.

New initiatives such as the engagement of highly skilled diasporas in the development of the innovative capacity of their country of origin are increasingly being developed to mitigate the effects of a possible brain drain (Vincent-Lancrin, 2008; OECD, 2008).

15.4 SUMMARY AND CONCLUSION

Why invest and develop capacity in education and tertiary education as part of capacity development? This paper has argued that there were several reasons for that.

A first is that capacity development relies on learning, at individual level, organizational and sectoral level. While learning does not need formal education to take place, formal education is obviously an engine for learning and for knowledge adoption and diffusion. Primary and secondary education contribute to teach the knowledge and practices that can lead to a more efficient use of water, as well as to slowly change or challenge some societal practices that can hinder capacity development.

A second reason is that education and tertiary education, when its quality is adequate, can contribute to economic development and thus give some countries the resources to access sophisticated technologies and to improve their infrastructure, for example in water.

A third reason to invest in tertiary education at the same time as basic education lies in its contribution to the design of capacity-building strategies, to the construction of an information base for monitoring its progress and its quality, but also to the development and absorption of cutting-edge sectoral knowledge.

Many developing countries still need to make efforts to develop an educational capacity comparable to that of developed countries and that would allow them to produce, diffuse and absorb critical knowledge to their capacity development strategies.

The shape of this investment can vary. Besides the development of a domestic educational capacity, using foreign capacity through cross-border education or the mobility of people is a way to develop capacity. This is true in general as well as for specific sectors such as water. More generally, an important means of capacity development in a specific sector, for example water, relies on curricular and pedagogical changes as well as on the development of vocational education and training.

NOTE

The views expressed are the author's and are not necessarily those of the OECD and its member countries.

REFERENCES

Aghion, P. and Howitt, P. (1998), *Endogenous Growth Theory*, MIT Press, Cambridge, MA.

Arndt, C. (2000), "Technical co-operation", in F. Tarp (ed.), *Foreign Aid and Development. Lessons Learnt and Directions for the Future*, Routledge, London, pp. 155–177.

Banerjee, A. and Duflo, E. (2008), "Mandated Empowerment. Handing Antipoverty Policy Back to the Poor?", *Annals of the New York Academy of Sciences*, 1136, 333–341, http://econ-www.mit.edu/files/3284

Barro, R.J. (1991), "Economic Growth in a Cross-section of Countries", *Quarterly Journal of Economics*, Vol. 106, pp. 407–443.

Bayoumi, T., Coe, D. and Helpman, E. (1999), "R&D Spillovers and Global Growth", *Journal of International Economics*, Vol. 47, pp. 399–428

Benhabib, J. and Spiegel, M. (1994), "The Role of Human Capital in Economic Development: Evidence from Cross-country Data", *Journal of Monetary Economics*, Vol. 34, pp. 143–173.

Bolger, J. (2000), "Capacity Development: Why, What and How", CIDA, Policy branch, Capacity Development Occasional Series, Vol. 1, N°1.

CKNet-INA. (2008), "General project data." WRIM Capacity Building project.

Coe, D.T. and Helpman, E. (1995), "International R&D Spillovers", *European Economic Review*, Vol. 39, pp. 859–887.

Duflo, E. and Kremer, M. (2005), "Use of Randomization in the Evaluation of Development Effectiveness" in Pitman, G., Feinstein, O. and Ingram, G. (eds.) *Evaluating Development Effectiveness*, New Brunswick, NJ: Transaction Publishers, 205–232.

Finn, M. (2005), "Stay rates of foreign doctorate recipients from US universities, 2003", Oak Ridge Institute for Science and Education.

Fuente (de la), A. and Ciccone, A. (2002), "Le capital humain dans une économie mondiale fondée sur la connaissance", European Commission, Brussels.

Gemmell, N. (1996), "Evaluating the Impacts of Human Capital Stocks and Accumulation on Economic Growth: Some New Evidence", *Oxford Bulletin of Economics and Statistics*, Vol. 58(1), pp. 9–28.

Grossman, G. and Helpman, E. (1991), *Innovation and Growth in the Global Economy*, MIT Press, Cambridge, MA.

Hall, R.E. and Jones, C.I. (1999), "Why do Some Countries Produce so Much More Output per Worker than Others?", *Quarterly Journal of Economics*, February, pp. 83–116.

Hanushek, E. and Wössmann, L. (2008), "The Role of Cognitive Skills in Economic Development", *Journal of Economic Literature*, 46(3), 607–668.

Hanushek, E., Lavy, V. and Hitomi, K. (2008), "Do Students Care about School Quality? Determinants of Dropout Behavior in Developing Countries", *Journal of Human Capital*, 2(1), 69–105.

Hanushek, E.A. and Kimko, D.D. (2000), "Schooling, Labor-force Quality, and the Growth of Nations", *American Economic Review*, Vol. 90(5), pp. 1184–1208.

Krueger, A.B. and Lindahl, M. (2001), "Education for Growth: Why and for Whom?", *Journal of Economic Literature*, Vol. 34 (December), pp. 1101–1136.

Lucas, R. (1988), "On the Mechanics of Economic Development", *Journal of Monetary Economics*, Vol. 22, pp. 3–42.

Mankiw, N.G., Romer, D. and Weil, D. (1992), "A Contribution to the Empirics of economic growth", *Quarterly Journal of Economics*, Vol. 107, pp. 407–437.

Mollinga, P., Bolding, A. Eds. (2004). The politics of irrigation reform. Ashgate, UK.

National Science Board (2004), *Science and Engineering Indicators 2004*. Two volumes. (Arlington, National Science Foundation). http://www.nsf.gov/statistics/seind04/pdfstart.htm

OECD (2000), *The Well-Being of Nations*, OECD, Paris.

OECD (2001), *Strategies for Sustainable Development: Guidance for Development Co-operation*, The DAC Guidelines, OECD, Paris.

OECD (2004a), *Education at a Glance. OECD Indicators 2004*, OECD, Paris.

OECD (2004b), *Internationalisation and Trade in Higher Education. Opportunities and Challenges*, OECD, Paris.

OECD (2006), *Working towards Good Practice. The Challenge of Capacity Development*, DAC Guidelines, OECD, Paris.

OECD (2007), *Understanding the social outcomes of learning*, OECD Publishing, Paris.

OECD-World Bank (2007), *Cross-border tertiary education: a way towards capacity development*, OECD Publishing, Paris.

OECD (2008), *The Global Competition for Talent. Mobility of the Highly Skilled*, OECD Publishing, Paris.

Pritchett, L. (2001), "Where has all the Education Gone?", *World Bank Review*, Vol. 15(3), pp. 367–391.

Romer, P.M. (1989), "Human Capital and Growth: Theory and Evidence", NBER Working Paper No. 3173.

Romer, P.M. (1993), "Two Strategies for Economic Development: Using Ideas and Producing Ideas", *Proceedings of the World Bank Annual Research Conference 1992*, supplement to the *World Bank Economic Review*, March, pp. 63–91.

Schultz, T.P. (1999), "Health and Schooling Investments in Africa", *Journal of Economic Perspectives*, Vol. 13(3), pp. 67–88.

Sen, A. (1999), *Development as Freedom*, Alfred Knopf, New York.

Sen, A. and Williams, B. (eds.) (1982), *Utilitarianism and Beyond*, Cambridge.

Sianesi, B. and Van Reenen, J. (2003), "The Returns to Education: Macroeconomics", *Journal of Economic Surveys*, Vol. 17(2), pp. 157–200.

Suhardiman, D. (2008), *Bureaucratic design: The paradox of irrigation management transfer in Indonesia*. Wageningen UR. Prom./coprom.: Vincent, L.F. & Mollinga, P.P, – Wageningen, P. xvi + 285.

Suter, B. and Jandl, M. (2006), *Comparative Study on Policies towards Foreign Graduates. Study on Admission and Retention Policies towards Foreign Students in Industrialised Countries*, International Centre for Migration Policy Development (ICMPD), Vienne, www.icmpd.org/

Thorbecke, E. (2000), "The Development Doctrine and Foreign Aid 1950–2000", in F. Tarp (ed.), *Foreign Aid and Development. Lessons Learnt and Directions for the Future*, Routledge, London, pp. 17–47.

UNESCO (2008), *Overcoming inequality: why governance matters*, Education for All Global Monitoring Report 2009, Unesco Publishing and Oxford University Press, Paris.

United Nations Development Programme (UNDP) (2003), *Ownership, Leadership and Transformation. Can we do Better for Capacity Development?*, edited by Carlos Lopes and Thomas Theisohn, Earthscan, London.

United Nations Development Programme (UNDP)/Global Environment Facitlity (GEF) (2003), Capacity development indicators, UNDP/GEF Resource Kit No. 4, www.undp.org/gef/undp-gef_monitoring_evaluation/ sub_me_policies_procedures.html

Vincent-Lancrin, S. (2006), "What is Changing in Academic Research? Trends and Futures Scenarios", *European Journal of Education*, Vol. 41(2).

Vincent-Lancrin, S. (2008), "Student mobility, internationalization of higher education and skilled migration", *World Migration Report 2008*, International Organization for Migrations, Geneva, 105–126.

World Bank (2002), *Constructing Knowledge Societies: New Challenges for Tertiary Education*, World Bank, Washington D.C.

Chapter 16

CAPACITY DEVELOPMENT IN AFRICA:
Lessons of the past decade

Mark Nelson & Ajay Tejasvi
World Bank Institute

ABSTRACT

The record of capacity development in Africa over the past decade is mixed. Yet the experience is also full of lessons learned and approaches tested that provide a treasure trove of evidence for building a more successful and sustainable approach. Some of the lessons are for the international donor community, whose practices of aid management have often contributed to this record of disappointing outcomes. Other lessons are for African countries, who have frequently failed to take the lead and accept ownership and responsibility for their own development strategies. These lessons focus not only on overall leadership and country-level incentive systems, but also on sector-level practices. To be sure, the experiences in Africa over the past decade have resulted in some improvements and changes that suggest the next decade may offer new hope. But a number of challenges remain. One challenge is to resolve the tension between short-term output-driven frameworks and longer-term adaptive approaches that have shown to be more effective. Another challenge is to create enabling conditions under which African governments engage with a wider variety of stakeholders from civil society in the design and implementation of capacity-development initiatives.

16.1 INTRODUCTION

In recent years, the concept of capacity development has received unprecedented attention as a critical goal of international development policy. Though often vaguely defined and suggesting different things to different people, the term capacity development has been widely used and accepted by many development professionals as a way to describe some of the critical roadblocks to reaching the Millennium Development Goals and stimulating sustainable growth in the developing world. Capacity development is a stated development objective for nearly two-thirds of World Bank projects and is the focus of much of the Bank's non-lending work. Other donors – Belgium, Denmark, Canada, Japan, the Netherlands, Sweden, and the U.S. for example – have reported similar or even higher percentages of their official development assistance devoted to enhancing capacity. Globally, estimates suggest that around a third of the $100 billion in annual international development assistance is directed at capacity development.

Nowhere has the capacity development issue received more attention than in Africa. Capacity development is often seen as the missing ingredient to development efforts, the magic bullet that could make the difference between development and stagnation or decline. Yet, donor and developing-country efforts to address this "capacity gap" have had mixed results. In Africa, in particular, capacity development has been an area of generous spending with mostly lackluster outcomes. "The Bank's traditional tools – technical assistance and training – have often proved ineffective in helping to build sustainable public sector capacity," states the World Bank's 2005 Independent Evaluation Group Report on Capacity Building in Africa. Other reviews like the report of the World Bank Task Force on Capacity Development in Africa and numerous independent reports and reviews have had similar findings.[1] While some countries in Africa have performed relatively well, Figure 16.1 shows that on average, sub-Saharan Africa's performance on some key indicators of government capacity have stagnated over the last decade.

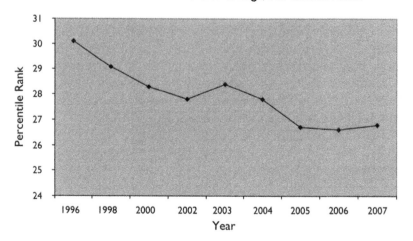

Figure 16.1 One view of Africa's capacity development challenge.[2]

Yet, the experience in Africa is also full of lessons learned and approaches tested that provide a treasure trove of evidence for building a more successful and sustainable approach to capacity development. Some of those lessons are for the international donor community, whose practices of aid management have often helped to create this record of disappointing outcomes. Other lessons are for African countries, who have often failed to take the lead and accept ownership of and responsibility for their own development strategies. These lessons focus not only on overall leadership and country-level incentive systems, but also on sector-level practices – both successes and failures. These experiences in Africa over the past decade have resulted in some improvements and changes that suggest the next decade may offer new hope. Yet, current practices of aid management point to a still-unresolved tension between technocratic reforms driven by results frameworks and short-term, quantifiable outcomes (e.g. the number of vaccinations given or kilometers of roads built), and the need for longer-term and more adaptive aid management approaches that focus on the cross-cutting sustainable capacities needed for carrying out country-led development.

The aim of this article is to review the record of capacity development in Africa over the past decade and examine some of the lessons that have emerged from the sometimes ill-fated experience. This paper will rely on a number of studies and reports that have shaped the background of the Paris Declaration and the Accra Action Agenda as well as case studies on specific country and sector experiences, and a 2007 review of the World Bank's portfolio.[3]

16.2   JUST WHAT IS CAPACITY DEVELOPMENT?

Capacity development is a broad concept that refers to "the ability of people, organizations and society as a whole to manage their affairs successfully." (OECD/DAC, 2006). It has evolved over the past years from a narrow preoccupation with training and technical assistance to include an understanding of a more multi-faceted and complex process. This broader understanding includes the enabling environment in which people and organizations operate, the formal and informal norms and values that affect behaviors. It is thus related to the functioning of complex systems within countries and organizations, and is rarely improved by the injection of supply-driven training programs or technical assistance alone. Capacity development requires an approach that takes into account the political economy of reform and demand-side pressures. It is, therefore, closely intertwined with the governance agenda, and benefits from efforts to improve institutions, laws, incentives, transparency and leadership. "Capacity is about skills, performance and governance," says the Bank's Task Force on Capacity Development in Africa (2005). The international development community has consistently overestimated its ability to build capacity in the absence of national commitment, local ownership and reasonably good governance.

One way to illustrate this understanding of capacity development is shown in Figure 16.2 below, with the enabling environment (cf. e.g. Chapter 1 in this Volume) on the Y-axis and skills and resources on the X-axis. Moving from point A to point B is capacity development, implying that moving along one of these axes alone is insufficient. Indeed, the overwhelming amount of spending on capacity development has been concentrated on the skills building and resources side of the equation. And while this may help some individuals and organizations, it rarely results in sustainable outcomes.

The pathways of moving up this capacity-development slope are rarely linear and may take many different trajectories. As depicted in Figure 16.3 below,

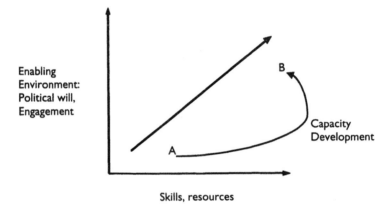

Figure 16.2 Capacity development = Skills + Will, though the path is rarely linear.

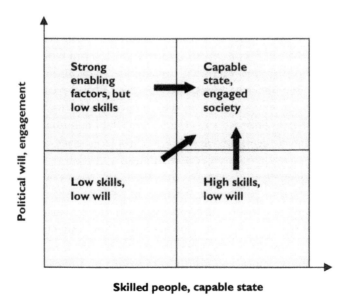

Figure 16.3 Multiple paths to a capable state and engaged society.

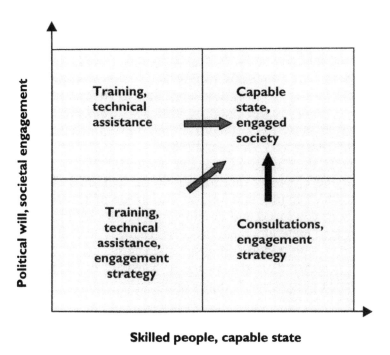

Figure 16.4 From diagnosis to intervention.

countries with both low skills and weak enabling factors start in the lower, southwest quadrant. Other countries may start with stronger initial conditions, a government with a strong determination to develop or a vibrant civil society demanding results. Figure 16.4 illustrates some examples of interventions that may be appropriate for moving towards the northeast quadrant, given the country's initial conditions.

Over the past decade, evaluations and reviews of capacity development in Africa have repeatedly focused on the failure to account for the governance aspects of the capacity problem. The World Bank Task Force on Capacity Development in Africa found that governance aspects were so important that it decided to entitle its 2005 report *Building Effective States, Forging Engaged Societies*. Projects that approach the problem with this in mind tend to be designed differently, with much more up-front engagement with key stakeholders and attempts to strengthen broader country ownership of the activities.

One major, path-breaking study by the European Center for Development Policy Management has gone much further in describing the systemic nature of capacity development (cf. Chapter 9 in this Volume). Drawn from 16 country case studies and extensive international consultations with practitioners, theorists and policy-makers, *Capacity, Change and Performance* looks at capacity development as a process of change, and draws on the theory and practice of "complex, adaptive systems" and "systems thinking."

These non-linear ways of considering the multiple interrelationships of individuals, organizations and institutions that are usually needed for sustainable change pose a challenge to the typical left-to-right results formulas of inputs to output to outcomes. "Complexity theory tells us that small initial changes can have huge effects and vice-versa. Nonlinear patterns of behavior can escalate micro-interventions into large, system-wide changes. This could mean that big system change could be instigated from the bottom through small intervention as well as pushed from the top through larger, more complex activities." (Baser and Morgan, 2008)

More recently, the Capacity Collective reinforced the argument for a concept of capacity development that focuses on the dynamics of change and on the systems in which change takes place. This group of academics and practitioners from both developed and developing countries identified the need for "all actors engaged in development processes to focus on change and adaptive management that is rooted in endogenous strengths, needs, aspirations, and expectations ... rather than always seeing [capacity development] from an exogenous, deficit perspective."(Taylor and Clarke, 2008)

16.3 EVALUATIONS AND REVIEWS OF CAPACITY DEVELOPMENT IN AFRICA

Several reviews of capacity development over the past decade have found major shortcomings in the approaches, particularly in Africa. The World Bank, the African Development Bank, the African Capacity Building Foundation and many bilateral organizations have come to a similar set of conclusions about the problems and the possible solutions.

It is important, of course, to differentiate among the many variations on this record within the diverse African continent. Some countries – Tanzania or Botswana, to name two examples – have made substantial progress, and some of their efforts have helped define "good practice" examples to be used by others. Other countries, particularly the post-conflict and fragile states concentrated in central Africa, continue to lag or decline. Within countries, substantial variations can be seen among sectors and regions. On the whole, however, Africa continues to be the place where the biggest capacity challenges continue to reside.

The reviews of capacity development in Africa focus on the shortcomings of donor practices and the failure to find ways to make programs be led and owned by Africans themselves. Among the key findings are the following:

- Failure to systematically include capacity development in country-led development strategies
- Failure to develop effective, participatory diagnostic processes that engage not only governments but users of the end-products or services, and to track progress and outcomes

- Too much focus on one-off training programs and technical assistance (TA) through the use of external consultants and too little on South-South learning and partnerships
- Complicated technocratic approaches that pay little attention to the political economy of reform, and issues of country ownership and leadership
- Over-reliance on supply-driven gap filling TA instead of helping developing partners learn sustainable "do-it-yourself" methodologies
- Lack of attention to demand-side pressures and the role of civil society organizations in helping define strategies.

In a review of lending in the Africa region, a World Bank Institute team found that capacity development is a major and possibly growing focus of projects. The study showed that there was an increase in the number of programs that included capacity development as part of their development objective to 75 percent in fiscal year 2007, up from 62 percent in fiscal year 2006 (see Table 16.1). However, the review showed that, in terms of implementation of these goals, Bank projects still relied heavily on training and consultants as the main instruments (see Table 16.2).

The World Bank's Independent Evaluation Group's (IEG) review of Bank support for Capacity Building in Africa in 2005 concluded that the World Bank should put capacity development at the center of its relations with its African clients and ensure that the support is country-owned, results-oriented, and

Table 16.1 Is capacity included in the development objective? (Projects in the African Region).

Fiscal year 2006		Fiscal year 2007	
Yes	62%	Yes	75%
No	34%	No	14%
Unclear	4%	Unclear	11%

Source: World Bank Institute desk study, 2008 (unpublished)

Table 16.2 Capacity development inputs in world bank projects in Africa.

	Fiscal year 2006	Fiscal year 2007
Equipment	20%	14%
Study tours	2%	7%
Consultants	32%	50%
Training	36%	54%
Knowledge sharing	12%	29%

Source: World Bank Institute desk study, 2008 (unpublished)

evidence-based. (Committee on Development Effectiveness 2005-0016, 2005). Though the Bank has made its support for capacity development in Africa more relevant by extending its traditional focus beyond building organizations and individual skills to strengthening institutions and demand for improved public services, most support for capacity development in country programs remained fragmented – designed and managed on a project-by-project basis. This made it difficult to capture cross-sectoral issues and opportunities, and to learn lessons across operations (Worldbank OED, 2005).

The challenges of capacity development varied markedly across countries and sectors. The report noted that though the Bank was moving toward better customization of its capacity development approaches to country conditions, it also needed to develop sector-specific guidance on diagnosing capacity needs and evaluating capacity-building measures. The review also found that the World Bank did not apply the same rigorous business practices to its capacity development work when compared to other areas. The main tools, notably technical assistance and training, were not being effectively used; and the range of instruments, notably programmatic support, economic and sector work (analytical and advisory services), and activities of the World Bank Institute were not being fully utilized. The report also found that most activities lacked standard quality assurance processes at the design stage and were not being routinely tracked, monitored, and evaluated.

IEG recommended that the Bank needed not only to strengthen its knowledge base, but transform traditional capacity-building tools to improve results. It should also ensure that guidelines and processes are in place for self- and independent evaluation of Bank interventions. Sector-specific guidance and diagnostic tools for capacity development were also needed, along with more focus on capacity within the Bank's Country Assistance Strategies. Finally, the Bank was also told to reassess what role training should play in its capacity-building support, how it should be provided, and what the respective roles of a central training unit and regional programs in any future support for training activities should be.

The Operations Evaluation Department of the African Development Bank (AfDB), in its 2008 evaluation of its training arm, the Joint Africa Institute, repeated many of the same concerns about more recent approaches to capacity development. The OED found that the AfDB lacked a clear and comprehensive policy for capacity development; its policy relied too heavily on training and failed to rely on careful needs assessment or ongoing monitoring and evaluation.

Another broad set of criticisms have long been leveled against the practices of bilateral donors whose tied aid and supply-driven technical assistance have tended to ignore both the supply side (i.e. the local suppliers and consultants who can deliver capacity services) as well as the demand side of reforms. Even countries that have used less tied aid, such as Denmark, have found shortcomings in the way development assistance was conceived and delivered.

In a review of its technical assistance, DANIDA, the Danish aid agency found that instead of generating independence in partner institutions, technical assistance "increased their sense of dependence, thus weakening the commitment and ownership of national actors." The TA created conditions that encouraged the circumventing of local procedures and systems, and drained local organizations of their most talented staff. TA also led to the adoption of "wrong policies," and "too complex and expensive solutions"(Danida, 2005).

16.4   STRENGTHENING SOUTH-SOUTH LEARNING AND NETWORKS

One of the major shifts that has taken place over the past decade in Africa is a growing demand from developing countries to learn from the experiences of their peers. This has changed the focus of capacity development from a process of transferring know-how from the North to facilitating interactions among peers who have gone through major development challenges.

Various African networks are emerging as an important delivery mechanism for these interactions (Bloom et al, 2008). Communities of practice are being tapped to support efforts at the regional, country and sector levels. Professional networks of Southern experts from the public sector, private sector and civil society are promoting unmediated interactions among themselves to identify and develop new and innovative solutions to their problems. Southern organizations and regional knowledge networks are positioning themselves to play a leadership role in strengthening capacity. This could then lay the foundations for a more predictable, long-term support to create effective and innovative regional capacity development initiatives.

Grant-giving regional intermediaries like PACT and the African Capacity Building Foundation (ACBF) are supporting regional and sub-regional knowledge networks. The World Bank Institute has also identified networks as central to its capacity development strategy. The World Bank Task Force on Capacity Development in Africa noted that networks across the world should be carefully yet systematically engaged whenever and wherever they can constructively help governments, private firms, and citizens improve the delivery of essential services and aid in the development of the country as a whole (World Bank Task Force on Capacity Development in Africa, 2005, pp. 99–101).

Networks and South-South partnerships like the ACBF and the New Partnership for Africa's Development (NEPAD) support and sustain knowledge generation and capacity development in several important ways. Studies show that regional networks provide a critical mass of professional peer reviews not available at the national level, thus sustaining peer pressure for learning and excellence as well as alleviating professional isolation. They are an effective mechanism for keeping in touch with the rapidly changing frontier of knowledge through contact with the rest of the world and information sharing.

Networks are a medium for experience sharing and a mechanism for drawing good practices from specific policy and knowledge contexts, making them an important resource for collective knowledge. In particular, networks may be an important way to replace the supply-driven, one-off training programs with more effective approaches to provide specialized skills-building that would not be available at the national level (World Bank Task Force on Capacity Development in Africa, 2005, pp 68–69). Knowledge and practitioner networks are changing the way information and shared wisdom travel through the world and are helping level the playing field for practitioners from the South (Tejasvi, 2007).

16.5 THE PARIS DECLARATION, THE ACCRA AGENDA FOR ACTION (AAA) AND THE NEW CONSENSUS

The international community has, over the past five years, begun to try to operationalize the findings of the various evaluations and to elevate capacity development as central to the challenge of improving the effectiveness of aid. In successive international gatherings, and starting especially with the Paris Declaration in 2005, donors and development partners have vowed to change the basic business model that defines the way capacity development assistance is delivered.

This has included not only signing commitments to operationalize the well-rehearsed good practices that have been enumerated above, but also to adopt a more inclusive and open approach that involves civil society as a key partner and stakeholder in capacity development.

Most recently, at the Third High Level Forum on Aid Effectiveness in Accra, more than 100 countries and major development organizations agreed to a set of reforms aimed at incorporating these lessons. Pushed by a large contingent of civil society organizations, the ministers signed what is considered the most ambitious commitment to capacity development ever. While the proof will be in the implementation, the final AAA contains important innovations in the way international aid is delivered. The final document contains a number of new elements:

- New language focused on reducing the use of policy conditionality in favor of mutually agreed conditions linked to a country's national development strategy
- New numerical targets for program-based approaches and use of country systems

- New language on South-South cooperation that was demanded by Brazil and other middle-income countries
- New commitments on transparency, audit, parliamentary scrutiny, mutual assessments of performance and the use of "credible, independent evidence" to assess both partner country and donor performance.

The last bullet point above was a victory for the very large presence of civil society and an articulate and organized parliamentary delegation that was very effective in getting its arguments into the Accra meeting. This meeting had the largest ever civil society presence at an international aid policy meeting, and their influence was conspicuous at every event. In preparation for the event, an advisory group conducted far-reaching global consultations and analytical work, with more than 5000 civil society representatives from 3500 organizations. This resulted in the publication of a set of case studies, a synthesis of findings, and a set of recommendations that were broadly endorsed at the roundtable.

After Accra, civil society representatives said they will now expect much more information from donors and recipients about aid programs to help them with their monitoring efforts and to help them better coordinate their own aid programs. One study presented in Accra estimated that civil society organizations themselves contribute up to $25 billion a year in development assistance.

Participants acknowledged that while civil society organizations are widely accepted as critical to the development process, much work still remains to be done. "Governments have tended to see the role of civil society as one of contestation," said Emmanuel Akwetey, Director of the Institute for Democratic Governance in Ghana, who noted that certain key reforms, such as right-to-information laws, have stagnated in many countries. "The greatest obstacle is the perception that civil society is someone out in the street protesting who cannot understand anything," he commented during the Accra meeting as recorded by one of the authors. Governments need to develop their capacity and skills to engage more effectively with an increasingly demanding public, he added.

The Accra meeting also brought into focus the tension between the accountability of donors to Northern taxpayers for short-term results and the need for longer-term, more adaptive approaches that respond to changes as they happen and set new goals as the path unwinds. One of the more promising parts of the capacity development story is the growing realization among donors that their role has to change in order to improve the outcomes from capacity development initiatives. To address the often competing demands of Northern taxpayers and developing country aid recipients, the AAA promises more transparency and joint accountability mechanisms and says that "developing countries and donors will jointly review and strengthen existing international accountability mechanisms, including peer reviews with participants of developing countries."[4]

Recognizing the growing consensus on the conceptual framework that helps define capacity development and how it can be best managed within various development contexts, the challenge ahead is now one of implementation.[5] Country-level practitioners have begun to create a more professional capacity development practice that is informed by evidence and experience. "There's no longer the excuse that we don't know what we need to do, what works, what doesn't," says an official of the United Nations Development Program. "We have moved beyond that now."

16.6   CRITICAL PARTNERSHIPS AND NETWORKS

Aided by practitioner networks like the Learning Network on Capacity Development (LenCD), donors are looking to link up with relevant partners on substantive issues, connecting regional and national networks to the global policy debate and growing knowledge base in this area. LenCD is an open network on capacity development that links many initiatives globally and supports an evolving community of practice. LenCD emerged as a consequence of informal networking linked to several streams of research, workshops and conferences. Since its conception in June 2004, LenCD has helped establish a collective learning process that now spans many countries and official development agencies.

The Organization for Economic Cooperation and Development's Development Assistance Committee (OECD-DAC) is also changing its approach to capacity development (CD). After many years of considering capacity development as a sub-theme of its governance work, the OECD-DAC decided in 2007 to create a small unit inside the DAC secretariat to coordinate and promote awareness about capacity development among the many work streams within DAC structures. Areas of new work on capacity development include fragile states, procurement, the environment and other such topics. The new coordinator will try to make sure that lessons learned in one area are known and shared in the others, and that the groups systematically incorporate these findings into their policy advice.

The DAC has also proposed the idea of an alliance among the various organizations working on CD, in particular, to bring the organizations in the developed world in contact with those in the developing world. The proposed alliance would be a Southern-led forum that, over the next three years, would coordinate efforts for relevant meetings, including ministerial gatherings and other events convened at the international, regional and sub-regional levels. Side events, knowledge fairs and other learning opportunities would also be organized. The alliance will encourage the development of specific commitments by southern parties to collaborate on CD issues including strategies, measuring and monitoring capacity, South-South cooperation and other matters deemed of high priority.

16.7 MAKING CAPACITY DEVELOPMENT WORK AT THE COUNTRY AND SECTOR LEVELS

Beneath the surface of the promising developments underway at the global policy level remains the challenge of how to abide by these global commitments, which tend to focus on broader, country-level incentive systems and leadership, and at the same time get concrete results at the level where most development assistance resources are spent – at the sector level. It is at the sector level where technocratic, supply-driven development approaches, foreign consultants and tied aid remain the main tools of the trade.

It its 2005 report, the World Bank's Independent Evaluation Group commissioned six separate case studies of capacity development in Africa and found a consistent series of messages about sector-level approaches. The case studies found that much work in sectors has been pursued by technical specialists who propose overly complicated solutions and pay insufficient attention to the political and enabling conditions. More time needs to be spent engaging with end-users of services and coming up with approaches that enjoy stronger demand and support, not only from the government but from the broader community of stakeholders, as well.

The IEG case studies reviewed the effectiveness of programs in Benin, Ethiopia, Ghana, Malawi, Mali and Mozambique and looked at four major sectors: roads, health, education and public financial management. The results varied from sector to sector. Sectors that have relatively few transactions with the public, such as roads, have been relatively successful in building sustainable capacities. Sectors that involve many transactions and the maintenance of large, complex infrastructures of professionals – such as schools and the health sector – have proved more challenging.

Several reasons emerged for the relative success of the roads sector:

- High demand from stakeholders for construction and maintenance of roads
- Strong buy-in and ownership from the partner countries
- Use of peer-to-peer learning approaches and adaptive management of changes using information from peer experience
- Supported targeted interventions, a redefinition of roles of the various stakeholders and the creation of accountability through the inclusion of the road-users in planning and decision-making.

In the education and health sectors, the study found that a clearer, more participatory articulation of the capacity development strategies would improve the design of the interventions. The evidence suggested that the efforts in these sectors were fragmented and poorly coordinated with the country's national development agenda.

In Ghana, for example, the reforms in the education sector sought to increase learning achievements and enrollment in primary schools. This was to be achieved through increasing the amount and quality of instructional and learning time in primary schools. Head teachers were identified as critical change agents. They were expected to improve supervision in the schools to reduce teacher absenteeism and monitor the performance of teachers to bring about the expected improvements in learning and teaching outcomes. An assumption made during project design was that the provision of houses to head teachers would create an incentive for them to perform their supervisory functions effectively and reduce teacher absenteeism. It was thought that houses would also create an incentive for motivated teachers to accept placement in deprived communities. But poor diagnoses on the issues, based on insufficient consultation with teachers and other stakeholders, resulted in the failure of this housing incentive (which made up 18.5 percent of project costs) to serve its intended purpose.[6]

In Ethiopia, the World Bank's role in the roads sector has been that of a lead donor with substantial influence at all levels of the capacity development programs. In the health and education sectors, by contrast, the Bank has attempted to fit its support into a sector program framework without earmarking funds for specific inputs or activities. Though the Bank has undertaken quality analytical work including limited capacity needs assessments in the four sectors (i.e. roads, health, education and public financial management), the analysis was rarely reflected in the design of subsequent operations, nor was there strong country ownership of this analytical work.[7]

Some of the recommendations IEG and others have made about sector-level capacity development include the following:

- **Participatory diagnostics** that involve all relevant stakeholders: Projects need to be understandable to a wide range of users of services and take into account local concerns about the difficulties of making structural or technical reforms.
- **Clarity of objectives and quality of design:** Goals should be widely discussed and realistic, and the design should take into consideration existing capacities and constraints.
- **Ownership:** Since the weak outcomes of the programs in the health, education and public financial management sectors reflected the lack of client commitment and ownership, donors need to spend more time facilitating consensus building and understanding the common objectives of relevant stakeholders.
- **Long-term engagement:** Outside technical assistance can be useful in helping establish good organizational practices and procedures, but this type of assistance needs to be provided over a longer-term than is typically the case.

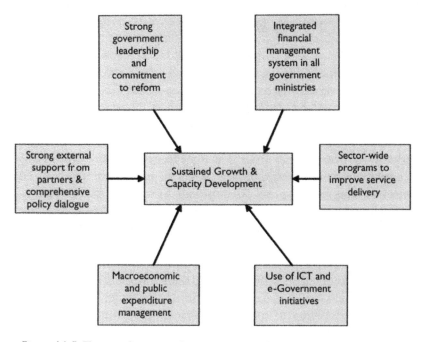

Figure 16.5 Tanzania's comprehensive approach to capacity development.

Success at the sector level can be bolstered by a broader, more comprehensive approach across sectors. In Tanzania, for example, strong government leadership and coordination, at both political and technocratic levels, were the hallmarks of the initiatives for capacity development over the past decade. This leadership and cross-sectoral mainstreaming created an enabling environment that improved the ability of the sectors to delve deeper into their capacity-specific issues. The pooling of development partners' resources through basket funds and budget support was another feature of these reform programs. At the same time, the availability of substantial and predictable donor support over an extended period and the comprehensive policy dialogue were also seen as critical. Tanzania saw the benefits of the strategy it adopted: macroeconomic and fiscal stability, and a reduction in poverty and illiteracy. While challenges remain, Tanzania's country-led approach to capacity development (Figure 16.5) is an inspiration to other developing countries in the region.

16.8 CONCLUSION

Francis Fukuyama points out that if we really want to increase the institutional capacity of a developing country, we need to change the metaphor that describes what we hope to achieve. "We are not arriving in the country with girders, bricks, cranes, and construction blueprints, ready to hire natives to help build the factory that we have designed," he writes. "Instead, we should be arriving with resources to motivate the natives to design their own factory and to help them figure out how to build and operate it themselves. Every bit of technical assistance that displaces a comparable capability on the part of the local society should be regarded as a two-edged sword and treated with great caution. Above all, the outsiders need to avoid the temptation to speed up the process by running the factory themselves' (Fukuyama, 2004).

This critical message on the need for country leadership and ownership seems to be slowly taking hold. The international forums in Paris and Accra have helped to bolster the push for developing-country leadership. Both donors and their developing country partners have committed to changing the business model for capacity development to make it more country-driven and sensitive to the demands of stakeholders. African nations, meanwhile, are increasingly setting their own pace, and creating pan-African institutions and networks that are gradually playing a more important role.

The challenge that remains for African nations is to broaden the coalitions of stakeholders involved in strengthening their national capacities. Capacity development efforts are not just an affair of the state, but involve society at-large. For these efforts to take hold, African governments need to engage more effectively with civil society organizations, parliamentarians, the media and private sector. A greater share of international development assistance needs to be applied to supporting peer networks that have been shown to be highly effective in strengthening the professionalism and competence of their members. This becomes all the more important for fragile states, where the state's ability to deliver services to its people is often lacking.

The critical role of more effective dialogue and diagnosis around capacity development work is just as evident at the sector level as it is in broader, national strategies. Grounding sector-level capacity development in a more systematic and structured dialogue with relevant stakeholders requires more time and planning. The evidence referenced in this paper makes it all the more apparent that such dialogue is critical to success. Rather than relying on more and higher levels of imported technical expertise, countries should insist on building local institutions, coalitions for reform and learning networks with other countries that have experienced similar challenges.

NOTES

1   World Bank Task Force on Capacity Development in Africa, 2005. *Building Effective States: Forging Engaged Societies*, Washington DC, World Bank, pp. 18–22. For an example of NGO views, see Coopération internationale pour le développement et la solidarité (CIDSE), 2005. "Long Due Reform?: The International Monetary Fund, the World Bank, and Global Economic Governance 60 Years Later." CIDSE Position Paper.

2   Kaufmann D., A. Kraay, and M. Mastruzzi 2008: "Governance Matters VII: Governance Indicators for 1996–2007." *www.govindicators.org The indicators*

quoted here aggregate the views on the quality of governance provided by a large number of enterprise, citizen and expert survey respondents in industrial and developing countries. These data are gathered from a number of survey institutes, think tanks, non-governmental organizations, and international organizations.

3 Among the recent World Bank reviews of this subject are the Independent Evaluation Group's review of capacity building in Africa (2005); the Quality Assurance Group's assessment of the Bank's non-lending technical assistance (July 2005); the report of the Task Force on Capacity Development in Africa (September 2005); a policy note on project implementation units and their impact on capacity (October 2005); the Capacity Development Management Action Plan for the Africa Region (September 2006); a broader Bank-wide study on technical assistance (October 2006); and a variety of completed and ongoing sector- and subject-specific conferences and studies.

  This work by the Bank was pursued parallel to a major effort by the OECD/DAC to generate an ongoing multi-stakeholder discussion about capacity development resulting in the publication of *The Challenge of Capacity Development: Working towards Good Practice* (September 2006) and the emergence of the Learning Network on Capacity Development (LenCD), a multi-stakeholder group of donor agencies, practitioners and experts who banded together to try to solve the capacity puzzle.

4 Accra Agenda for Action, September 4, 2008.

5 To review the consensus position among donors, see OECD/DAC, 2006, "The Challenge of Capacity Development: Working Towards Good Practice." Paris, OECD.

6 See Independence Evaluation Group, 2005. "An Independent Review of World Bank Support to Capacity Building in Africa: The Case of Ghana", World Bank, Washington, D.C.

7 See Independence Evaluation Group, 2005. "An Independent Review of World Bank Support to Capacity Building in Africa: The Case of Ethiopia", World Bank, Washington, D.C.

REFERENCES

Baser, H. and Morgan, P. 2008, *Capacity, Change and Performance: Study Report.* Maastricht, European Center for Development Policy Management, pp. 18–19.

Bloom, E. et al. 2008, "Strengthening Networks: Using Organizational Network Analysis to Promote Network Effectiveness, Scale, and Accountability," Capacity Development Brief Issue No. 28, World Bank Institute, Washington, D.C. Available at www.worldbank.org/capacity

Committee on Development Effectiveness 2005-0016, 2005, *World Bank Support for Capacity Building in Africa.*, The World Bank, Washington, D.C.

Danish International Development Assistance, 2005, "Technical Assistance in Danish Bilateral Aid – Policy Paper." Copenhagen, DANIDA, p. 5.

Fukuyama, F. 2004, State-Building: Governance and World Order in the 21st Century, Cornell University Press, New York, p. 88.

OECD/DAC, 2006, "The challenge of capacity development: working towards good practice." Paris, OECD, p.12.

Taylor, P. and Clarke, P. 2008, *Capacity for a Change: Document Based on Outcomes of the Capacity Collective Workshop.* Institute of Development Studies, Sussex, p. 25.

Tejasvi, A. 2007, "South-South Capacity Development: The Way to Grow?" *Capacity Development Brief* No. 20. World Bank Institute, p. 2. Available at www.worldbank.org/capacity

World Bank Operations Evaluations Department (now called Independent Evaluation Group), 2005, *Capacity Building in Africa: An OED Evaluation of World Bank Support,* World Bank, Washington D.C.

World Bank Task Force on Capacity Development in Africa, 2005, *Building Effective States – Forging Engaged Societies.* Washington, DC, World Bank, p. 25.

Chapter 17

CAPACITY ENHANCEMENT INDICATORS:
A review of the literature

Yemile Mizrahi
World Bank Institute, The World Bank Washington, D.C., USA

ABSTRACT

Since the international development conferences in Doha, Johannesburg and Monterrey, capacity enhancement has acquired a central place as a driver of sustainable development. The World Bank Institute's (WBI) responsibility is to keep the focus sharp on capacity enhancement as a core feature of the Bank's development business.

The World Bank Institute Evaluation Group (IEG) commissioned a literature review to identify indicators used to operationalize and measure "capacity" in capacity enhancement programs supported by a number of organizations.

The stark conclusion this review paper brings us to is that despite the importance accorded to the concept, little effort has gone into concretely defining what "capacity enhancement" means. To a large extent, the difficulty emerges from a vague understanding of the term "capacity," and even less clarity about the results to be expected from capacity enhancement efforts. The paper suggests that the analytical framework and results orientation of capacity enhancement programs can be strengthened considerably by asking the questions: *capacity for whom?* And, *capacity for what?* Although a general agreement is emerging with respect to the *levels* at which capacity enhancement endeavors can be directed – individuals, organizations, and institutions – the capacity-related outcomes will be amenable to measurement only if the outcomes expected are concretely defined. The author also encourages us to think of capacity building as a process and, therefore, to define interim benchmarks.

17.1　INTRODUCTION

A consensus exists in the development community, both among practitioners and academics, that strengthening capacity is fundamental for development. The transference of resources from rich to poor countries, although important, is not sufficient to improve the performance of public and private organizations in developing countries. It is equally critical to enhance the capacity of these organizations to use, manage, and deploy these resources so that they are able to accomplish their strategic objectives (Horton, 2001). The latter involves not only the ability to identify needs and acquire the missing resources, but also the ability to design adequate incentives and create the opportunities to use these resources effectively and efficiently. This often requires the introduction of substantial institutional and organizational reforms and the ability to manage the changes that necessarily accompany this process. Poor institutional contexts, particularly weak bureaucracies and corruption, constitute a serious constraint on the ability of technical cooperation to contribute to capacity development (Browne, 2002).

Although the question of capacity enhancement has received extensive attention in the development literature, and the topic has been approached from different analytic perspectives, little agreement exists about how to define, operationalize, and measure capacity and capacity enhancement. This paper is an attempt to synthesize the literature to draw some conclusions about the measurement of capacity enhancement.

The paper provides a brief background on the concept of capacity enhancement, highlights the difficulties in measuring capacity enhancement, summarizes the capacity enhancement indicators used in development projects, and, finally, suggests an analytic framework for designing capacity enhancement indicators. A large body of literature was reviewed for this document, but only the most relevant pieces are cited in the bibliography.

17.2　BACKGROUND

Definitions of the terms **capacity** and **capacity enhancement** abound in the literature. Cohen (1993) believes the term capacity building (or enhancement) has been used too broadly and inconsistently to the point where it has lost its analytic power and utility. He claims that **capacity** needs to be narrowly defined as "individual ability, competence to carry out a specific task" and that capacity enhancement must, therefore, focus on increasing the abilities of specific types of personnel within an organization. This narrow definition, however, does not "travel" well to the other analytic dimensions of capacity and capacity enhancement: the organizational and institutional levels.[1] The UNDP (United Nations Development Programme) may provide the most analytically useful and less controversial definition. Capacity refers to the "ability to perform functions, solve problems and set and achieve objectives." This definition recognizes that national capacity is not just the sum total of individual capacities; that the concept is richer and more complex, it "weaves individual strengths into a stronger and more resilient fabric ... If countries and societies want to develop capacities, they must do more than expand individual skills. They also have to create the opportunities and incentives for people to use and expand individual skills" (Fukuda-Parr, Lopes, and Malik, 2002).

The term **capacity enhancement** adds a time dimension to this definition of capacity. Although conceptually related, capacity enhancement refers to the acquisition of these abilities within a certain period of time. Capacity enhancement projects thus refer to the necessary resources and conditions that are needed to develop these abilities. Capacity enhancement is a process and it is measured in degrees. Therefore, to become operational, capacity enhancement requires the establishment of some benchmarks along a low-high continuum.

In addition to the difficulties of defining these terms operationally, most authors writing on capacity and capacity enhancement have eluded questions of measurement and the identification of indicators. As Morgan (1997) correctly notes, "measures exist at the input and output end of the spectrum, and many indicators can be found which address service delivery and performance outcomes. There remains however a 'black box' in the middle of the indicator spectrum to do with capacity development which remains vague and unclear".

This vagueness is puzzling because despite the consensus about the importance of strengthening the capacity of developing countries' governments and societies, most analyses of capacity enhancement projects implemented during the 1980s concluded that the majority of these projects failed (Berg, 1993; Jaycox, 1993; David, 2001; Browne, 1999). In the 1990s some financial institutions became highly critical of these projects because "extensive investments apparently produced little in terms of the increased capacity of public sector officials or organizations to perform efficiently, effectively, and responsively" (Grindle, 1997). Yet, without adequate instruments to measure, monitor, and evaluate capacity enhancement, it is difficult to understand specifically what aspects/elements of capacity enhancement projects failed, to identify partial successes, and to design better, more effective and feasible projects.

---

[1] The European Centre for Development Policy Management (ECDPM) moves a step forward in the analytic dimension and defines "organizational capacity" as "an organization's potential to perform – its ability to successfully apply its skills and resources toward the accomplishments of its goals and the satisfaction of its stakeholders' expectations" (ECDPM, 2003).

Reflecting on previous capacity enhancement projects, a large part of the current literature on capacity enhancement focuses on new ways to design and implement technical assistance projects. Critical of the previous experience, this literature develops a "new paradigm" for capacity enhancement (or technical cooperation) which stresses the importance of country ownership, shifts its focus from the transference of knowledge to the acquisition of knowledge, and acknowledges the existence of local capacities (Browne, 2002; Morgan, 1997; Horton, 2001; Lopes and Theisohn, 2003).[2]

However, fewer efforts have been devoted to the particular question of measurement.

17.3 DIFFICULTIES IN MEASURING CAPACITY ENHANCEMENT

In large part, the difficulty of measuring capacity enhancement is that by definition, capacity enhancement is a process, rather than a final outcome or an output (the results of capacity) which are more easily identified and quantified. Moreover, enhancement may lead to different degrees of capacity.

Measurements of capacity enhancement may be qualitative in nature and involve a time frame, since capacity is strengthened over time. More importantly, capacity enhancement involves a complex process of learning, adaptation, and attitudinal change at the individual, organizational, and institutional levels. Benchmarks used to assess degrees or levels of capacity are often based on subjective evaluations and partial or incomplete information. Identifying indicators and measurement tools that grasp these complexities and address these different levels of analysis is much more challenging and difficult than identifying indicators that measure outputs or outcomes.

Moreover, focusing on capacity enhancement is also less "glamorous" than focusing on results or outcomes (Morgan, 1997). This also explains why fewer efforts have been devoted to measuring this process and the capacity that leads to performance. Of course, one could indirectly measure capacity enhancement by measuring performance outcomes. The logic is that if an organization or institution has greater capacity to perform its functions, performance will likely improve. However, capacity and performance are not synonymous, and failure to distinguish between these concepts can lead to misleading conclusions.

17.3.1 CAPACITY VERSUS PERFORMANCE

There are several reasons why performance indicators are not appropriate for measuring capacity:

---

[2] The motto of this new paradigm is "scan globally, reinvent locally" (Fukuda-Parr, Lopes, and Malik, 2002).

First, while performance may be a good indicator of adequate or good capacity, it does not yield insights into which aspects of capacity are particularly good, or which may be weakening. The personnel within a particular organization, for example, may have adequate levels of skills and yet the organization may be failing in its performance. Analyzing declining levels of performance, however, cannot reveal much about capacity gaps, for it may be that this gap is not at the skill level, but at a higher level of management.

Furthermore, performance indicators do not reveal what aspect of capacity is responsible for a better or failed performance. Weak performance indicators tell us little about the origins or causes of these results. Capacity enhancement projects may not be successful in generating better performance indicators or more satisfactory outputs, yet without adequately disaggregating capacity and finding indicators and benchmarks to measure capacity enhancement through its different analytic dimensions, it is difficult to assess what aspects of the process are failing, where additional support is required, and whether capacity enhancement projects are even realistic or feasible. Weak performance can be attributed to the lack of skilled personnel, to the unclear definition of roles and responsibilities within an organization, to the lack of adequate financial support, to the weakness of the regulatory framework, or to a combination of all these factors. Understanding these different analytic dimensions and designing measurements to evaluate progress at each level is important for designing better and more effective capacity enhancement projects.

Second, like in many other development programs, capacity enhancement programs may be only partially successful. Yet partial success is difficult to recognize if the criteria for evaluating these programs is solely based on performance outcomes. Measuring the "process" of capacity enhancement and developing benchmarks is thus critical for allowing the analyst to recognize partial and incomplete results. The prevalent frustration with many capacity enhancement programs stems in large part from the failure to recognize partial success. Confronted with what was perceived as "total failure," many projects attempted to start from scratch every time a new project was introduced.

Identifying partial successes lends not only to a more balanced judgment, but also to the adoption of more gradual, piecemeal, and realistic development strategies that take as a starting point "existing local capacity." The latter has been identified by the UNDP as a critical element in the new "paradigm" of capacity development (Browne, 2002).

Third, an institution or organization can improve its performance indicators, but nothing guarantees that this level of performance can be sustained over time. Unlike performance indicators, indicators of capacity and capacity enhancement indicators provide information about sustainability by revealing information about the extent of institutionalization or routinization of reforms introduced to enhance capacity. Technical assistance projects may have an initial positive impact on performance results, but as soon as the funding of these projects ends or foreign experts leave the country, performance indicators deteriorate. Unlike indicators of performance, indicators of capacity enhancement tell us something about the extent of "country ownership," a critical element for the sustainability of any capacity enhancement project. More on this is written in the sections below.

Finally, the relationship between capacity enhancement and performance is by no means direct and linear. The performance of governments, businesses, or civil society organizations is affected by a multiplicity of factors, above and beyond capacity enhancement. A severe economic crisis, for example, can have a substantial impact in the growth of poverty rates, regardless of the capacity of public officials to design and implement better poverty reduction strategies. Rapid economic growth, on the other hand, can have a greater impact on reducing poverty rates than the enhancement of government's long term capacity to deal with macro-economic stability. Similarly, low HIV rates may not accurately reveal the government's capacity to respond, should the problem emerge at a later stage. Finally, a business may be successful in a closed economy protected from competition, regardless of its capacity to produce quality products.

17.3.2   PRECONDITIONS FOR ENHANCING CAPACITY

In a recent publication that reflects on previous capacity enhancement projects, the UNDP recognizes that "successful and sustainable capacity development has remained an elusive goal" and that despite the training of thousands of people, "development undertakings have constantly faced lack of necessary skills and weak institutions." (Fokuda-Parr, Lopes, and Malik, 2002, p. 3). In large part, this is due to the lack of "ownership" of most capacity enhancement projects, the strong dependency on the donor community and foreign experts, and the tendency to concentrate on the training of individual skills without consideration of the larger organizational and institutional context.

It is clear that a necessary precondition for enhancing the capacity of governments, public or private organizations, or firms, is the willingness and commitment of key actors within this institutions to introduce reforms geared at improving performance outputs. An organization may have the technical capacity to accomplish a particular task, yet without a strong commitment of this organization's leadership, this organization may lack the adequate resources and/or the appropriate regulatory framework to accomplish these tasks.

17.4   CAPACITY ENHANCEMENT INDICATORS

The literature on capacity and capacity enhancement is extensive but mostly vague when it comes to operationalization of concepts, identification of measurement tools, and definition of indicators. While a consensus exists that capacity involves something more than the sum total of individual capacities, and that capacity enhancement projects therefore need to consider the broader institutional and organizational framework in which individuals operate, there is very little agreement on how to assess, monitor and measure capacity and capacity enhancement in the absence of specific developmental or sectoral objectives.

Furthermore, even when most authors acknowledge that capacity and capacity enhancement needs to be approached on a three dimensional analytical framework, authors usually tend to focus more on one analytic dimension more than another. Thus, for example, in his study of building technical capacity in the public sector, Cohen (1993) focuses more on building functional capacities (skills), at the individual level. In contrast, Peterson's analysis of bureaucracies in Africa (1997) focuses on the organizational dimension, and Boknick's study of fiscal discipline in Zambia (1997) focuses on the institutional level of analysis.

In a 2002 publication, the UNDP recognizes that despite years of training through technical assistance projects, weak institutions and poor skills remain unyielding constraints to development. While this document stresses that these projects failed due in large part to lack of "ownership," it devotes most of its attention to "finding new solutions to old problems," but offers little in terms of suggesting more concrete indicators to evaluate the effectiveness or progress of capacity enhancement projects. A more recent publication, however, grants greater attention to defining capacity and capacity enhancement in more operational terms.[3]

Building on the experiences and lessons of previous capacity enhancement projects in different countries, the main goal of this UNDP document is to offer practical and useful advice for practitioners and decision makers in developing countries and the international donor community. To disaggregate the term capacity, the document identifies "key capacities" that could be expected from "an empowered and capable individual, organization, or society in molding its own destiny" (Lopes and Theisohn, 2003).[4]

---

[3] In the UNDP language, capacity enhancement is described as "capacity development." They prefer this term to that of "capacity building" because it acknowledges existing local capacities and it connotes a long term process that covers many crucial stages and that ensures national ownership and sustainability. The concept "capacity development" encompasses organizations and institutions that lie entirely outside the public sector: private enterprise and civil society organizations in particular. Capacity development involves human resource development, institution building, and capacities in the society as a whole. (UNDP, 2002, Introduction).

[4] Although this list is less vague than the term "capacity enhancement" or "capacity development," the items on the list still need to be operationalized and linked to particular institutional and organizational contexts to be subjected to measurement.

Box 17.1 UNDP's core capacities.

UNDP's core capacities:

1 The capacity to set objectives
2 The capacity to develop strategies
3 The capacity to draw action plans
4 The capacity to develop and implement appropriate policies
5 The capacity to develop regulatory and legal frameworks
6 The capacity to build and manage partnerships
7 The capacity to foster an enabling environment for civil society
8 The capacity to mobilize and manage resources
9 The capacity to implement action plans
10 The capacity to monitor progress

Source: Lopes and Theisohn, 2003

While stopping short of defining general capacity enhancement indicators, the book recommends some tools and techniques to conduct capacity assessments and to identify capacity gaps at the individual, organizational, and societal levels.

This is without a doubt a step forward in the definition of an analytic framework and the identification of measurement tools to analyze and assess capacity enhancement. However, the tools and techniques proposed to analyze capacity still remain abstract and vague. They serve better as general guiding principles for thinking about capacity enhancement than as useful measurement tools. For example, to assess the capacity of an organization to fulfill its functions, the book recommends that one should know if "the institutional processes such as planning, quality management, monitoring and evaluation, work effectively." Similarly, to assess the capacity of individuals, one should know if "the incentives are sufficient to promote excellence" (See annex 1 for the full list). However, how should one evaluate whether planning works effectively? What are the benchmarks? How should we know if the incentives are sufficient or not to promote excellence? These elements need to be defined before we can build more concrete indicators.

A more ambitious attempt to analyze capacity enhancement in the public sector through its various levels of analysis is Merilee S. Grindle edited volume, "Getting Good Government." This edited volume presents an analytic framework, or conceptual map, to analyze capacity and provides concrete case studies of capacity enhancement projects in various parts of the world. Contributors to this volume agree that "good government is advanced when skilled and professional public officials undertake to formulate and implement their policies, when bureaucratic units perform their assigned tasks effectively, and when fair and authoritative rules for economic and political interaction are regularly observed and enforced" (Grindle, 1997). A central contribution of this volume is the recognition that enhancing the capacity of governments to perform efficiently, effectively and responsibly requires addressing the different dimensions of governance: developing human resources, strengthening organizations, and reforming (or creating) institutions.

Hilderbrand and Grindle's chapter provides the most comprehensive and substantial analytic framework to assess capacity and capacity enhancement (capacity building in their words). By disaggregating capacity by different levels of analysis, this framework helps to identify capacity gaps and tools for designing more effective projects. Like most recent studies on capacity enhancement, these authors argue that training individuals and transferring technology is not sufficient for enhancing capacity, for individuals do not perform in a vacuum. Performance of individuals depends on the larger organizational and institutional framework. Moreover, they argue that unlike many authors who stress that civil servants improve their performance if they are adequately trained, well paid and have well defined responsibilities, Grindle and Hilderbrand persuasively argue that although these elements are important, good performance depends much more on improved management (Grindle, 1997). Taking a more systematic approach, they define five levels of analysis that affect capacity and should guide capacity building interventions:

Box 17.2 Hilderbrand and grindle's analytical framework.

Hilderbrand and Grindle's analytic framework:

1   **The action environment,** or the political, social, and economic context in which governments carry out their activities (rate of economic growth; degree of political conflict; human resource profile of the country)

2   **The institutional context of the public sector,** which includes such factors as the rules and procedures set for government operations and public officials, the financial resources the government has to carry out its activities, the responsibilities that government assumes for development initiatives.

3   **The task network** refers to the organizations involved in accomplishing any given task. The degree of communication and coordination among these organizations and the extent to which organizations are able to carry their responsibilities effectively.

4   **Organizations** are the building blocks of the task network. Factors such as the structure, processes, resources, and management styles affect how organizations define their goals, structure their work, provide incentives structures, and establish authority relations.

5   **Human Resources** refers to the level of skills and the retention of skilled personnel within organizations.

Source: Hilderbrand and Grindle, 1997

Using this analytic framework, the cases reviewed in this volume provide concrete examples of successes and failures of various capacity enhancement projects. Moreover, some of the case studies identify concrete indicators to assess the outcomes of the particular project under review. However, aside from proposing general guiding principles, the volume does not develop specific indicators, benchmarks or measurement tools that can be used across regions.

Taking a similar analytical perspective but devoting more attention to developing a methodology for carrying out systematic institutional capacity analysis, Alain Tobelem (Tobelem, 1992) defines different indicators to assess capacity gaps. Tobelem's document is intended as an "operation manual" to conduct institutional capacity analysis. Tobelem's methodological approach focuses more on disaggregating and operationalizing capacity, rather than building benchmarks for assessing capacity enhancement. Nevertheless, his framework is extremely useful for identifying indicators of capacity through its different analytic dimensions.

Although he does not use the term "ownership," Tobelem stresses that a government's commitment and political will to engage in a capacity engagement project is a necessary precondition. In his words "when a government, an administration, or an entity does not intend to change anything – when it does not want to do anything more or anything better – the related installed institutional capacity is by definition sufficient and therefore does not need to be assessed" (Tobelem, 1992).

Once this level of commitment or "ownership" has been evaluated, Tobelem suggests five analytic perspectives to assess institutional capacity that very much resemble those proposed by Grindle (1997) in the volume

---

Box 17.3 Tobelem's analytic dimensions to assess capacity.

Tobelem's analytic dimensions to assess capacity:

1 **Rules of the game** (institutional background), which includes governance, constitution, legislation, regulations and rules.
2 **Inter-institutional Relationships,** refers to the number and the extent of coordination of the different institutional entities (or organizations) in charge of a particular function or task.
3 **Internal Organization,** which includes the roles, the mandates, the distribution of functions, the internal relationship flows, the management style, and the resources of an organization.
4 **Personnel Policy and Reward System** refers to the existing civil service regulations.
5 **Skills,** which includes the personnel's knowledge and skill levels to accomplish their functions.

Source: Tobelem, 1992

referred to above. However, in addition to describing each one of these analytic perspectives, Tobelem defines a set of indicators to measure capacity and assess capacity gaps.

Tobelem provides a list of indicators and many of these indicators remain vague unless they make explicit reference to concrete development objectives and particular regional contexts. In abstraction, they can serve as an overall framework which can then be used to operationalize concepts and identify indicators.

An example of capacity indicators that relate to a specific development objective is the Paris21 Task Team on Statistical Capacity Building. Recognizing the centrality of statistical information for the formulation of development policies, particularly in the area of poverty reduction, this team developed a set of quantitative and qualitative indicators of statistical capacity to help identify capacity gaps and to track the progress of countries building statistical capacity (Laliberte, 2002). The process of defining these indicators took three years, from 1999 to 2002.[5] These indicators are particularly applicable to countries "that are statistically challenged, that have major deficiencies in available statistics and require sizable statistical capacity building, including fundamental changes to improve statistical operations and that cannot develop their statistical capacity without external assistance." (Laliberte, 2002).[6]

The quantitative indicators measure the performance of data-producing agencies by providing information on the depth and breath of statistical activities: the financing, staff, number of data sources, and diversity of statistical outputs.[7] Quantitative indicators focus on the statistics produced and can be used to assess if the statistical agency has attained the goal of delivering its products (Laliberte, 2002). These indicators, however, do not provide information about whether the data is effectively used by governmental agencies or other users. Nor do they reveal whether the data is produced in an efficient manner. This information is supplied by the qualitative indicators, which are applied to the statistical data sets.[8] Qualitative indicators reveal information on effectiveness and efficiency by taking into consideration the broader

---

[5] The Paris21 Consortium is a partnership of national, regional, and international statisticians, policymakers, development professionals, and other users of statistics. This Consortium was launched in 1999 and its purpose is to promote, influence, and facilitate capacity-building activities and the better use of statistics. Its founding organizers are the UN, OECD, World Bank, IMF, and EC.

[6] Other sector specific indicators probably exist, but nothing was found with reference to particular development projects.

[7] To limit the reporting burden, The Paris21 Task Team suggests that only three representative agencies be assessed (Laliberte, 2002).

[8] Like in the case of quantitative indicators, the Paris21 Task Team suggest to limit these data sets to three representative statistical outputs (they suggest GDP, population, and household income/expenditure data sets, which represent the economic, demographic, and social domains respectively. Laliberte, 1997).

environment in which the statistical agency operates. They show if the legal environment facilitates the production of statistics; if the culture is amenable to quality work; if the integrity and professionalism are protected and transparency measures are in place; if the data produced follows international methodological standards; if measures are in place to maintain the relevancy of products; and if the characteristics of the statistics produced fit the user's needs (Laliberte, 2002, p. 15).[9] Qualitative indicators are structured according to six criteria that are relevant for statistical operations:

Box 17.4 Paris21.

Paris21 Statistical capacity building criteria used to build qualitative indicators:

1   Institutional prerequisites
2   Integrity
3   Methodological Soundness
4   Accuracy and Reliability
5   Serviceability
6   Accessibility

In addition to describing the quantitative and qualitative concepts and developing indicators for each different aspect of the statistical operation process, the Paris21 Task Team defined benchmarks to assess these indicators according to a four-level range. Level 4 refers to optimal conditions for data production, and level 1 to least favourable conditions. Developing benchmarks related to a particular objective (in this case, strengthening statistical capacity of an agency) allows the analyst to assess not only capacity gaps, but also to analyze the process of capacity enhancement.

Like much of the recent literature on capacity enhancement, the Paris21 Team recognizes that capacity enhancement depends on something more than the development of skills and the acquisition technical equipment. Although their indicators are not disaggregated by levels of analysis (individual, organizational, institutional), these different dimensions are implicit in their definition.

Another example of an attempt to develop indicators to assess particular capacity enhancement projects is Cohen and Wheeler's study on training and retention of public sector bureaucracies in Africa (Cohen and Wheeler, 1997). This study assesses the impact of six externally funded capacity building projects that focused on training public sector economists, planners, statisticians, and financial managers.

---

[9] The Paris21 and Tobelem's approaches were combined to build the indicators of statistical capacity building. See Section V of this document.

Table 17.1 Paris21 Benchmarks descriptions.

Paris21 Benchmarks descriptions for data-related indicators (An example) Indicator: Effective coordination of statistics.

Level 4	Level 3	Level 2	Level 1
1 Legal or other formal arrangements/procedures clearly specify the responsibilities for coordination of statistical work and promotion of statistics standards, and this is implemented effectively through: 2 Development of a coordinated national program of statistical activities; identification of data gaps in meeting users' needs; elimination of duplication of statistical effort 3 Promotion of standard frameworks, concepts, classifications, and methodologies throughout the dataproducing agencies	1 Legal or other formal arrangements/procedures allocate responsibility for coordination of statistical work, but this is not fully effective in practice 2 There is some (but not significant) data gaps and/or duplication of statistical effort 3 Standard frameworks, etc. are promoted but there are some instances of noncompliance	1 Legal or other formal arrangements/procedures do not allocate responsibility for coordination of statistical work, and coordination does not occur. 2 There is significant data gaps in certain area and/or duplication of statistical effort (statistical outputs produced by different agencies may lack consistency and coherence) 3 Standard framework etc. are not actively promoted and there is significant noncompliance	1 There is no legal or other formal arrangements/procedures that specify responsibility for coordination of statistical work. 2 There is significant data gaps and duplication of statistical effort. 3 Standard frameworks, etc. are not promoted and are generally not observed. Dataproducing agencies may produce and use statistical outputs that are in conflict with those produced by others.

Source: Laliberte, 2002

Cohen and Wheeler recognize that although many of these projects have been widely perceived as failures, there are few systematic studies to assess the success and failures of training and staff retention programs. Although their analysis focuses at the skill (or human resources) level of analysis, they recognize that analyzing the capacity of the public sector requires a broader analytic perspective that takes into consideration management styles and other organizational aspects.

To assess the success of training projects aimed at building human resource capacity in the public sector, they develop the indicators of box 17.5:

Box 17.5 Indicators.

Indicators of capacity enhancement in African Public Sectors:

1   Retention rates of trained personnel in the targeted ministry
2   Retention of trained personnel in other government ministries or agencies
3   Attrition rates (how long trained personnel stay in the public sector)
4   Decline of expatriate experts
5   Profile of those who leave the public sector: whether the best and brightest stayed in or left the public sector.

Source: Cohen and Wheeler, 1992

Using these indicators, these authors found that retention rates were much higher than expected, even though there is a wide perception that public servants are badly paid, lack equipment, work in a demoralized environment and suffer from poor management (Cohen and Wheeler, 1997). The questions then are, why so many trained individuals stayed in their jobs and why despite such high retention rates, bureaucracies in these African countries continue to perform poorly. The answer is that for many public officials, their government job represented only a minor component of their salary. The combination of low salaries and weak management allowed these people to keep their public sector employment while seeking jobs outside. Trained people stayed in the public sector, but they were under-performing. "The opportunity to use office hours and equipment to significantly augment official salaries through private-income earning activities provides a major incentive to remain in the civil service" (Cohen and Wheeler, 1997). Paradoxically, weak management, lack of clarity of roles and responsibilities, duplication of functions, and absence of performance evaluations, increased public officials' incentives to remain in the public service. Lack of accountability and motivation within the public sector allowed these officials to use their jobs as safety nets while devoting their time and talent to other more profitable and rewarding occupations.

The conclusion of this study is that quantitative indicators such as skill retention rates and attrition rates are not sufficient to assess government capacity. More qualitative indicators of management styles and civil service rules are required in assessing the capacity of an organization. Although Cohen and Wheeler suggest different organizational aspects that need to be taken into consideration in assessing capacity (pay levels, team work, supportive supervision, development of a more attractive scheme of service, transparent and timely promotions, etc), they do not define indicators to assess organizational capacity.

In an article entitled "The design and use of capacity development indicators," Peter Morgan (1997) offers a framework for thinking about capacity indicators and provides some operational guidelines for their design. He stresses the importance of understanding that measuring capacity is different from measuring performance or outcomes; capacity indicators reveal something about the efforts that are necessary to improve organizational performance.

Morgan contends that designing indicators is an activity that has to be "demystified". In his words, "indicator factories in funding agencies now produce lists and lists of indicators for many different sectors. These are then tacked on to development projects and inserted into approval documents and contracts with little empirical evidence of their benefit or impact ... Yet at the same time, there needs to be more attention paid to he use and design of indicators as one part of the broader process of the strategic management of capacity development (Morgan, 1997, p. 3)." Rather than providing yet another list of indicators, Morgan suggests criteria for how to design indicators that are more effective and realistic and that reveal information about the "process" of building or strengthening capacities.

Although Morgan recognizes that there are no generic indicators of organizational capacity development, and that indicators need to relate to the specific development objectives (capacity for what?) and the actors for whom it is aimed

---

Box 17.6 Organizational capacity.

Morgan's boiler plate principles of organizational capacity:

1   The organization can learn and adapt to changing circumstances. It has a self-renewing capacity.

2   The organization can form productive relationships with outside groups or organizations as part of a broader effort to achieve its objectives.

3   The organization has an effective program for the recruitment, development and retention of staff that can adequately perform its critical functions.

4   The organization has some ability to legitimize its existence.

5   The organization has a structure, technology, and set of procedures that enable the staff to carry out the critical functions.

6   The organization has a culture, a set of values and an organizational motivation that values and rewards performance.

7   The organization has the ability, the resources and the autonomy to focus on a manageable set of objectives over a reasonable period of time.

Source: Morgan, 1997

(capacity for whom?), he still identifies some "boiler plate principles" that can be used to assess the capacity of any organization to fulfill its functions.

As is evident from this list, the issues addressed here by Morgan refer to the organizational level or dimension discussed by other authors. In abstraction, however, the elements of this list too are vague to be used as indicators. They need to be adapted to particular development goals and to concrete institutional and organizational contexts.

Morgan gives some examples of capacity development indicators that refer to particular development goals and that are targeted to particular agencies or institutions. Some of these examples are reproduced in Annex 4. Although Morgan's indicators are more specific about the goals and the agencies involved in particular development projects, they do not make explicit reference to the different analytic dimensions of capacity enhancement. This limits their use in assessing the capacity of an institution or an organization to fulfill their assigned functions in a sustainable manner.

In reviewing the literature on capacity enhancement, five conclusions emerge:

a Consensus exists that analysis of capacity and capacity enhancement should be approached through three different levels of analysis: The individual (human skills), the organizational, and the institutional. The recognition of the relationship between these three dimensions or levels is fundamental in the emerging paradigm for capacity enhancement among the international donor community. Unlike the past, where most capacity enhancement projects centered around strengthening human skills through training, today there is a recognition that the broader social, economic, and political context needs to be taken into account for any project to have a feasible possibility of success. Teachers and trainers can transfer information effectively, but trained individuals need a facilitating environment to apply their acquired knowledge. To have more analytic value, indicators of capacity enhancement have to be defined for these different analytical dimensions.

b Capacity enhancement indicators acquire operational value when they refer to concrete development objectives and the actors towards which capacity enhancement projects are directed. In abstraction, indicators lose analytic utility. Thus, to build indicators it is essential to address two central questions: capacity for what? And capacity for whom? Indicators of capacity of a statistical agency for example, will be different from indicators of organizational capacity of public bureaucracies.

c Capacity enhancement is a dynamic process of learning and adaptation. To gauge this process, indicators require the definition of benchmarks or norms that allow the analyst to assess different *levels* of capacity along a continuum. Defining these benchmarks may be difficult because they are often based on subjective perceptions and are not always value-free. However, some minimum level of consensus among experts is needed for benchmarks to have any utility as measurement tools.

d   Capacity enhancement depends first and foremost on the existence of political will and commitment on the part of the recipients. A teacher's success highly depends on the will and motivation of the student to learn. Country ownership and motivation are therefore the single most important determinants of effectiveness of capacity enhancement projects. Evaluating the extent of political will and motivation may be difficult, but it is essential to assess whether this element exists before any project is launched. The definition of the indicators, therefore, will need to be sensitive to the country's sense of ownership and the leaders' capacity and will to change in that direction.

e   Finally, most authors agree that aside from political will (or country ownership), capacity enhancement projects require "champions" of reform. Like in all reform process, capacity enhancement projects generate winners and losers. It is essential not only to minimize the losers and maximize the winners, but in most cases, the success of any reform depends on good leadership, or as Stephen Peterson (1994) argues, on the existence of "saints" (government reformers willing to introduce reforms and confront potential opposition). Although leadership is difficult to measure, capacity enhancement indicators need to be sensitive to this element for in many cases, regardless of the quality of technical assistance projects, the success or failure of capacity enhancement projects depends on good leadership.

17.5   DESIGNING CAPACITY ENHANCEMENT INDICATORS – AN EXAMPLE

To illustrate how the analytic framework can be used to design indicators of capacity enhancement, this document provides a hypothetical example of a project designed to enhance the capacity of a country's statistical agency to collect and analyze data. Strengthening this country's statistical capacity is regarded as a crucial step for designing and implementing better social policies, and particularly, better anti-poverty policies. Measuring and monitoring poverty requires stronger methodological and analytical tools. The latter strongly relies on the quality and coverage of poverty data. Reducing poverty and inequality is the overall goal of the project, but the latter requires good anti-poverty policies, which in turn depends on the existence of good analytical and measurement tools. See Table 17.2.

Taking strengthening the country's statistical agency as an immediate goal, Table 17.3 suggests a variety of indicators designed to measure capacity enhancement for measuring, monitoring, and analyzing poverty in this country. These indicators were developed by combining Tobelem's approach, who disaggregates capacity by different analytical dimensions, and the Paris21 Task Team who give particular attention to elements related to methodological soundness, integrity, and reliability of statistical data sources.

Table 17.2 Capacity enhancement for poverty reduction strategies.

Overall goal: poverty reduction	Intermediate goal: improving poverty reduction policies	Immediate goal: improving the methodology for measuring and analyzing poverty; improve the quality and coverage of poverty statistics
• Reduction of poverty levels (reduction of number of poor) • Improvement in the distribution of income • Creation of jobs; compensation of workers who lose their jobs; retraining • Improvement of quality of basic services: health, education, nutrition • Alleviation of extreme poverty, protection of the most vulnerable • Strengthening the targeting of social assistance • Inclusion of excluded social sector (regionally and ethnic considerations)	• Improving poverty assessments using more reliable and consistent statistical information and based on an improved methodology • Improve analytic skills to forecast impact of social reform policies on poverty • Improve targeting methods and delivery methods (based on definition of poverty line, number of poor, identification of populations at risk) • Introducing new methods and tools for monitoring and evaluating the poverty impact of policies and programs; assess the short term impact of enterprise restructuring • Improve understanding of tradeoffs and costs of different policy instruments and policies	• Improve the methodology of poverty measurement, improve the household budget survey (HBS) and a system of administrative statistics for enhanced poverty monitoring and policy-oriented analytic work • Generate consensus on the number and characteristics of the poor; the poverty line, and welfare indicators • Improve quality, consistency, reliability and coverage (particularly regional) of statistics on the poor • Publicize and disseminate data to other agencies within the government • Guarantee open access to the data • Promote a dialogue on poverty issues

Table 17.3 Statistical capacity building indicators.

DIMENSIONS	SUB-DIMENSION	INDICATOR	OPERATIONALIZATION
Institutional	1. Constitutional framework	Collection of information and preservation of information guaranteed by law	*Legislation guarantees access of information and protects confidentiality. *There are prescribed penalties for breach of confidentiality
	2. Regulatory framework	Statistical agency has clear mandate to collect data; functions are clearly assigned to agency	*Legislation gives data producing agencies full responsibility to compile and disseminate statistics
		Autonomy of operations	*Statistical Agency's authority to collect and process data; to hire and fire personnel, and to plan and budget its activities is legally recognized.
		Integrity	*Legal and institutional provisions protect professional independence (the choice and tenure of senior management; protection from political interference *Choices of sources and statistical guidelines and techniques are based on statistical criteria without outside interference.
	3. Legitimacy	Agencies within the government agree on mandate of Country's statistical agency	*Legal procedures clearly specify the responsibilities for promotion and coordination of statistical work. *Statistical agency is generally regarded by others as a professional, independent, objective and valued organization
	4. Ownership	Government commits resources to Statistical Agency	*Government funding as proportion of donor funding
		Government agencies share statistical information and are deeply involved in the project	*Existence of a task force or coordinating committee

(Continued)

Table 17.3 (Continued)

DIMENSIONS	SUB-DIMENSION	INDICATOR	OPERATIONALIZATION
Inter-Institutional	1. Effective coordination of statistics	Centralization of data collection / Central-local coordination of data collection	*Development of coordinated national program of statistic activities *Elimination of duplication of statistical effort *Promotion of standard frameworks, concepts, classifications, methodologies throughout the data producing agencies
	2. Agreement exists between entities having to relate formally or informally	Flow of information between agencies	*Existence of committees, task forces, coordination agencies
Organizational	1. Adequacy of buildings and equipment	Information and communication technology equipment	*Main frame, internal network, internet dissemination, PCs in use are adequate to perform tasks *Office buildings for data producing agencies are sound
	2. Distribution of Functions	Comprehensiveness of existing functions as compared with requirements	*Functions within the organization are well defined.
		Relevance of existing functions as compared with requirements	*Little duplication/overlap of functions
	3. Management style	Organization focus on quality/mission	*Motivation of personnel; understanding of mission
		Planning, monitoring and evaluation measures implemented	*Existence of a long-term statistical program *Management information systems are used regularly for: strategic plans, annual work programs, establishing and monitoring budget, performance of major projects, costs of inputs, records of staff participation in agency activities, individual staff performance and records *In response to changing priorities, management has flexibility to redirect resources between statistical projects *Evaluations of statistical activities are conducted periodically *Evaluation of project activities by external funding agencies are well coordinated with data producing agency
		Culture of professional and ethical	*Management promotes strong focus on quality/results oriented

(Continued)

Table 17.3 (Continued)

DIMENSIONS	SUB-DIMENSION	INDICATOR	OPERATIONALIZATION
	4. Civil Service Rules	standards	*Quality reviews conducted periodically; problems and suggestions for improvement acted upon *Manuals of operations exist and are kept up to date
		Procedures exist for hiring, firing, and promoting public officials	*Salary levels and work conditions are adequate and competitive
		Pay levels based on merit/competence	*Hiring is conducted on the basis of merit, not clientelism or patrimonialism
		Transparency/Accountability	*Information about operations and statistical process is available to the public; statistical outputs are available to the public
	5. Budgeting	Financial sustainability	*Share of Statistical agency in national accounts *Allocation of resources between projects within the agency and between head/regional offices is consistent with workloads and user priorities
	6. Methodological soundness	International/regional standards implemented	Current international accepted concepts are used and adjusted *International classifications and correspondence tables are used to link macro aggregates with micro data
		Source of data used	*Household surveys, census; administrative sources
		Source of data adequacy	*Data collected captures required information; targeted population is included
	7. Accuracy and Reliability		*Survey design is sound and covers all geographic regions and ethnic groups
		Response monitoring	*Response rates to surveys and censuses are consistently monitored *Active program of understanding variation of response rates across different types of respondents *Statistically valid methods are used for imputing and adjusting for non-response
		Validation of administrative data	*Effective continuing contact is maintained with the administrative authority that provides the source data

(Continued)

Table 17.3 (Continued)

DIMENSIONS	SUB-DIMENSION	INDICATOR	OPERATIONALIZATION
		Validation of intermediate and final outputs	*Administrative data is used for statistical purposes *An active process is in place to check comparability and internal consistency of data from individual census and surveys questionnaires *Assessment and validation of intermediate data and statistical output are carried out *Revision studies are undertaken regularly to assess reliability of preliminary data
	8. Serviceability	User consultation	*Users and other experts systematically consult data
		Timeliness of statistical outputs	*Statistical outputs are released on time
		Periodicity of statistical outputs	*Statistical outputs are released according to a time-program *Advance notice of changes in process and publication of annual report on activities.
	9. Accessibility	Effective dissemination	*Statistical outputs are released simultaneously to all users and are produced in different media *Statistical outputs are well designed and clear to follow *Seasonal and other analytical series are provided where appropriate
		Updated metadata	*A full range of information on underlying concepts, definitions, classifications, methodology, data sources, is documented, available and freely accessible to users *Catalogs of data products are widely available
Individual	1. Staff level and expertise adequacy	Intellectual/Technical skills needed to implement tasks	The number of staff is sufficient to handle ongoing statistical activities Staff is educated to the levels required
	2. Skill retention	Rate of turnover	Staff turnover is manageable
	3. Continuous training	Skills are updated	Existence of on the job training programs

NOTE

This document originates from: Capacity Enhancement Indicators: Review of the Literature, Yemile Mizrahi, WBI Evaluation Studies, No. EG03-72, World Bank Institute, The World Bank Washington, D.C., 2003. The original report was prepared for the World Bank Institute (WBI) under the overall guidance of Marlaine Lockheed, Manager, Evaluation Group. The team was led by Nidhi Khattri. The findings, interpretations, and conclusions expressed in this paper are entirely those of the authors and do not necessarily represent the views of the World Bank Group. WBI Evaluation Studies are available on line at http://www.worldbank.org/wbi/evaluation/puball.htm

REFERENCES

Achibache, B., Belindas, M., Dinc, M., Eele, G. and Swanson, E. 2002, *"Strengthening Statistical Systems."* Chapter 5. *Poverty Reduction Strategy Papers Sourcebook.* World Bank.

Berg, E.J. 1993, *Rethinking Technical Cooperation: Reforms for Capacity Building in Africa.* Washington DC: United Nations Development Program/Development Alternatives Inc.

Bolnick, B.R. 1997, *"Establishing Fiscal Discipline: The Cash Budget in Zambia,"* in Merilee, S. Grindle, ed. Getting Good Government, Harvard University Press.

Browne, S. 1999, *Beyond Aid: From Patronage to Partnership.* Aldershot: Ashgate.

Browne, S. ed. 2002, *Developing Capacity Through Technical Cooperation.* Country Experiences. UNDP. Earthscan Publications.

Cohen, J.M. and Wheeler, J.R. 1997, *"Training and Retention in African Public Sectors,"* in Merilee S. Grindle, ed. Getting Good Government. Harvard University Press.

Cohen, J.M. 1993, *"Building Sustainable Public Sector Managerial, Professional, and Technical Capacity: A Framework for Analysis and Intervention,"* Harvard Institute for International Development, Harvard University. Development Discussion Papers, n. 473.

David, I.P. 2001, *"Why Statistical Capacity Building Technical Assistance Projects Fail?,"* International Association for Official Statistics. Manuscript.

ECDPM. 2003, *Evaluating Capacity Development.* Issue 17, April. www.capacity.org

Fukuda-Parr, S., Lopes, C. and Malik, K. 2002, *Capacity for Development.* New Solutions to Old Problems. United Nations Development Program and Earthscan Publications.

Grindle, M.S. ed. 1997, *Getting Good Government.* Capacity Building in the Public Sectors of Developing Countries. Harvard Institute for International Development. Harvard University Press.

Grindle, M.S. and Hilderbrand, M.E. 1995, *"Building sustainable capacity in the public sector: what can be done?,"* Public Administration and Development, Vol. 15, 441–463.

Horton, D. 2001, *Learning About Capacity Development Through Evaluation.* Perspectives and Observations from a Collaborative Network of National and International Organizations and Donor Agencies. The Hague: International Service for National Agricultural Research.

Horton, D. 2002, *"Planning, Implementing, and Evaluating Capacity Development,"* International Service for National Agricultural Research. Briefing paper, N. 50, July.

Jaycox, E.V.K. 1993, *"Capacity Building: The Missing Link in African Development,"* Transcript of address to the African-American Institute Conference, African Capacity Building: Effective and Enduring Partnerships, Reston, VA.

Laliberte, L. 2002, *"Statistical Capacity Building Indicators. Final Report."* Paris21 Task Team on Statistical Capacity Indicators. September.

Lopes, C. and Theisohn, T. 2003, *"Can We Do Better? Insights for Capacity Development."* Unpublished Manuscript. UNDP.

Morgan, P. 1997, *"The Design and Use of Capacity Development Indicators."* Paper prepared for the Policy Branch of CIDA.

Morgan, P. 2001, *"Technical Cooperation. Success and Failure: An overview,"* unpublished paper submitted to the UNDP.

Peterson, S.B. 1994, *"Saints, Demons, Wizards and Systems: Why Information Technology Reforms Fail or Under perform in Public Bureaucracies in Africa."* Harvard Institute for Institutional Development, Harvard University. Development Discussion Paper, n. 48.

Peterson, S.B. 1997, *"Hierarchy versus Networks: Alternative Strategies for Building Organizational Capacity in Public Bureaucracies in Africa,"* in Merilee S. Grindle, ed. Getting Good Government. Harvard University Press.

Tobelem, A. "Institutional Capacity Analysis and Development System (ICADS)".

Author index

Printed and bound by CPI Group (UK) Ltd, Croydon, CR0 4YY

23/10/2024

01778251-0002